An Introduction to Digital Audio

For Chrissie

An Introduction to Digital Audio

Second Edition

John Watkinson

Focal Press OXFORD AMSTERDAM BOSTON LONDON NEW YORK PARIS
SAN DIEGO SAN FRANCISCO SINGAPORE SYDNEY TOKYO

Focal Press
An imprint of Elsevier Science
Linacre House, Jordan Hill, Oxford OX2 8DP
225 Wildwood Avenue, Woburn, MA 01801-2041

First published 1994
Reprinted 1995, 1998, 1999
Second Edition 2002

British Library Cataloguing in Publication Data
A catalogue record for this book is available from the British Library

Library of Congress Cataloguing in Publication Data
Watkinson, John.
An introduction to digital audio/John Watkinson – 2nd ed.
p.cm.
ISBN 0 240 51643 5 (alk. paper)
1. Sound – Recording and reproducing – Digital techniques.   I. Title.

TK7881.4 W3834 2002
621.389'3 – dc21   2002026543

ISBN 0 240 51643 5

For information on all Focal Press publications visit our website at:
www.focalpress.com

Composition by Genesis Typesetting, Rochester, Kent
Printed and bound in Great Britain by Biddles Ltd, *www.biddles.co.uk*

Contents

Preface to the second edition

Digital audio has had a short but spectacular history, suddenly emerging from laboratories as the economics of LSI chips and data recording allowed ideas to become affordable equipment. That sudden emergence caused problems of its own, not least the dilemma of those who were trying to use the new equipment armed only with experience of the traditional analog audio.

Today digital audio has become the norm, and many users will not have been exposed to traditional analog equipment. In both cases a useful solution is access to a text that explains the foundations of the subject in an understandable format.

More years ago than I care to remember, *The Art of Digital Audio* was written to fill that need and today exists in its third edition. Given the progress in the art, this is now a large volume and exceeds the requirements, to say nothing of the budget, of many readers, leading to a need for a smaller book. This is that book.

Although a more affordable volume, this book does not make the mistake of simplifying the material to the point of being misleading. Instead the topics are carefully selected and explained with the same precision as the larger book.

This second edition recognizes changes in the subject such as the dominance of disk-based recording and the introduction of computer networks. Many of the explanations have been refined by my experience of teaching the subject.

John Watkinson
Burghfield Common, England

Introducing digital audio

1.1 Audio as data

The most exciting aspects of digital technology are the tremendous possibilities which were not available with analog technology. Many processes which are difficult or impossible in the analog domain are straightforward in the digital domain. Once audio is in the digital domain, it becomes data, and only differs from generic data in that it needs to be reproduced with a certain timebase.

The worlds of digital audio, digital video, communication and computation are closely related, and that is where the real potential lies. The time when audio was a specialist subject which could evolve in isolation from other disciplines has gone. Audio has now become a branch of information technology (IT); a fact which is reflected in the approach of this book.

Systems and techniques developed in other industries for other purposes can be used to store, process and transmit audio, video or both at once. IT equipment is available at low cost because the volume of production is far greater than that of professional audiovisual equipment. Disk drives and memories developed for computers can be put to use in such products. Communications networks developed to handle data can happily carry audiovisual data over indefinite distances without quality loss.

As the power of processors increases, it becomes possible to perform under software control processes which previously required dedicated hardware. This allows a dramatic reduction in hardware cost. Inevitably the very nature of audiovisual equipment and the ways in which it is used is changing along with the manufacturers who supply it. The computer industry is competing with traditional manufacturers, using the economics of mass production.

Tape is a linear medium and it is necessary to wait for the tape to wind to a desired part of the recording. In contrast, the head of a hard disk drive can access any stored data in milliseconds. This is known in computers as direct access and in audio production as non-linear access. As a result the non-linear editing workstation based on hard drives has eclipsed the use of tape for editing.

Digital broadcasting uses coding techniques to eliminate the interference, fading and multipath reception problems of analog broadcasting. At the same time, more efficient use is made of available bandwidth. The hard drive-based consumer audio recorder gives the consumer more power.

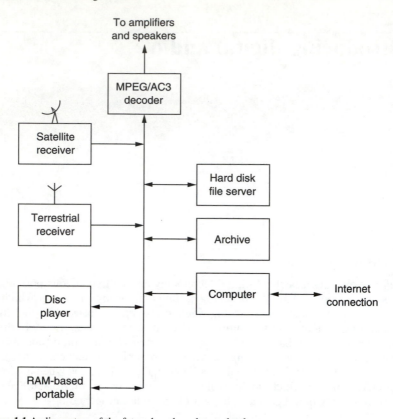

Figure 1.1 Audio system of the future based on data technology.

Figure 1.1 shows what the home audio system of the future may look like. MPEG-compressed signals may arrive in real time by terrestrial or satellite broadcast, via the Internet, or as the soundtrack of media such as DVD. Media such as Compact Disc supply uncompressed data for higher quality. The heart of the system is a hard drive-based server. This can be used to time shift broadcast programs, to skip commercial breaks or to assemble requested audio material transmitted in non-real time at low bit rates. If equipped with a web browser, the server may explore the web looking for material which is of the same kind the user normally wants. As the cost of storage falls, the server may download this material speculatively.

For portable use, the user may download compressed audio files into memory-based devices which act as audio players yet have no moving parts. On playback the bitstream is recovered from memory, decoded and converted typically to a signal which can drive headphones.

Ultimately digital technology will change the nature of broadcasting out of recognition. Once the viewer has non-linear storage technology and electronic program guides, the traditional broadcaster's transmitted schedule is irrelevant. Increasingly consumers will be able to choose what is played and when, rather than the broadcaster deciding for them. The broadcasting of conventional commercials will cease to be effective when viewers have the technology to skip them. Anyone with a web site which can stream audio data can become a broadcaster.

1.2 What is an audio signal?

An analog audio signal is an electrical waveform which is a representation of the velocity of a microphone diaphragm. Such a signal is two-dimensional in that it carries a voltage changing with respect to time. In analog systems, these waveforms are conveyed by some infinite variation of a continuous parameter. In a recorder, distance along the medium is a further, continuous, analog of time. It does not matter at what point a recording is examined along its length, a value will be found for the recorded signal. That value can itself change with infinite resolution within the physical limits of the system.

Those characteristics are the main weakness of analog signals. Within the allowable bandwidth, *any* waveform is valid. If the speed of the medium is not constant, one valid waveform is changed into another valid waveform; a problem which cannot be detected in an analog system and which results in wow and flutter. In addition, a voltage error simply changes one valid voltage into another; noise cannot be detected in an analog signal. Noise might be suspected, but how is one to know what proportion of the received signal is noise and what is the original? If the transfer function of a system is not linear, distortion results, but the distorted waveforms are still valid; an analog system cannot detect distortion. Again distortion might be suspected, but it is impossible to tell how much of the energy at a given frequency is due to the distortion and how much was actually present in the original signal.

It is a characteristic of analog systems that degradations cannot be separated from the original signal, so nothing can be done about them. At the end of a system a signal carries the sum of all degradations introduced at each stage through which it passed. This sets a limit to the number of stages through which a signal can be passed before it is useless. Alternatively, if many stages are envisaged, each piece of equipment must be far better than necessary so that the signal is still acceptable at the end. The equipment will naturally be more expensive.

Digital audio is simply an alternative means of carrying an audio waveform. Although there are a number of ways in which this can be done, there is one system, known as pulse code modulation (PCM), which is in virtually universal use.[1] Figure 1.2 shows how PCM works. Instead of being continuous, the time axis is represented in a discrete, or stepwise manner. The audio waveform is not carried by continuous representation, but by measurement at regular intervals. This process is called sampling and the frequency with which samples are taken is called the sampling rate or sampling frequency F_s. Each sample still varies infinitely as the original waveform did. To complete the conversion to PCM, each sample is then represented to finite accuracy by a discrete number in a process known as quantizing.

At the ADC (analog-to-digital convertor), every effort is made to rid the sampling clock of jitter, or time instability, so every sample is taken at an exactly even time step. Clearly, if there is any subsequent timebase error, the instants at which samples arrive will be changed and the effect can be detected. If samples arrive at some destination with an irregular timebase, the effect can be eliminated by temporarily storing the samples in a memory and reading them out using a

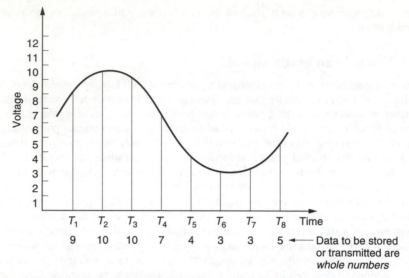

Figure 1.2 In pulse code modulation (PCM) the analog waveform is measured periodically at the sampling rate. The voltage (represented here by the height) of each sample is then described by a whole number. The whole numbers are stored or transmitted rather than the waveform itself.

stable, locally generated clock. This process is called timebase correction and all properly engineered digital audio systems will use it.

Those who are not familiar with digital principles often worry that sampling takes away something from a signal because it appears not to be taking notice of what happened between the samples. This would be true in a system having infinite bandwidth, but no analog signal can have infinite bandwidth. All analog signal sources from microphones and so on have a resolution or frequency response limit, as indeed do devices such as loudspeakers and human hearing. When a signal has finite bandwidth, the rate at which it can change is limited, and the way in which it changes becomes predictable. When a waveform can only change between samples in one way, it is then only necessary to convey the samples and the original waveform can be unambiguously reconstructed from them. A more detailed treatment of the principle will be given in Chapter 4.

As stated, each sample is also discrete, or represented in a stepwise manner. The magnitude of the sample, which will be proportional to the voltage of the audio signal, is represented by a whole number. This process is known as quantizing and results in an approximation, but the size of the error can be controlled until it is negligible. The link between quality and sample resolution is explored in Chapter 4. The advantage of using whole numbers is that they are not prone to drift.

If a whole number can be carried from one place to another without numerical error, it has not changed at all. By describing audio waveforms numerically, the original information has been expressed in a way which is more robust.

Essentially, digital audio carries the sound numerically. Each sample is a numerical analog of the voltage at the corresponding instant in the sound.

1.3 Why binary?

Arithmetically, the binary system is the simplest numbering scheme possible.

Figure 1.3(a) shows that there are only two symbols: 1 and 0. Each symbol is a binary digit, abbreviated to *bit*. One bit is a datum and many bits are data. Logically, binary allows a system of thought in which statements can only be true or false.

What is binary?

(a) Mathematically:
     The simplest numbering scheme possible, there are only two symbols:

                              1 and **0**

     Logically:
     A system of thought in which there are only two states:

                          True and **False**

(b) Binary information is **not** subject to misinterpretation

                  Black   **White**
                  In   **Out**
                  Guilty   **Innocent**

(c) Variables or non-binary terms:

                  Somewhat   **Undecided**
                  Probably   **Not proven**
                  Grey   **Under par**

Figure 1.3 Binary digits (a) can only have two values. At (b) are shown some everyday binary terms, whereas (c) shows some terms which cannot be expressed by a binary digit.

The great advantage of binary systems is that they are the most resistant to misinterpretation. In information terms they are *robust*. Figure 1.3(b) shows some binary terms and (c) some non-binary terms for comparison. In all real processes, the wanted information is disturbed by noise and distortion, but with only two possibilities to distinguish, binary systems have the greatest resistance to such effects.

Figure 1.4(a) shows an ideal binary electrical signal is simply two different voltages: a high voltage representing a true logic state or a binary 1 and a low voltage representing a false logic state or a binary 0. The ideal waveform is also shown at (b) after it has passed through a real system. The waveform has been considerably altered, but the binary information can be recovered by comparing the voltage with a threshold which is set half-way between the ideal levels. In this way any received voltage which is above the threshold is considered a 1 and any voltage below is considered a 0. This process is called slicing, and can reject significant amounts of unwanted noise added to the signal. The signal will be carried in a channel with finite bandwidth, and this limits the slew rate of the signal; an ideally upright edge is made to slope.

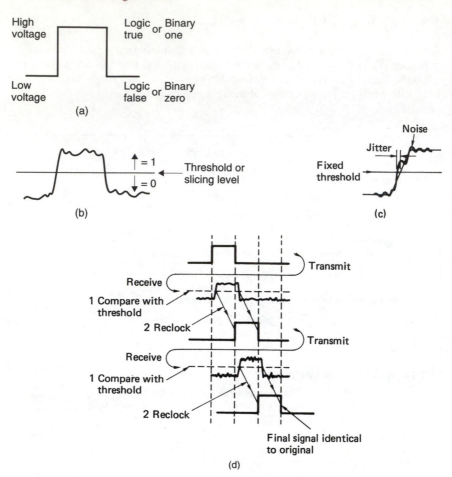

Figure 1.4 An ideal binary signal (a) has two levels. After transmission it may look like (b), but after slicing the two levels can be recovered. Noise on a sliced signal can result in jitter (c), but reclocking combined with slicing makes the final signal identical to the original as shown in (d).

Noise added to a sloping signal (c) can change the time at which the slicer judges that the level passed through the threshold. This effect is also eliminated when the output of the slicer is reclocked. Figure 1.4(d) shows that however many stages the binary signal passes through, the information is unchanged except for a delay. Of course, an excessive noise could cause a problem. If it had sufficient level and an appropriate polarity, noise could force the signal to cross the threshold and the output of the slicer would then be incorrect. However, as binary has only two symbols, if it is known that the symbol is incorrect, it need only be set to the other state and a perfect correction has been achieved. Error correction really is as trivial as that, although determining which bit needs to be changed is somewhat harder.

Figure 1.5 shows that binary information can be represented by a wide range of real phenomena. All that is needed is the ability to exist in two states. A switch can be open or closed and so represent a single bit. This switch may control the

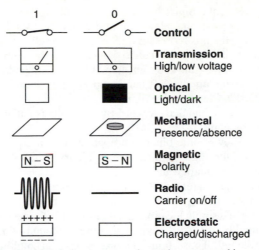

Figure 1.5 A large number of real phenomena can be used to represent binary data.

voltage in a wire which allows the bit to be transmitted. In an optical system, light may be transmitted or obstructed. In a mechanical system, the presence or absence of some feature can denote the state of a bit. The presence or absence of a radio carrier can signal a bit. In a random access memory (RAM), the state of an electric charge stores a bit.

Figure 1.5 also shows that magnetism is naturally binary as two stable directions of magnetization are easily arranged and rearranged as required. This is why digital magnetic recording has been so successful: it is a natural way of storing binary signals.

The robustness of binary signals means that bits can be packed more densely onto storage media, increasing the performance or reducing the cost. In radio signalling, lower power can be used.

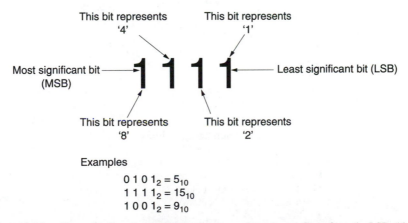

Figure 1.6 In a binary number, the digits represent increasing powers of two from the LSB. Also defined here are MSB and wordlength. When the wordlength is eight bits, the word is a byte. Binary numbers are used as memory addresses, and the range is defined by the address wordlength. Some examples are shown here.

In decimal systems, the digits in a number (counting from the right, or least significant end) represent ones, tens, hundreds and thousands, etc. Figure 1.6 shows that in binary, the bits represent one, two, four, eight, sixteen, etc. A multi-digit binary number is commonly called a word, and the number of bits in the word is called the wordlength. The right-hand bit is called the least significant bit (LSB) whereas the bit on the left-hand end of the word is called the most significant bit (MSB). Clearly more digits are required in binary than in decimal, but they are more easily handled. A word of eight bits is called a byte, which is a contraction of 'by eight'.

Figure 1.6 also shows some binary numbers and their equivalent in decimal. The radix point has the same significance in binary: symbols to the right of it represent one half, one quarter and so on.

Binary words can have a remarkable range of meanings. They may describe the magnitude of a number such as an audio sample or an image pixel or they may specify the address of a single location in a memory. In all cases the possible range of a word is limited by the wordlength. The range is found by raising two to the power of the wordlength. Thus a four-bit word has sixteen combinations, and could

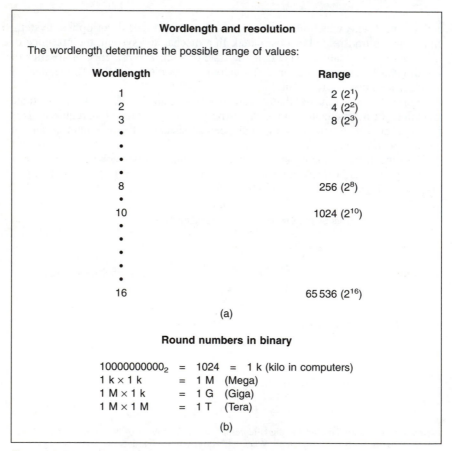

Wordlength and resolution

The wordlength determines the possible range of values:

Wordlength	Range
1	$2\ (2^1)$
2	$4\ (2^2)$
3	$8\ (2^3)$
•	
•	
•	
•	
8	$256\ (2^8)$
•	
10	$1024\ (2^{10})$
•	
•	
•	
•	
•	
16	$65\,536\ (2^{16})$

(a)

Round numbers in binary

10000000000_2 = 1024 = 1 k (kilo in computers)
1 k × 1 k = 1 M (Mega)
1 M × 1 k = 1 G (Giga)
1 M × 1 M = 1 T (Tera)

(b)

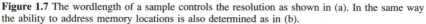

Figure 1.7 The wordlength of a sample controls the resolution as shown in (a). In the same way the ability to address memory locations is also determined as in (b).

address a memory having sixteen locations. A sixteen-bit word has 65 536 combinations. Figure 1.7(a) shows some examples of wordlength and resolution.

The capacity of memories and storage media is measured in bytes, but to avoid large numbers, kilobytes, megabytes and gigabytes are often used. A ten-bit word has 1024 combinations, which is close to one thousand. In digital terminology, 1 K is defined as 1024, so a kilobyte of memory contains 1024 bytes. A megabyte (1 MB) contains 1024 kilobytes and would need a twenty-bit address. A gigabyte contains 1024 megabytes and would need a thirty-bit address. Figure 1.7(b) shows some examples.

1.4 Why digital?

There are two main answers to this question, and it is not possible to say which is the most important, as it will depend on one's standpoint.

(a) The quality of reproduction of a well-engineered digital audio system is independent of the medium and depends only on the quality of the conversion processes and of any compression scheme.
(b) The conversion of audio to the digital domain allows tremendous opportunities which were denied to analog signals.

Someone who is only interested in sound quality will judge the former the most relevant. If good-quality convertors can be obtained, all the shortcomings of analog recording and transmission can be eliminated to great advantage. An extremely good signal-to-noise ratio is possible, coupled with very low distortion. Timing errors between channels can be eliminated, making for accurate stereo images. One's greatest effort is expended in the design of convertors, whereas those parts of the system which handle data need only be workmanlike. When a digital recording is copied, the same numbers appear on the copy: it is not a dub, it is a clone. If the copy is undistinguishable from the original, there has been no generation loss. Digital recordings can be copied indefinitely without loss of quality. This is, of course, wonderful for the production process, but when the technology becomes available to the consumer the issue of copyright becomes of great importance.

In the real world everything has a cost, and one of the greatest strengths of digital technology is low cost. When the information to be recorded consists of discrete numbers, they can be packed densely on the medium without quality loss. Should some bits be in error because of noise or dropout, error correction can restore the original value. Digital recordings take up less space than analog recordings for the same or better quality. Digital circuitry costs less to manufacture because more functionality can be put in the same chip.

Digital equipment can have self-diagnosis programs built-in. The machine points out its own failures so the the cost of maintenance falls. A small operation may not need maintenance staff at all; a service contract is sufficient. A larger organization will still need maintenance staff, but they will be fewer in number and their skills will be oriented more to systems than devices.

1.5 Some digital audio processes outlined

Whilst digital audio is a large subject, it is not necessarily a difficult one. Every process can be broken down into smaller steps, each of which is relatively easy to follow. The main difficulty with study is to appreciate where the small steps fit into the overall picture. Subsequent chapters of this book will describe the key processes found in digital technology in some detail, whereas this chapter illustrates why these processes are necessary and shows how they are combined in various ways in real equipment. Once the general structure of digital devices is appreciated, the following chapters can be put in perspective.

Figure 1.8(a) shows a minimal digital audio system. This is no more than a point-to-point link which conveys analog audio from one place to another. It consists of a pair of convertors and hardware to serialize and de-serialize the samples. There is a need for standardization in serial transmission so that various devices can be connected together. The standards for digital interfaces are described in Chapter 7.

Analog audio entering the system is converted in the analog-to-digital convertor (ADC) to samples which are expressed as binary numbers. A typical sample would have a wordlength of sixteen bits. The sample is connected in

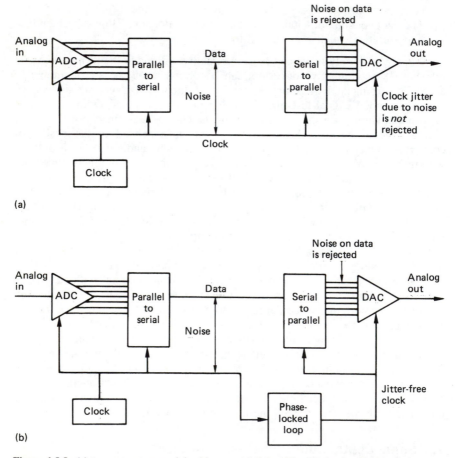

Figure 1.8 In (a) two convertors are joined by a serial link. Although simple, this system is deficient because it has no means to prevent noise on the clock lines causing jitter at the receiver. In (b) a phase-locked loop is incorporated, which filters jitter from the clock.

parallel into an output register which controls the cable drivers. The cable also carries the sampling rate clock. The data are sent to the other end of the line where a slicer rejects noise picked up on each signal. Sliced data are then loaded into a receiving register by the clock, and sent to the digital-to-analog convertor (DAC), which converts the sample back to an analog voltage.

As Figure 1.4 showed, noise can change the timing of a sliced signal. Whilst this system rejects noise which threatens to change the numerical value of the samples, it is powerless to prevent noise from causing jitter in the receipt of the sample clock. Noise on the clock means that samples are not converted with a regular timebase and the impairment caused will be audible.

The jitter problem is overcome in Figure 1.8(b) by the inclusion of a phase-locked loop which is an oscillator that synchronizes itself to the *average* frequency of the clock but which filters out the instantaneous jitter.

The system of Figure 1.8 is extended in Figure 1.9 by the addition of some random access memory (RAM). The operation of RAM is described in Chapter 3. What the device does is determined by the way in which the RAM address is controlled. If the RAM address increases by one every time a sample from the ADC is stored in the RAM, an audio recording can be made for a short period until the RAM is full. The recording can be played back by repeating the address sequence at the same clock rate but reading the memory into the DAC. The result is generally called a sampler. If the memory capacity is increased, the device can be used for general recording. RAM recorders are replacing dictating machines and the tape recorders used by journalists. In general they will be restricted to a fairly short playing time because of the high cost of memory in comparison with other storage media.

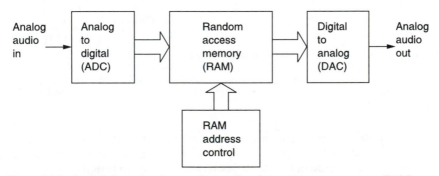

Figure 1.9 In the digital sampler, the recording medium is a random access memory (RAM). Recording time available is short compared with other media, but access to the recording is immediate and flexible as it is controlled by addressing the RAM.

Using compression, the playing time of a RAM-based recorder can be extended. For unchanging sounds such as test signals and station IDs, read only memory (ROM) can be used instead as it is non-volatile.

1.6 Time compression and expansion

Data files such as computer programs are simply lists of instructions and have no natural time axis. In contrast, audio and video data are sampled at a fixed rate and need to be presented to the viewer at the same rate. In audiovisual systems the audio also needs to be synchronized to the video. Continuous bitstreams at a fixed bit rate are difficult for generic data recording and transmission systems to handle. Such systems mostly work on blocks of data which can individually be addressed and/or routed. The bit rate may be fixed at the design stage at a value which may be too low or too high for the audio or video data to be handled.

The solution is to use time compression or expansion. Figure 1.10 shows a RAM which is addressed by binary counters which periodically overflow to zero and start counting again, giving the RAM a ring structure. If write and read addresses increment at the same speed, the RAM becomes a fixed data delay as the addresses retain a fixed relationship. However, if the read address clock runs at a higher frequency but in bursts, the output data are assembled into blocks with spaces in between. The data are now time compressed. Instead of being an unbroken stream which is difficult to handle, the data are in blocks with convenient pauses in between them. In these pauses numerous processes can take place. A hard disk might move its heads to another track. In all types of recording and transmission, the time compression of the samples allows time for synchronizing patterns, subcode and error-correction words to be inserted.

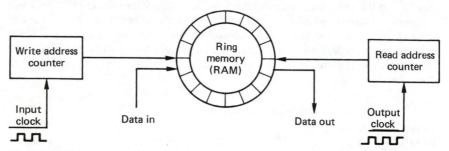

Figure 1.10 If the memory address is arranged to come from a counter which overflows, the memory can be made to appear circular. The write address then rotates endlessly, overwriting previous data once per revolution. The read address can follow the write address by a variable distance (not exceeding one revolution) and so a variable delay takes place between reading and writing.

Subsequently, any time compression can be reversed by time expansion. This requires a second RAM identical to the one shown. Data are written into the RAM in bursts, but read out at the standard sampling rate to restore a continuous bitstream. In a recorder, the time-expansion stage can be combined with the timebase correction stage so that speed variations in the medium can be eliminated at the same time. The use of time compression is universal in digital recording and widely used in transmission. In general the *instantaneous* data rate in the channel is not the same as the original rate although clearly the *average* rate must be the same.

Where the bit rate of the communication path is inadequate, transmission is still possible, but not in real time. Figure 1.11 shows that the data to be transmitted will have to be written in real time on a storage device such as a disk drive, and the drive will then transfer the data at whatever rate is possible to another drive at the receiver. When the transmission is complete, the second drive can then provide the data at the corrrect bit rate.

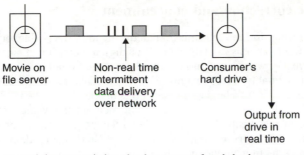

Figure 1.11 In non-real-time transmission, the data are transferred slowly to a storage medium which then outputs real-time data. Recordings can be downloaded to the home in this way.

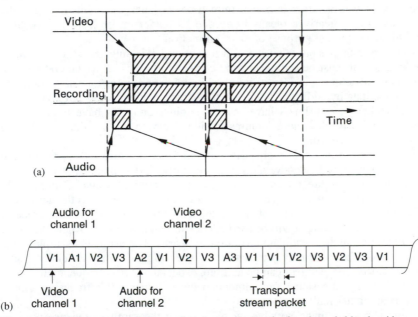

Figure 1.12 (a) Time compression is used to shorten the length of track needed by the video. Heavily time-compressed audio samples can then be recorded on the same track using common circuitry. In MPEG, multiplexing allows data from several TV channels to share one bitstream (b).

In the case where the available bit rate is higher than the correct data rate, the same configuration can be used to copy an audio data file faster than in real time. Another application of time compression is to allow several streams of data to be carried along the same channel in a technique known as *multiplexing*. Figure 1.12 shows some examples. At (a) multiplexing allows audio and video data to be recorded on the same heads in a digital video recorder such as DVC. At (b), several radio or television channels are multiplexed into one MPEG transport stream.

1.7 Error correction and concealment

All practical recording and transmission media are imperfect. Magnetic media, for example, suffer from noise and dropouts. In a digital recording of binary data, a bit is either correct or wrong, with no intermediate stage. Small amounts of noise are rejected, but inevitably, infrequent noise impulses cause some individual bits to be in error. Dropouts cause a larger number of bits in one place to be in error. An error of this kind is called a burst error. Whatever the medium and whatever the nature of the mechanism responsible, data are either recovered correctly or suffer some combination of bit errors and burst errors. In optical disks, random errors can be caused by imperfections in the moulding process, whereas burst errors are due to contamination or scratching of the disk surface.

The audibility of a bit error depends upon which bit of the sample is involved. If the LSB of one sample was in error in a detailed musical passage, the effect would be totally masked and no-one could detect it. Conversely, if the MSB of one sample was in error during a pure tone, no-one could fail to notice the resulting click. Clearly a means is needed to render errors from the medium inaudible. This is the purpose of error correction.

In binary, a bit has only two states. If it is wrong, it is only necessary to reverse the state and it must be right. Thus the correction process is trivial and perfect. The main difficulty is in identifying the bits which are in error. This is done by coding the data by adding redundant bits. Adding redundancy is not confined to digital technology, airliners have several engines and cars have twin braking systems. Clearly the more failures which have to be handled, the more redundancy is needed.

In digital recording, the amount of error which can be corrected is proportional to the amount of redundancy, and it will be shown in Chapter 6 that within this limit, the samples are returned to exactly their original value. Consequently *corrected* samples are undetectable. If the amount of error exceeds the amount of redundancy, correction is not possible, and, in order to allow graceful degradation, concealment will be used. Concealment is a process where the value of a missing sample is estimated from those nearby. The estimated sample value is not necessarily exactly the same as the original, and so under some circumstances concealment can be audible, especially if it is frequent. However, in a well-designed system, concealments occur with negligible frequency unless there is an actual fault or problem.

Concealment is made possible by rearranging the sample sequence prior to recording. This is shown in Figure 1.13 where odd-numbered samples are separated from even-numbered samples prior to recording. The odd and even sets of samples may be recorded in different places on the medium, so that an uncorrectable burst error affects only one set. On replay, the samples are recombined into their natural sequence, and the error is now split up so that it results in every other sample being lost in two different places. In those places, the waveform is described half as often, but can still be reproduced with some loss of accuracy. This is better than not being reproduced at all even if it is not perfect. Most tape-based digital audio recorders use such an odd/even distribution for concealment. Clearly if any errors are fully correctable, the distribution is a waste of time; it is only needed if correction is not possible.

The presence of an error-correction system means that the audio quality is independent of the medium/head quality within limits. There is no point in trying

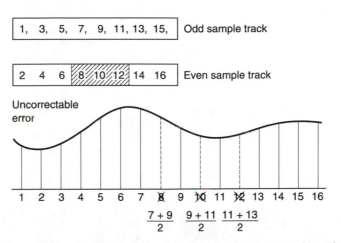

Figure 1.13 In cases where the error correction is inadequate, concealment can be used provided that the samples have been ordered appropriately in the recording. Odd and even samples are recorded in different places as shown here. As a result an uncorrectable error causes incorrect samples to occur singly, between correct samples. In the example shown, sample 8 is incorrect, but samples 7 and 9 are unaffected and an approximation to the value of sample 8 can be had by taking the average value of the two. This interpolated value is substituted for the incorrect value.

to assess the health of a machine by listening to the audio, as this will not reveal whether the error rate is normal or within a whisker of failure. The only useful procedure is to monitor the frequency with which errors are being corrected, and to compare it with normal figures.

Digital systems such as broadcast channels, optical disks and magnetic recorders are prone to burst errors. Adding redundancy equal to the size of expected bursts to every code is inefficient. Figure 1.14(a) shows that the efficiency of the system can be raised using interleaving. Sequential samples from the ADC are assembled into codes, but these are not recorded/transmitted in their natural sequence. A number of sequential codes are assembled along rows in a memory. When the memory is full, it is copied to the medium by reading down columns.

Subsequently, the samples need to be de-interleaved to return them to their natural sequence. This is done by writing samples from tape into a memory in columns, and when it is full, the memory is read in rows. Samples read from the memory are now in their original sequence so there is no effect on the information. However, if a burst error occurs as is shown shaded on the diagram, it will damage sequential samples in a vertical direction in the de-interleave memory. When the memory is read, a single large error is broken down into a number of small errors whose size is exactly equal to the correcting power of the codes and the correction is performed with maximum efficiency.

An extension of the process of interleave is where the memory array has not only rows made into codewords but also columns made into codewords by the addition of vertical redundancy. This is known as a product code. Figure 1.14(b) shows that in a product code the redundancy calculated first and checked last is

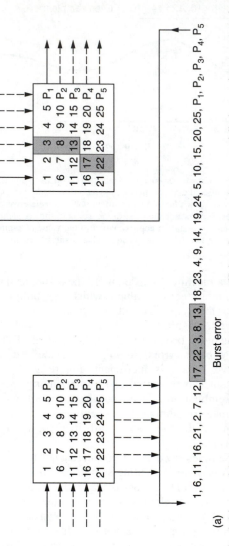

(a)

Burst error

1, 6, 11, 16, 21, 2, 7, 12, 17, 22, 3, 8, 13, 18, 23, 4, 9, 14, 19, 24, 5, 10, 15, 20, 25, P_1, P_2, P_3, P_4, P_5

Figure 1.14 (a) Interleaving is essential to make error-correction schemes more efficient. Samples written sequentially in rows into a memory have redundancy P added to each row. The memory is then read in columns and the data sent to the recording medium. On replay the non-sequential samples from the medium are de-interleaved to return them to their normal sequence. This breaks up the burst error (shaded) into one error symbol per row in the memory, which can be corrected by the redundancy P.

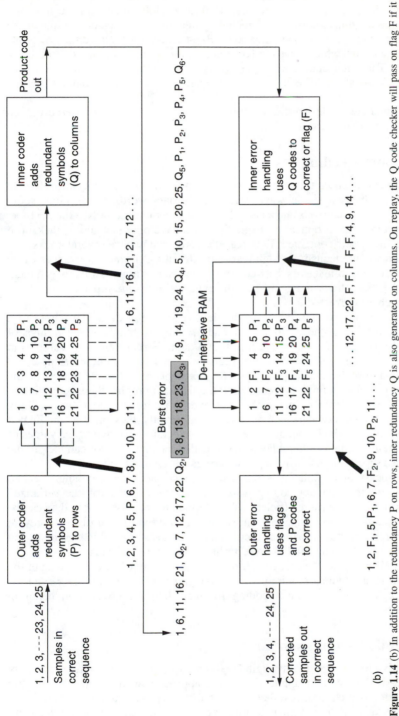

Figure 1.14 (b) In addition to the redundancy P on rows, inner redundancy Q is also generated on columns. On replay, the Q code checker will pass on flag F if it finds an error too large to handle itself. The flags pass through the de-interleave process and are used by the outer error correction to identify which symbol in the row needs correcting with P redundancy. The concept of crossing two codes in this way is called a product code.

called the outer code, and the redundancy calculated second and checked first is called the inner code. The inner code is formed along tracks on the medium. Random errors due to noise are corrected by the inner code and do not impair the burst-correcting power of the outer code. Burst errors are declared uncorrectable by the inner code which flags the bad samples on the way into the de-interleave memory. The outer code reads the error flags in order to locate the erroneous data. As it does not have to compute the error locations, the outer code can correct more errors.

The interleave, de-interleave, time-compression and timebase-correction processes inevitably cause delay.

1.8 Channel coding

In most recorders used for storing digital information, the medium carries a track which reproduces a single waveform. Clearly data words representing audio samples contain many bits and so they have to be recorded serially, a bit at a time. Some media, such as optical or magnetic disks, have only one active track, so it must be totally self-contained. Tape-based recorders may have several tracks read or written simultaneously. At high recording densities, physical tolerances cause phase shifts, or timing errors, between tracks and so it is not possible to read them in parallel. Each track must still be self-contained until the replayed signal has been timebase corrected.

Recording data serially is not as simple as connecting the serial output of a shift register to the head. In digital audio, samples may contain strings of identical bits. For example, silence in digital audio is represented by samples in which all the bits are zero. If a shift register is loaded with such a sample and shifted out serially, the output stays at a constant level for the period of the identical bits, and nothing is recorded on the track. On replay there is nothing to indicate how many bits were present, or even how fast to move the medium. Clearly, serialized raw data cannot be recorded directly, they must be modulated into a waveform which contains an embedded clock irrespective of the values of the bits in the samples. On replay a circuit called a data separator can lock to the embedded clock and use it to separate strings of identical bits.

The process of modulating serial data to make them self-clocking is called channel coding. Channel coding also shapes the spectrum of the serialized waveform to make it more efficient. With a good channel code, more data can be stored on a given medium. Spectrum shaping is used in optical disks to prevent the data from interfering with the focus and tracking servos, and in hard disks and in certain tape formats to allow rerecording without erase heads.

Channel coding is also needed to broadcast digital signals where shaping of the spectrum is an obvious requirement to avoid interference with other services.

The techniques of channel coding for recording are covered in detail in Chapter 6.

1.9 Audio compression

In its native form, high quality digital audio requires a high data rate, which may be excessive for certain applications. One approach to the problem is to use compression which reduces that rate significantly with a moderate loss of subjective quality. The human hearing system is not equally sensitive to all

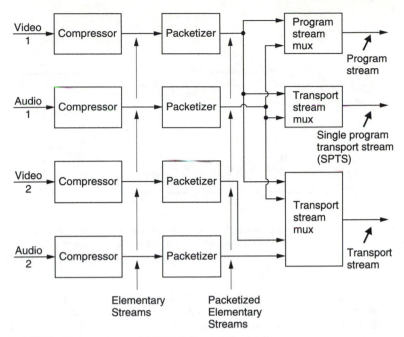

Figure 1.15 The bitstream types of MPEG-2. See text for details.

frequencies, so some coding gain can be obtained by using fewer bits to describe the frequencies which are less audible.

Whilst compression may achieve considerable reduction in bit rate, it must be appreciated that compression systems reintroduce the generation loss of the analog domain to digital systems.

One of the most popular compression standards for audio and video is known as MPEG. Figure 1.15 shows that the output of a single MPEG compressor is called an *elementary stream*. In practice audio and video streams of this type can be combined using multiplexing. The *program stream* is optimized for recording and is based on blocks of arbitrary size. The *transport stream* is optimized for transmission and is based on blocks of constant size.

It should be appreciated that many successful products use non-MPEG compression.

Compression and the corresponding decoding are complex processes and take time, adding to existing delays in signal paths. Concealment of uncorrectable errors is also more difficult on compressed data.

1.10 Disk-based recording

The magnetic disk drive was perfected by the computer industry to allow rapid random access to data, and so it makes an ideal medium for editing. As will be seen in Chapter 9, the heads do not touch the disk, but are supported on a thin air film which gives them a long life but which restricts the recording density. Thus disks cannot compete with tape for archiving, but for work such as Compact Disc production they have no equal.

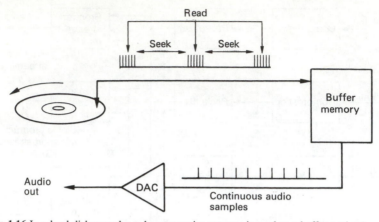

Figure 1.16 In a hard disk recorder, a large-capacity memory is used as a buffer or timebase corrector between the convertors and the disk. The memory allows the convertors to run constantly despite the interruptions in disk transfer caused by the head moving between tracks.

The disk drive provides intermittent data transfer owing to the need to reposition the heads. Figure 1.16 shows that disk-based devices rely on a quantity of RAM acting as a buffer between the real-time audio environment and the intermittent data environment.

Figure 1.17 shows the block diagram of an audio recorder based on disks and compression. The recording time and sound quality will not compete with full bandwidth tape-based devices, but following acquisition the disks can be used directly in an edit system, allowing a useful time saving in ENG (electronic news gathering) applications.

Development of the optical disk was stimulated by the availability of low-cost lasers. Optical disks are available in many different types, some which can only be recorded once, some which are erasable. These will be contrasted in Chapter

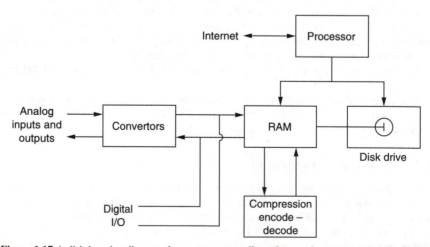

Figure 1.17 A disk-based audio recorder can capture audio and transmit compressed audio files over the Internet.

11. Optical disks have in common the fact that access is generally slower than with magnetic drives and it is difficult to obtain high data rates, but most of them are removable and can act as interchange media.

1.11 Rotary-head digital recorders

The rotary-head recorder has the advantage that the spinning heads create a high head-to-tape speed offering a high bit rate recording without high linear tape speed. Whilst mechanically complex, the rotary-head transport has been raised to a high degree of refinement and offers the highest recording density and thus lowest cost per bit of all digital recorders.

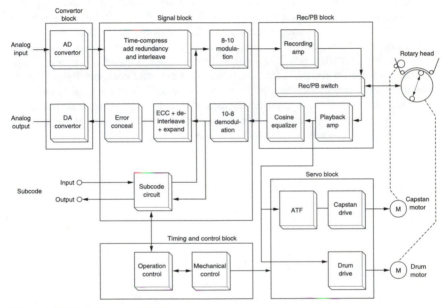

Figure 1.18 Block diagram of DAT

Figure 1.18 shows a representative block diagram of a rotary head machine. Following the convertors, a compression process may be found. In an uncompressed recorder, there will be distribution of odd and even samples for concealment purposes. An interleaved product code will be formed prior to the channel coding stage which produces the recorded waveform. On replay the data separator decodes the channel code and the inner and outer codes perform correction as in section 1.7. Following this the data channels are recombined and any necessary concealment will take place. Any compression will be decoded prior to the output convertors.

1.12 Digital audio broadcasting

Although it has given good service for many years, analog broadcasting is an inefficient use of bandwidth. Using compression, digital modulation and error-correction techniques, acceptable sound quality can be obtained in a fraction of the bandwidth of analog. Pressure on spectrum use from other uses such as cellular telephones will only increase and this may result in rapid changeover to digital broadcasts.

In addition to conserving spectrum, digital transmission is (or should be) resistant to multipath reception and gives consistent quality throughout the service area. Resistance to mutipath means that omnidirectional antennae can be used, an essential for mobile reception.

1.13 Networks

Communications networks allow transmission of data files whose content or meaning is irrelevant to the transmission medium. These files can therefore contain digital audio. Production systems can be based on high bit rate networks instead of traditional routing techniques. Contribution feeds between broadcasters and station output to transmitters no longer requires special-purpose links. Audio delivery is also possible on the Internet. As a practical matter, most Internet users suffer from a relatively limited bit rate and compression will have to be used until greater bandwidth becomes available. Whilst the quality does not compare with that of traditional broadcasts, this is not the point. Internet audio allows a wide range of services which traditional broadcasting cannot provide and phenomenal growth is expected in this area.

Reference

1. Devereux, V.G., Pulse code modulation of video signals: 8 bit coder and decoder. *BBC Res. Dept. Rept.*, **EL-42**, No.25 (1970)

Some audio principles

2.1 The physics of sound

Sound is simply an airborne version of vibration. The air which carries sound is a mixture of gases. In gases, the molecules contain so much energy that they break free from their neighbours and rush around at high speed. As Figure 2.1(a) shows, the innumerable elastic collisions of these high-speed molecules produce pressure on the walls of any gas container. If left undisturbed in a container at a constant temperature, eventually the pressure throughout would be constant and uniform.

Sound disturbs this simple picture. Figure 2.1(b) shows that a solid object which moves *against* gas pressure increases the velocity of the rebounding molecules, whereas in (c) one moving *with* gas pressure reduces that velocity. The average velocity and the displacement of all the molecules in a layer of air near to a moving body is the same as the velocity and displacement of the body. Movement of the body results in a local increase or decrease in pressure of some kind. Thus sound is both a pressure and a velocity disturbance.

Despite the fact that a gas contains endlessly colliding molecules, a small mass or *particle* of gas can have stable characteristics because the molecules leaving are replaced by new ones with identical statistics. As a result acoustics seldom needs to consider the molecular structure of air and the constant motion can be neglected. Thus when particle velocity and displacement is considered, this refers

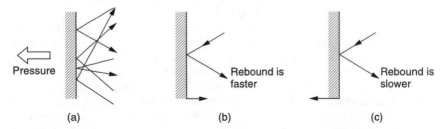

Figure 2.1 (a) The pressure exerted by a gas is due to countless elastic collisions between gas molecules and the walls of the container. (b) If the wall moves against the gas pressure, the rebound velocity increases. (c) Motion with the gas pressure reduces the particle velocity.

to the average values of a large number of molecules. In an undisturbed container of gas the particle velocity and displacement will both be zero everywhere.

When the volume of a fixed mass of gas is reduced, the pressure rises. The gas acts like a spring; it is compliant. However, a gas also has mass. Sound travels through air by an interaction between the mass and the compliance. Imagine pushing a mass via a spring. It would not move immediately because the spring would have to be compressed in order to transmit a force. If a second mass is connected to the first by another spring, it would start to move even later. Thus the speed of a disturbance in a mass/spring system depends on the mass and the stiffness. Sound travels through air without a net movement of the air.

The speed of sound is proportional to the square root of the absolute temperature. On earth, temperature changes with respect to absolute zero (−273°C) also amount to around one per cent except in extremely inhospitable places. The speed of sound experienced by most of us is about 1000 feet per second or 344 metres per second.

2.2 Wavelength

Sound can be due to a one-off event known as percussion, or a periodic event such as the sinusoidal vibration of a tuning fork. The sound due to percussion is called transient whereas a periodic stimulus produces steady-state sound having a frequency f.

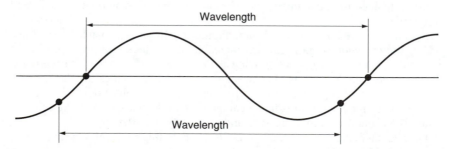

Figure 2.2 Wavelength is defined as the distance between two points at the same place on adjacent cycles. Wavelength is inversely proportional to frequency.

Because sound travels at a finite speed, the fixed observer at some distance from the source will experience the disturbance at some later time. In the case of a transient sound caused by an impact, the observer will detect a single replica of the original as it passes at the speed of sound. In the case of the tuning fork, a periodic sound source, the pressure peaks and dips follow one another away from the source at the speed of sound. For a given rate of vibration of the source, a given peak will have propagated a constant distance before the next peak occurs. This distance is called the wavelength lambda. Figure 2.2 shows that wavelength is defined as the distance between any two identical points on the whole cycle. If the source vibrates faster, successive peaks get closer together and the wavelength gets shorter. Figure 2.2 also shows that the wavelength is inversely proportional to the frequency. It is easy to remember that the wavelength of 1000 Hz is a foot (about 30 cm).

2.3 Periodic and aperiodic signals

Sounds can be divided into these two categories and analysed both in the time domain in which the waveform is considered, or in the frequency domain in which the spectrum is considered. The time and frequency domains are linked by transforms of which the best known is the Fourier transform. Transforms will be considered further in Chapter 3.

Figure 2.3(a) shows that an ideal periodic signal is one which repeats after some constant time has elapsed and goes on indefinitely in the time domain. In the frequency domain such a signal will be described as having a fundamental frequency and a series of harmonics or partials which are at integer multiples of the fundamental. The timbre of an instrument is determined by the harmonic structure. Where there are no harmonics at all, the simplest possible signal results which has only a single frequency in the spectrum. In the time domain this will be an endless sine wave.

Figure 2.3(b) shows an aperiodic signal known as white noise. The spectrum shows that there is equal level at all frequencies, hence the term 'white' which is analogous to the white light containing all wavelengths. Transients or impulses may also be aperiodic. A spectral analysis of a transient (c) will contain a range

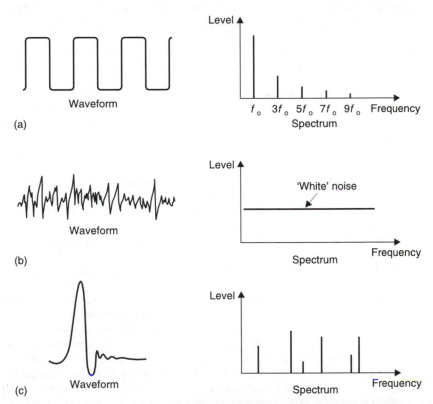

Figure 2.3 (a) Periodic signal repeats after a fixed time and has a simple spectrum consisting of fundamental plus harmonics. (b) Aperiodic signal such as noise does not repeat and has a continuous spectrum. (c) Transient contains an anharmonic spectrum.

of frequencies, but these are not harmonics because they are not integer multiples of the lowest frequency. Generally the narrower an event in the time domain, the broader it will be in the frequency domain and vice versa.

2.4 Sound and the ear

Experiments can tell us that the ear only responds to a certain range of frequencies within a certain range of levels. If sound is defined to fall within those ranges, then its reproduction is easier because it is only necessary to reproduce those levels and frequencies which the ear can detect.

Psychoacoustics can describe how our hearing has finite resolution in both the time and frequency domains such that what we perceive is an inexact impression. Some aspects of the original disturbance are inaudible to us and are said to be masked. If our goal is the highest quality, we can design our imperfect equipment so that the shortcomings are masked. Conversely if our goal is economy we can use compression and hope that masking will disguise the inaccuracies it causes.

A study of the finite resolution of the ear shows how some combinations of tones sound pleasurable whereas others are irritating. Music has evolved empirically to emphasize primarily the former. Nevertheless we are still struggling to explain why we enjoy music and why certain sounds can make us happy and others can reduce us to tears. These characteristics must still be present in digitally reproduced sound.

The frequency range of human hearing is extremely wide, covering some ten octaves (an octave is a doubling of pitch or frequency) without interruption.

By definition, the sound quality of an audio system can only be assessed by human hearing. Many items of audio equipment can only be designed well with a good knowledge of the human hearing mechanism. The acuity of the human ear is finite but astonishing. It can detect tiny amounts of distortion, and will accept an enormous dynamic range over a wide number of octaves. If the ear detects a different degree of impairment between two audio systems in properly conducted tests, we can say that one of them is superior.

However, any characteristic of a signal which can be heard can in principle also be measured by a suitable instrument although in general the availability of such instruments lags the requirement. The subjective tests will tell us how sensitive the instrument should be. Then the objective readings from the instrument give an indication of how acceptable a signal is in respect of that characteristic.

The sense we call hearing results from acoustic, mechanical, hydraulic, nervous and mental processes in the ear/brain combination, leading to the term psychoacoustics. It is only possible briefly to introduce the subject here. The interested reader is referred to Moore[1] for an excellent treatment.

Figure 2.4 shows that the structure of the ear is divided into the outer, middle and inner ears. The outer ear works at low impedance, the inner ear works at high impedance, and the middle ear is an impedance matching device. The visible part of the outer ear is called the pinna which plays a subtle role in determining the direction of arrival of sound at high frequencies. It is too small to have any effect at low frequencies. Incident sound enters the auditory canal or meatus. The pipe-like meatus causes a small resonance at around 4 kHz. Sound vibrates the eardrum or tympanic membrane which seals the outer ear from the middle ear.

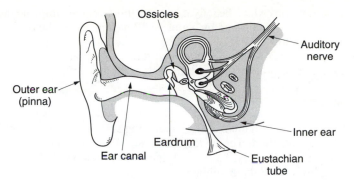

Figure 2.4 The structure of the human ear. See text for details.

The inner ear or cochlea works by sound travelling though a fluid. Sound enters the cochlea via a membrane called the oval window.

If airborne sound were to be incident on the oval window directly, the serious impedance mismatch would cause most of the sound to be reflected. The middle ear remedies that mismatch by providing a mechanical advantage. The tympanic membrane is linked to the oval window by three bones known as ossicles which act as a lever system such that a large displacement of the tympanic membrane results in a smaller displacement of the oval window but with greater force. Figure 2.5 shows that the malleus applies a tension to the tympanic membrane rendering it conical in shape. The malleus and the incus are firmly joined together to form a lever. The incus acts upon the stapes through a spherical joint. As the area of the tympanic membrane is greater than that of the oval window, there is a further multiplication of the available force. Consequently small pressures over the large area of the tympanic membrane are converted to high pressures over the small area of the oval window.

The middle ear is normally sealed, but ambient pressure changes will cause static pressure on the tympanic membrane which is painful. The pressure is relieved by the Eustachian tube which opens involuntarily while swallowing. The Eustachian tubes open into the cavities of the head and must normally be closed to avoid one's own speech appearing deafeningly loud.

The ossicles are located by minute muscles which are normally relaxed. However, the middle ear reflex is an involuntary tightening of the *tensor tympani*

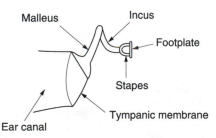

Figure 2.5 The malleus tensions the tympanic membrane into a conical shape. The ossicles provide an impedance-transforming lever system between the tympanic membrane and the oval window.

and *stapedius* muscles which heavily damp the ability of the tympanic membrane and the stapes to transmit sound by about 12 dB at frequencies below 1 kHz. The main function of this reflex is to reduce the audibility of one's own speech. However, loud sounds will also trigger this reflex which takes some 60–120 ms to occur, too late to protect against transients such as gunfire.

2.5 The cochlea

The cochlea, shown in Figure 2.6(a), is a tapering spiral cavity within bony walls which is filled with fluid. The widest part, near the oval window, is called the *base* and the distant end is the *apex*. Figure 2.6(b) shows that the cochlea is divided lengthwise into three volumes by Reissner's membrane and the basilar membrane. The *scala vestibuli* and the *scala tympani* are connected by a small aperture at the apex of the cochlea known as the *helicotrema*. Vibrations from the stapes are transferred to the oval window and become fluid pressure variations which are relieved by the flexing of the round window. Effectively the basilar membrane is in series with the fluid motion and is driven by it except at very low frequencies where the fluid flows through the helicotrema, bypassing the basilar membrane.

The vibration of the basilar membrane is sensed by the organ of Corti which runs along the centre of the cochlea. The organ of Corti is active in that it contains elements which can generate vibration as well as sense it. These are connected in a regenerative fashion so that the Q factor, or frequency selectivity of the ear, is higher than it would otherwise be. The deflection of hair cells in the organ of Corti triggers nerve firings and these signals are conducted to the brain by the auditory nerve. Some of these signals reflect the time domain, particularly during the transients with which most real sounds begin and also at low

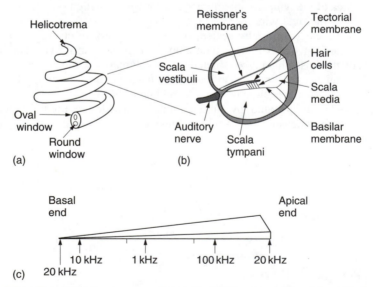

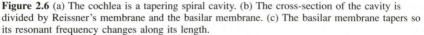

Figure 2.6 (a) The cochlea is a tapering spiral cavity. (b) The cross-section of the cavity is divided by Reissner's membrane and the basilar membrane. (c) The basilar membrane tapers so its resonant frequency changes along its length.

frequencies. During continuous sounds, the basilar membrane is also capable of performing frequency analysis.

Figure 2.6(c) shows that the basilar membrane is not uniform, but tapers in width and varies in thickness in the opposite sense to the taper of the cochlea. The part of the basilar membrane which resonates as a result of an applied sound is a function of the frequency. High frequencies cause resonance near to the oval window, whereas low frequencies cause resonances further away. More precisely the distance from the apex where the maximum resonance occurs is a logarithmic function of the frequency. Consequently tones spaced apart in octave steps will excite evenly spaced resonances in the basilar membrane. The prediction of resonance at a particular location on the membrane is called *place theory*. Essentially the basilar membrane is a mechanical frequency analyser.

Nerve firings are not a perfect analog of the basilar membrane motion. On continuous tones a nerve firing appears to occur at a constant phase relationship to the basilar vibration, a phenomenon called phase locking, but firings do not necessarily occur on every cycle. At higher frequencies firings are intermittent, yet each is in the same phase relationship.

The resonant behaviour of the basilar membrane is not observed at the lowest audible frequencies below 50 Hz. The pattern of vibration does not appear to change with frequency and it is possible that the frequency is low enough to be measured directly from the rate of nerve firings.

2.6 Mental processes

The nerve impulses are processed in specific areas of the brain which appear to have evolved at different times to provide different types of information. The time domain response works quickly, primarily aiding the direction-sensing mechanism and is older in evolutionary terms. The frequency domain response works more slowly, aiding the determination of pitch and timbre and evolved later, presumably as speech evolved.

The earliest use of hearing was as a survival mechanism to augment vision. The most important aspect of the hearing mechanism was the ability to determine the location of the sound source. Figure 2.7 shows that the brain can examine several possible differences between the signals reaching the two ears. At (a) a phase shift will be apparent. At (b) the distant ear is shaded by the head resulting in a different frequency response compared to the nearer ear. At (c) a transient sound arrives later at the more distant ear. The inter-aural phase, delay and level mechanisms vary in their effectiveness depending on the nature of the sound to be located. At some point a fuzzy logic decision has to be made as to how the information from these different mechanisms will be weighted.

There will be considerable variation with frequency in the phase shift between the ears. At a low frequency such as 30 Hz, the wavelength is around 11.5 metres and so this mechanism must be quite weak at low frequencies. At high frequencies the ear spacing is many wavelengths producing a confusing and complex phase relationship. This suggests a frequency limit of around 1500 Hz which has been confirmed by experiment.

At low and middle frequencies sound will diffract round the head sufficiently well that there will be no significant difference between the level at the two ears. Only at high frequencies does sound become directional enough for the head to shade the distant ear causing what is called an inter-aural intensity difference (IID).

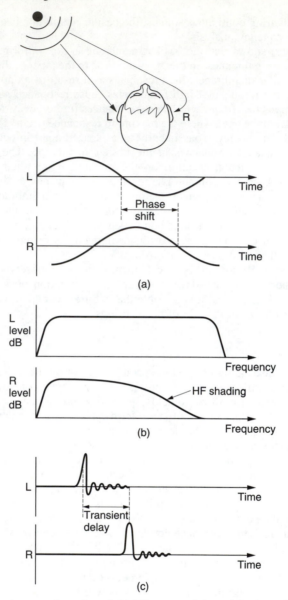

Figure 2.7 Having two spaced ears is cool. (a) Off-centre sounds result in phase difference. (b) Distant ear is shaded by head producing loss of high frequencies. (c) Distant ear detects transient later.

Phase differences are only useful at low frequencies and shading only works at high frequencies. Fortunately real-world noises and sounds are broadband and often contain transients. Timbral, broadband and transient sounds differ from tones in that they contain many different frequencies. Pure tones are rare in nature.

A transient has an unique aperiodic waveform which, as Figure 2.7(c) shows, suffers no ambiguity in the assessment of inter-aural delay (IAD) between two

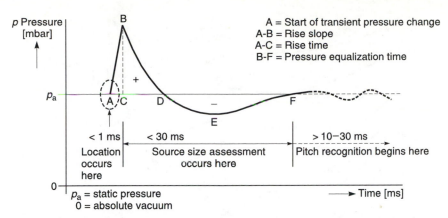

Figure 2.8 Real acoustic event produces a pressure step. Initial step is used for spatial location, equalization time signifies size of source. (Courtesy Manger Schallwandlerbau.)

versions. Note that a one-degree change in sound location causes a IAD of around 10 microseconds. The smallest detectable IAD is a remarkable 6 microseconds. This should be the criterion for spatial reproduction accuracy.

Transient noises produce a one-off pressure step whose source is accurately and instinctively located. Figure 2.8 shows an idealized transient pressure waveform following an acoustic event. Only the initial transient pressure change is required for location. The time of arrival of the transient at the two ears will be different and will locate the source laterally within a processing delay of around a millisecond.

Following the event which generated the transient, the air pressure equalizes. The time taken for this equalization varies and allows the listener to establish the likely size of the sound source. The larger the source, the longer the pressure-equalization time. Only after this does the frequency analysis mechanism tell anything about the pitch and timbre of the sound.

The above results suggest that anything in a sound reproduction system which impairs the reproduction of a transient pressure change will damage localization and the assessment of the pressure-equalization time. Clearly in an audio system which claims to offer any degree of precision, every component must be able to reproduce transients accurately and must have at least a minimum phase characteristic if it cannot be phase linear. In this respect digital audio represents a distinct technical performance advantage although much of this is later lost in poor transducer design, especially in loudspeakers.

2.7 Level and loudness

At its best, the ear can detect a sound pressure variation of only 2×10^{-5} Pascals r.m.s. and so this figure is used as the reference against which sound pressure level (SPL) is measured. The sensation of loudness is a logarithmic function of SPL and consequently a logarithmic unit, the deciBel, was adopted for audio measurement. The deciBel is explained in detail in section 2.12.

The dynamic range of the ear exceeds 130 dB, but at the extremes of this range, the ear is either straining to hear or is in pain. The frequency response of

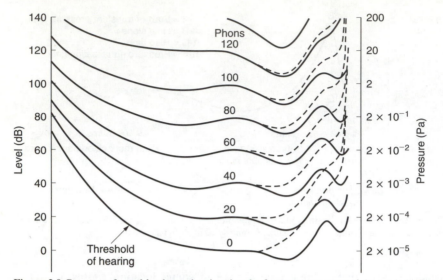

Figure 2.9 Contours of equal loudness showing that the frequency response of the ear is highly level dependent (solid line, age 20; dashed line, age 60).

the ear is not at all uniform and it also changes with SPL. The subjective response to level is called loudness and is measured in *phons*. The phon scale is defined to coincide with the SPL scale at 1 kHz, but at other frequencies the phon scale deviates because it displays the actual SPLs judged by a human subject to be equally loud as a given level at 1 kHz. Figure 2.9 shows the so-called equal loudness contours which were originally measured by Fletcher and Munson and subsequently by Robinson and Dadson. Note the irregularities caused by resonances in the meatus at about 4 kHz and 13 kHz.

Usually, people's ears are at their most sensitive between about 2 kHz and 5 kHz, and although some people can detect 20 kHz at high level, there is much evidence to suggest that most listeners cannot tell if the upper frequency limit of sound is 20 kHz or 16 kHz.[2,3] For a long time it was thought that frequencies below about 40 Hz were unimportant, but it is now clear that reproduction of frequencies down to 20 Hz improves reality and ambience.[4] The generally accepted frequency range for high-quality audio is 20 Hz to 20 000 Hz, although for broadcasting an upper limit of 15 000 Hz is often applied.

The most dramatic effect of the curves of Figure 2.9 is that the bass content of reproduced sound is disproportionately reduced as the level is turned down. This would suggest that if a sufficiently powerful yet high-quality reproduction system is available the correct tonal balance when playing a good recording can be obtained simply by setting the volume control to the correct level. This is indeed the case. A further consideration is that many musical instruments as well as the human voice change timbre with level and there is only one level which sounds correct for the timbre.

Audio systems with a more modest specification would have to resort to the use of tone controls to achieve a better tonal balance at lower SPL. A loudness control is one where the tone controls are automatically invoked as the volume is reduced. Although well meant, loudness controls seldom compensate accurately because they must know the original level at which the material was meant to be reproduced as well as the actual level in use.

A further consequence of level-dependent hearing response is that recordings which are mixed at an excessively high level will appear bass light when played back at a normal level. Such recordings are more a product of self-indulgence than professionalism.

Loudness is a subjective reaction and is almost impossible to measure. In addition to the level-dependent frequency response problem, the listener uses the sound not for its own sake but to draw some conclusion about the source. For example, most people hearing a distant motorcycle will describe it as being loud. Clearly at the source, it *is* loud, but the listener has compensated for the distance.

The best that can be done is to make some compensation for the level-dependent response using *weighting curves*. Ideally there should be many, but in practice the A, B and C weightings were chosen where the A curve is based on the 40-phon response. The measured level after such a filter is in units of dBA. The A curve is almost always used because it most nearly relates to the annoyance factor of distant noise sources.

2.8 Frequency discrimination

Figure 2.10 shows an uncoiled basilar membrane with the apex on the left so that the usual logarithmic frequency scale can be applied. The envelope of displacement of the basilar membrane is shown for a single frequency at (a). The vibration of the membrane in sympathy with a single frequency cannot be localized to an infinitely small area, and nearby areas are forced to vibrate at the same frequency with an amplitude that decreases with distance. Note that the envelope is asymmetrical because the membrane is tapering and because of frequency-dependent losses in the propagation of vibrational energy down the cochlea. If the frequency is changed, as in (b), the position of maximum displacement will also change. As the basilar membrane is continuous, the position of maximum displacement is infinitely variable allowing extremely good pitch discrimination of about one twelfth of a semitone which is determined by the spacing of hair cells.

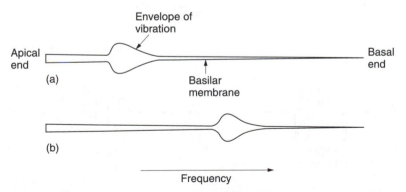

Figure 2.10 The basilar membrane symbolically uncoiled. (a) Single frequency causes the vibration envelope shown. (b) Changing the frequency moves the peak of the envelope.

In the presence of a complex spectrum, the finite width of the vibration envelope means that the ear fails to register energy in some bands when there is more energy in a nearby band. Within those areas, other frequencies are mechanically excluded because their amplitude is insufficient to dominate the local vibration of the membrane. Thus the Q factor of the membrane is responsible for the degree of auditory masking, defined as the decreased audibility of one sound in the presence of another. Masking is important because audio compression relies heavily on it.

The term used in psychoacoustics to describe the finite width of the vibration envelope is *critical bandwidth*. Critical bands were first described by Fletcher.[5] The envelope of basilar vibration is a complicated function. It is clear from the mechanism that the area of the membrane involved will increase as the sound level rises. Figure 2.11 shows the bandwidth as a function of level.

As will be seen in Chapter 3, transform theory teaches that the higher the frequency resolution of a transform, the worse the time accuracy. As the basilar membrane has finite frequency resolution measured in the width of a critical band, it follows that it must have finite time resolution. This also follows from the fact that the membrane is resonant, taking time to start and stop vibrating in response to a stimulus. There are many examples of this. Figure 2.12 shows the impulse response. Figure 2.13 shows the perceived loudness of a tone burst increases with duration up to about 200 ms due to the finite response time.

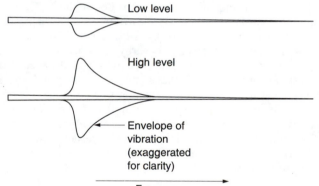

Figure 2.11 The critical bandwidth changes with SPL.

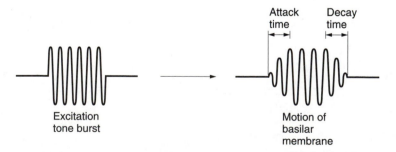

Figure 2.12 Impulse response of the ear showing slow attack and decay due to resonant behaviour.

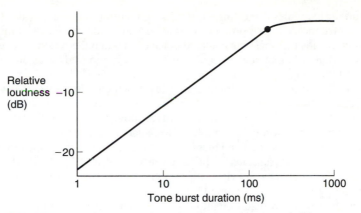

Figure 2.13 Perceived level of tone burst rises with duration as resonance builds up.

The ear has evolved to offer intelligibility in reverberant environments which it does by averaging all received energy over a period of about 30 ms. Reflected sound which arrives within this time is integrated to produce a louder sensation, whereas reflected sound which arrives after that time can be temporally discriminated and is perceived as an echo. Microphones have no such ability, which is why acoustic treatment is often needed in areas where microphones are used.

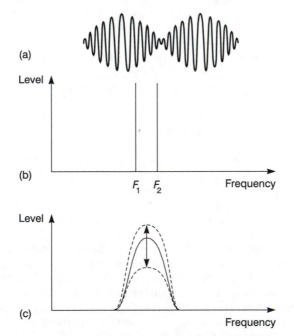

Figure 2.14 (a) Result of adding two sine waves of similar frequency. (b) Spectrum of (a) to infinite accuracy. (c) With finite accuracy only a single frequency is distinguished whose amplitude changes with the envelope of (a) giving rise to beats.

A further example of the finite time discrimination of the ear is the fact that short interruptions to a continuous tone are difficult to detect. Finite time resolution means that masking can take place even when the masking tone begins after and ceases before the masked sound. This is referred to as forward and backward masking.[6]

Figure 2.14 shows an electrical signal (a) in which two equal sine waves of nearly the same frequency have been linearly added together. Note that the envelope of the signal varies as the two waves move in and out of phase. Clearly the frequency transform calculated to infinite accuracy is that shown at (b). The two amplitudes are constant and there is no evidence of the envelope modulation. However, such a measurement requires an infinite time. When a shorter time is available, the frequency discrimination of the transform falls and the bands in which energy is detected become broader.

When the frequency discrimination is too wide to distinguish the two tones as in (c), the result is that they are registered as a single tone. The amplitude of the single tone will change from one measurement to the next because the envelope is being measured. The rate at which the envelope amplitude changes is called a *beat* frequency which is not actually present in the input signal. Beats are an artifact of finite frequency resolution transforms. The fact that human hearing produces beats from pairs of tones proves that it has finite resolution.

2.9 Frequency response and linearity

It is a goal in high-quality sound reproduction that the timbre of the original sound shall not be changed by the reproduction process. There are two ways in which timbre can inadvertently be changed, as Figure 2.15 shows. In (a) the spectrum of the original shows a particular relationship between harmonics. This signal is passed through a system (b) which has an unequal response at different frequencies. The result is that the harmonic structure (c) has changed, and with it the timbre. Clearly a fundamental requirement for quality sound reproduction is that the response to all frequencies should be equal.

Frequency response is easily tested using sine waves of constant amplitude at various frequencies as an input and noting the output level for each frequency.

Figure 2.16 shows that another way in which timbre can be changed is by non-linearity. All audio equipment has a transfer function between the input and the output which form the two axes of a graph. Unless the transfer function is exactly straight or *linear*, the output waveform will differ from the input. A non-linear transfer function will cause distortion which changes the distribution of harmonics and changes timbre.

At a real microphone placed before an orchestra a multiplicity of sounds may arrive simultaneously. The microphone diaphragm can only be in one place at a time, so the output waveform must be the sum of all the sounds. An ideal microphone connected by ideal amplification to an ideal loudspeaker will reproduce all of the sounds simultaneously by linear superimposition. However, should there be a lack of linearity anywhere in the system, the sounds will no longer have an independent existence, but will interfere with one another, changing one another's timbre and even creating new sounds which did not previously exist. This is known as *intermodulation*. Figure 2.17

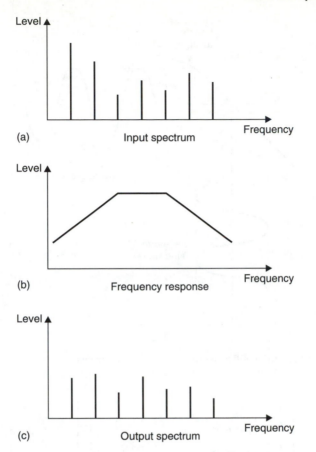

Figure 2.15 Why frequency response matters. Original spectrum at (a) determines timbre of sound. If original signal is passed through a system with deficient frequency response (b), the timbre will be changed (c).

shows that a linear system will pass two sine waves without interference. If there is any non-linearity, the two sine waves will intermodulate to produce sum and difference frequencies which are easily observed in the otherwise pure spectrum.

2.10 The sine wave

As the sine wave is such a useful concept it will be treated here in detail. Figure 2.18 shows a constant speed rotation viewed along the axis so that the motion is circular. Imagine, however, the view from one side in the plane of the rotation. From a distance only a vertical oscillation will be observed and if the position is plotted against time the resultant waveform will be a sine wave. Geometrically it is possible to calculate the height or displacement because it is the radius multiplied by the sine of the phase angle.

The phase angle is obtained by multiplying the angular velocity ω by the time t. Note that the angular velocity is measured in radians per second whereas frequency

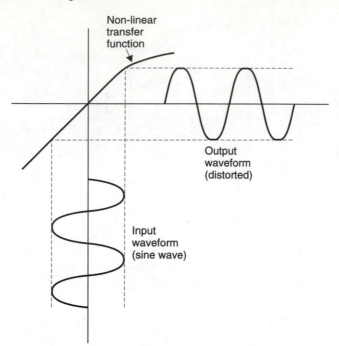

Figure 2.16 Non-linearity of the transfer function creates harmonies by distorting the waveform. Linearity is extremely important in audio equipment.

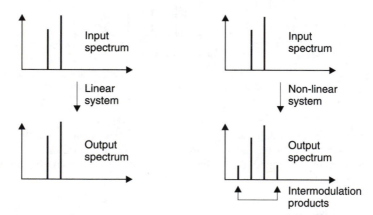

Figure 2.17 (a) A perfectly linear system will pass a number of superimposed waveforms without interference so that the output spectrum does not change. (b) A non-linear system causes inter-modulation where the output spectrum contains sum and difference frequencies in addition to the originals.

f is measured in rotations per second or Hertz (Hz). As a radian is unit distance at unit radius (about 57°) then there are 2π radians in one rotation. Thus the phase angle at a time **t** is given by $\sin\omega t$ or $\sin 2\pi ft$.

A second viewer, who is at right angles to the first viewer, will observe the same waveform but with different timing. The displacement will be given by the

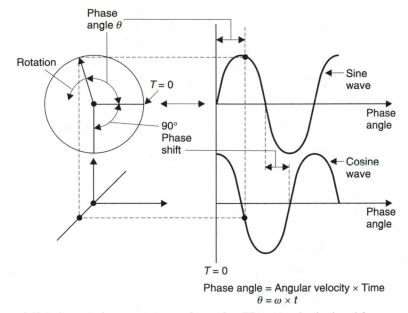

Figure 2.18 A sine wave is one component of a rotation. When a rotation is viewed from two places at right angles, one will see a sine wave and the other will see a cosine wave. The constant *phase shift* between sine and cosine is 90° and should not be confused with the time variant *phase angle* due to the rotation.

radius multiplied by the cosine of the phase angle. When plotted on the same graph, the two waveforms are *phase-shifted* with respect to one another. In this case the phase-shift is 90° and the two waveforms are said to be *in quadrature*. Incidentally the motions on each side of a steam locomotive are in quadrature so that it can always get started (the term used is quartering). Note that the *phase angle* of a signal is constantly changing with time whereas the *phase-shift* between two signals can be constant. It is important that these two are not confused.

The velocity of a moving component is often more important in audio than the displacement. The vertical component of velocity is obtained by differentiating the displacement. As the displacement is a sine wave, the velocity will be a cosine wave whose amplitude is proportional to frequency. In other words the displacement and velocity are in quadrature with the velocity lagging. This is consistent with the velocity reaching a minimum as the displacement reaches a maximum and vice versa. Figure 2.19 shows the displacement, velocity and acceleration waveforms of a body executing SHM. Note that the acceleration and the displacement are always anti-phase.

2.11 Root mean square measurements

Figure 2.20(a) shows that, according to Ohm's law, the power dissipated in a resistance is proportional to the square of the applied voltage. This causes no difficulty with direct current (DC), but with alternating signals such as audio it is harder to calculate the power. Consequently a unit of voltage for alternating signals

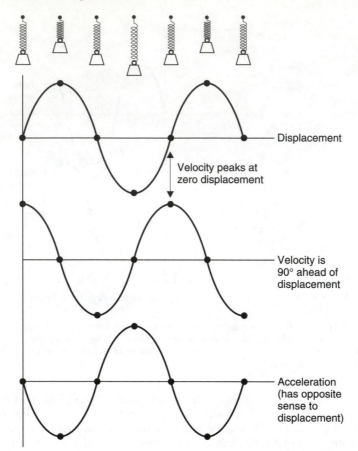

Figure 2.19 The displacement, velocity and acceleration of a body executing simple harmonic motion (SHM).

was devised. Figure 2.20(b) shows that the average power delivered during a cycle must be proportional to the mean of the square of the applied voltage. Since power is proportional to the square of applied voltage, the same power would be dissipated by a DC voltage whose value was equal to the square root of the mean of the square of the AC voltage. Thus the Volt rms (root mean square) was specified. An AC signal of a given number of Volts rms will dissipate exactly the same amount of power in a given resistor as the same number of Volts DC.

Figure 2.21(a) shows that for a sine wave the r.m.s. voltage is obtained by dividing the peak voltage V_{pk} by the square root of two. However, for a square wave (b) the r.m.s. voltage and the peak voltage are the same. Most moving coil AC voltmeters only read correctly on sine waves, whereas many electronic meters incorporate a true r.m.s. calculation.

On an oscilloscope it is often easier to measure the peak-to-peak voltage which is twice the peak voltage. The r.m.s. voltage cannot be measured directly on an oscilloscope since it depends on the waveform although the calculation is simple in the case of a sine wave.

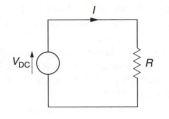

Ohm's law $V = IR$

Power $P = IV = \dfrac{V^2}{R}$ ◄—— Power goes as *square* of voltage

$\therefore V = \sqrt{RP}$

In the case of 600 Ω/1 mW:

$V = \sqrt{600 \times 0.001} = 0.7746$ V

(a)

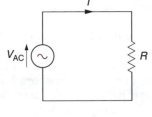

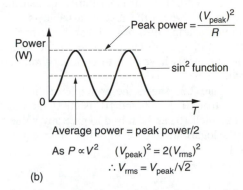

Peak power $= \dfrac{(V_{peak})^2}{R}$

sin² function

0

T

Average power = peak power/2

As $P \propto V^2$ $(V_{peak})^2 = 2(V_{rms})^2$

$\therefore V_{rms} = V_{peak}/\sqrt{2}$

(b)

Figure 2.20 (a) Ohm's law: the power developed in a resistor is proportional to the square of the voltage. Consequently, 1 mW in 600 Ω requires 0.775 V. With a sinusoidal alternating input (b), the power is a sine squared function which can be averaged over one cycle. A DC voltage which would deliver the same power has a value which is the square root of the mean of the square of the sinusoidal input.

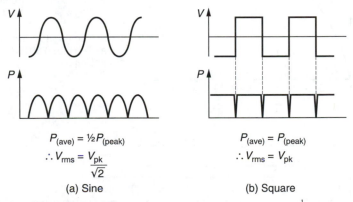

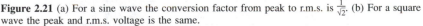

$P_{(ave)} = \frac{1}{2}P_{(peak)}$

$\therefore V_{rms} = \dfrac{V_{pk}}{\sqrt{2}}$

(a) Sine

$P_{(ave)} = P_{(peak)}$

$\therefore V_{rms} = V_{pk}$

(b) Square

Figure 2.21 (a) For a sine wave the conversion factor from peak to r.m.s. is $\frac{1}{\sqrt{2}}$. (b) For a square wave the peak and r.m.s. voltage is the same.

2.12 The deciBel

The first audio signals to be transmitted were on telephone lines. Where the wiring is long compared to the electrical wavelength (not to be confused with the acoustic wavelength) of the signal, a transmission line exists in which the distributed series inductance and the parallel capacitance interact to give the line a characteristic impedance. In telephones this turned out to be about 600Ω. In transmission lines the best power delivery occurs when the source and the load impedance are the same; this is the process of matching.

It was often required to measure the power in a telephone system, and one milliWatt was chosen as a suitable unit. Thus the reference against which signals could be compared was the dissipation of one milliWatt in 600Ω Figure 2.20(a) shows that the dissipation of 1 mW in 600Ω will be due to an applied voltage of 0.775 V r.m.s. This voltage became the reference against which all audio levels are compared.

The deciBel is a logarithmic measuring system and has its origins in telephony[7] where the loss in a cable is a logarithmic function of the length. Human hearing also has a logarithmic response with respect to sound pressure level (SPL). In order to relate to the subjective response audio signal level measurements have also to be logarithmic and so the deciBel was adopted for audio.

Figure 2.22 shows the principle of the logarithm. To give an example, if it is clear that 10^2 is 100 and 10^3 is 1000, then there must be a power between 2 and 3 to which 10 can be raised to give any value between 100 and 1000. That power is the logarithm to base 10 of the value. e.g. $\log_{10} 300 = 2.5$ approx. Note that 10^0 is 1.

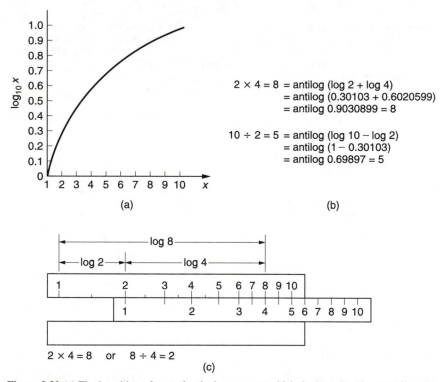

Figure 2.22 (a) The logarithm of a number is the power to which the base (in this case 10) must be raised to obtain the number. (b) Multiplication is obtained by adding logs, division by subtracting. (c) The slide rule has two logarithmic scales whose length can easily be added or subtracted.

Logarithms were developed by mathematicians before the availability of calculators or computers to ease calculations such as multiplication, squaring, division and extracting roots. The advantage is that, armed with a set of log tables, multiplication can be performed by adding and division by subtracting. Figure 2.22 shows some examples. It will be clear that squaring a number is performed by adding two identical logs and the same result will be obtained by multiplying the log by 2.

The slide rule is an early calculator which consists of two logarithmically engraved scales in which the length along the scale is proportional to the log of the engraved number. By sliding the moving scale two lengths can easily be added or subtracted and as a result multiplication and division is readily obtained.

The logarithmic unit of measurement in telephones was called the Bel after Alexander Graham Bell, the inventor. Figure 2.23(a) shows that the Bel was defined as the log of the *power* ratio between the power to be measured and some reference power. Clearly the reference power must have a level of 0 Bels since $\log_{10} 1$ is 0.

The Bel was found to be an excessively large unit for practical purposes and so it was divided into 10 deciBels, abbreviated dB with a small d and a large B and pronounced deebee. Consequently the number of dB is ten times the log of the power ratio. A device such as an amplifier can have a fixed power gain which is independent of signal level and this can be measured in dB. However, when

$$1 \text{ Bel} = \log_{10} \frac{P_1}{P_2} \qquad 1 \text{ deciBel} = 1/10 \text{ Bel}$$

$$\text{Power ratio (dB)} = 10 \times \log_{10} \frac{P_1}{P_2}$$

(a)

As power $\alpha\ V^2$, when using voltages:

$$\text{Power ratio (dB)} = 10 \log \frac{V_1^2}{V_2^2}$$

$$= 10 \times \log \frac{V_1}{V_2} \times 2$$

$$= 20 \log \frac{V_1}{V_2}$$

(b)

Figure 2.23 (a) The Bel is the log of the ratio between two powers, that to be measured and the reference. The Bel is too large so the deciBel is used in practice. (b) As the dB is defined as a power ratio, voltage ratios have to be squared. This is conveniently done by doubling the logs so the ratio is now multiplied by 20.

measuring the power of a signal, it must be appreciated that the dB is a ratio and to quote the number of dBs without stating the reference is about as senseless as describing the height of a mountain as 2000 without specifying whether this is feet or metres. To show that the reference is one milliWatt into 600 Ω the units will be dB(m). In radio engineering, the dB(W) will be found which is power relative to one Watt.

Although the dB(m) is defined as a power ratio, level measurements in audio are often done by measuring the signal voltage using 0.775 V as a reference in a circuit whose impedance is not necessarily 600 Ω. Figure 2.23(b) shows that as the power is proportional to the square of the voltage, the power ratio will be obtained by squaring the voltage ratio. As squaring in logs is performed by doubling, the squared term of the voltages can be replaced by multiplying the log by a factor of two. To give a result in deciBels, the log of the voltage ratio now has to be multiplied by 20.

Whilst 600 Ω matched impedance working is essential for the long distances encountered with telephones, it is quite inappropriate for analog audio wiring in a studio. The wavelength of audio in wires at 20 kHz is 15 km. Studios are built on a smaller scale than this and clearly analog audio cables are *not* transmission lines and their characteristic impedance is not relevant.

In professional analog audio systems impedance matching is not only unnecessary but also undesirable. Figure 2.24(a) shows that when impedance matching is required the output impedance of a signal source must artificially be raised so that a potential divider is formed with the load. The actual drive voltage must be twice that needed on the cable as the potential divider effect wastes 6 dB of signal level and requires unnecessarily high power supply rail voltages in equipment. A further problem is that cable capacitance can cause an undesirable HF roll-off in conjunction with the high source impedance.

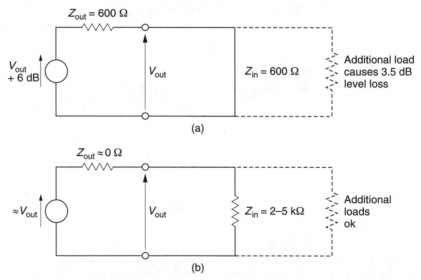

Figure 2.24 (a) Traditional impedance matched source wastes half the signal voltage in the potential divider due to the source impedance and the cable. (b) Modern practice is to use low-output impedance sources with high-impedance loads.

In modern professional analog audio equipment, shown in Figure 2.24(b), the source has the lowest output impedance practicable. This means that any ambient interference is attempting to drive what amounts to a short circuit and can only develop very small voltages. Furthermore shunt capacitance in the cable has very little effect. The destination has a somewhat higher impedance (generally a few kΩ to avoid excessive currents flowing and to allow several loads to be placed across one driver.

In the absence of a fixed impedance it is meaningless to consider power. Consequently only signal voltages are measured. The reference remains at 0.775 V, but power and impedance are irrelevant. Voltages measured in this way are expressed in dB(u); the most common unit of level in modern analog systems. Most installations boost the signals on interface cables by 4 dB. As the gain of receiving devices is reduced by 4 dB, the result is a useful noise advantage without risking distortion due to the drivers having to produce high voltages.

2.13 Audio level metering

There are two main reasons for having level meters in audio equipment: to line up or adjust the gain of equipment, and to assess the amplitude of the program material.

Line-up is often done using a 1 kHz sine wave generated at an agreed level such as 0 dB(u). If a receiving device does not display the same level, then its input sensitivity must be adjusted. Tape recorders and other devices which pass signals through are usually lined up so that their input and output levels are identical, i.e. their insertion loss is 0 dB. Line-up is important in large systems because it ensures that inadvertent level changes do not occur.

In measuring the level of a sine wave for the purposes of line-up, the dynamics of the meter are of no consequence, whereas on program material the dynamics matter a great deal. The simplest (and cheapest) level meter is essentially an AC voltmeter with a logarithmic response. As the ear is logarithmic, the deflection of the meter is roughly proportional to the perceived volume, hence the term volume unit (VU) meter.

In audio, one of the worst sins is to overmodulate a subsequent stage by supplying a signal of excessive amplitude. The next stage may be an analog tape recorder, a radio transmitter or an ADC, none of which respond favourably to such treatment. Real audio signals are rich in short transients which pass before the sluggish VU meter responds. Consequently the VU meter is also called the virtually useless meter in professional circles.

Broadcasters developed the peak program meter (PPM) which is also logarithmic, but which is designed to respond to peaks as quickly as the ear responds to distortion. Consequently the attack time of the PPM is carefully specified. If a peak is so short that the PPM fails to indicate its true level, the resulting overload will also be so brief that the ear will not hear it. A further feature of the PPM is that the decay time of the meter is very slow, so that any peaks are visible for much longer and the meter is easier to read because the meter movement is less violent. The original PPM as developed by the BBC was sparsely calibrated, but other users have adopted the same dynamics and added dB scales.

In broadcasting, the use of level metering and line-up procedures ensures that the level experienced by the listener does not change significantly from program to program. Consequently in a transmission suite, the goal would be to broadcast recordings at a level identical to that which was determined during production. However, when making a recording prior to any production process, the goal would be to modulate the recording as fully as possible without clipping as this would then give the best signal-to-noise ratio. The level could then be reduced if necessary in the production process.

References

1. Moore, B.C.J., *An Introduction to the Psychology of Hearing*, London: Academic Press (1989)
2. Muraoka, T., Iwahara, M. and Yamada, Y., Examination of audio bandwidth requirements for optimum sound signal transmission. *J. Audio Eng. Soc.*, **29**, 2–9 (1982)
3. Muraoka, T., Yamada, Y. and Yamazaki, M., Sampling frequency considerations in digital audio. *J. Audio Eng. Soc.*, **26**, 252–256 (1978)
4. Fincham, L.R., The subjective importance of uniform group delay at low frequencies. Presented at the 74th Audio Engineering Society Convention (New York, 1983), Preprint 2056(H-1)
5. Fletcher, H., Auditory patterns. *Rev. Modern Physics*, **12**, 47–65 (1940)
6. Carterette, E.C. and Friedman, M.P., *Handbook of Perception*, 305–319. New York: Academic Press (1978)
7. Martin, W.H., Decibel – the new name for the transmission unit. *Bell System Tech. J.* (January 1929)

3

Digital principles

3.1 Binary codes

In digital audio, binary numbers express the values of the samples which represent the original analog waveform. Digital audio recording consists of storing such numbers on a suitable medium, where the goal is to reproduce the numbers unchanged. However, if it is required to manipulate the audio waveform, this can be done in the digital domain by changing the sample values. To see how this can be done requires some knowledge of binary codes.

Figure 3.1 shows some binary numbers and their equivalent in decimal. The radix point has the same significance in binary: symbols to the right of it represent one half, one quarter and so on. Binary numbers easily become very long, and writing them by hand is tedious and error-prone. The octal and hexadecimal notations are both used for writing binary since conversion is so simple. Figure 3.1 also shows that a binary number is split into groups of three or four digits starting at the least significant end, and the groups are individually converted to octal or hexadecimal digits. Since sixteen different symbols are required in hex, the letters A–F are used for the numbers above nine.

The fixed number of bits in a PCM sample determines the extent of the quantizing range. In the sixteen-bit samples commonly used, there are 65 536 different numbers, each representing a different analog signal voltage. Care must be taken during conversion to ensure that the signal does not go outside the convertor range, or it will be clipped. In Figure 3.2 it will be seen that in a sixteen-bit pure binary system, the number range goes from 0000 hex, which represents the smallest voltage, through to FFFF hex, which represents the largest positive voltage. Effectively the zero voltage level of the analog waveform has been shifted so that the positive and negative voltages in a real audio signal may be expressed by binary numbers which are only positive. This approach is called offset binary and unfortunately it is unsuitable for audio signal processing in the digital domain.

Figure 3.3 shows that the level of the signal is measured by how far the waveform deviates from mid-range, around which attenuation, gain and mixing all take place. Digital audio mixing is achieved by adding sample values from two or more different sources, but the correct result will only be obtained if the quantizing intervals are of the same size and there are no offsets. In other cases the binary numbers are not proportional to the signal voltage.

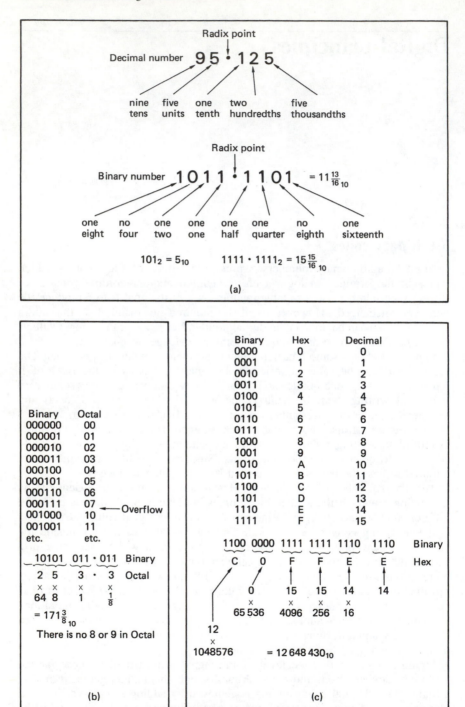

Figure 3.1 (a) Binary and decimal. (b) In octal, groups of three bits make one symbol 0–7. (c) In hex, groups of four bits make one symbol 0–F. Note how much shorter the number is in hex.

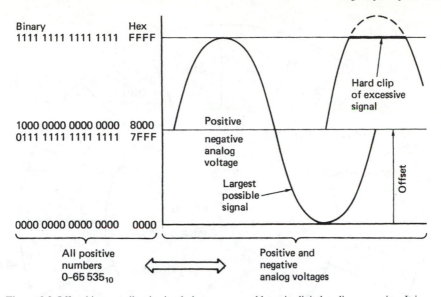

Figure 3.2 Offset binary coding is simple but causes problems in digital audio processing. It is seldom used.

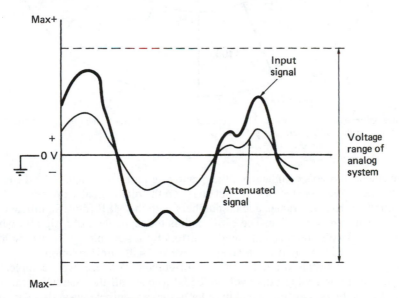

Figure 3.3 Attenuation of an audio signal takes place with respect to midrange.

In the two's complement system, the upper half of the pure binary number range has been redefined to represent negative quantities. If a pure binary counter is constantly incremented and allowed to overflow, it will produce all the numbers in the range permitted by the number of available bits, and these are shown for a four-bit example drawn around the circle in Figure 3.4. As a circle has no real beginning, it is possible to consider it to start wherever it is

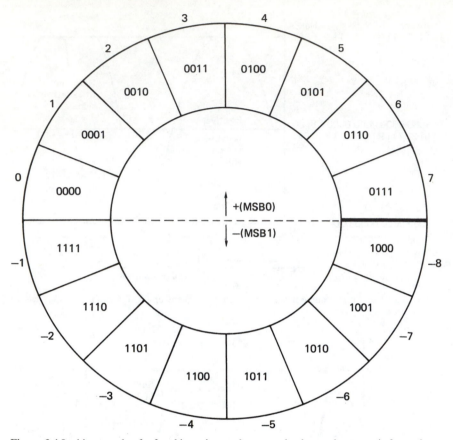

Figure 3.4 In this example of a four-bit two's complement code, the number range is from −8 to +7. Note that the MSB determines polarity.

convenient. In two's complement, the quantizing range represented by the circle of numbers does not start at zero, but starts on the diametrically opposite side of the circle. Zero is midrange, and all numbers with the MSB (most significant bit) set are considered negative. The MSB is thus the equivalent of a sign bit where 1 = minus. Two's complement notation differs from pure binary in that the most significant bit is inverted in order to achieve the half-circle rotation.

Figure 3.5 shows how a real ADC is configured to produce two's complement output. At (a) an analog offset voltage equal to one half the quantizing range is added to the bipolar analog signal in order to make it unipolar as at (b). The ADC produces positive only numbers at (c) which are proportional to the input voltage. The MSB is then inverted at (d) so that the all-zeros code moves to the centre of the quantizing range. The analog offset is often incorporated into the ADC as is the MSB inversion.

The two's complement system allows two sample values to be added, or mixed in audio parlance, and the result will be referred to the system midrange; this is analogous to adding analog signals in an operational amplifier. Figure 3.6 illustrates how adding two's complement samples simulates a bipolar mixing

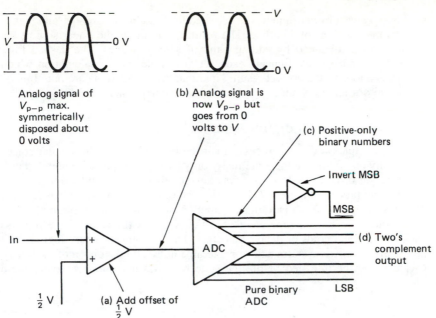

Figure 3.5 A two's complement ADC. At (a) an analog offset voltage equal to one-half the quantizing range is added to the bipolar analog signal in order to make it unipolar as at (b). The ADC produces positive-only numbers at (c), but the MSB is then inverted at (d) to give a two's complement output.

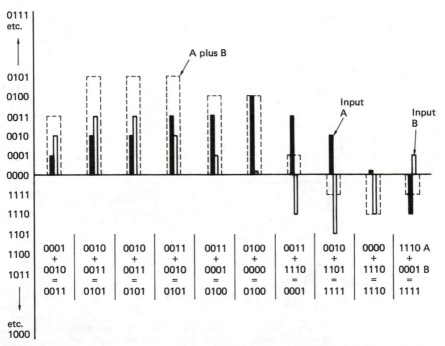

Figure 3.6 Using two's complement arithmetic, single values from two waveforms are added together with respect to midrange to give a correct mixing function.

process. The waveform of input A is depicted by solid black samples, and that of B by samples with a solid outline. The result of mixing is the linear sum of the two waveforms obtained by adding pairs of sample values. The dashed lines depict the output values. Beneath each set of samples is the calculation which will be seen to give the correct result. Note that the calculations are pure binary. No special arithmetic is needed to handle two's complement numbers.

3.2 Introduction to digital logic

However complex a digital process, it can be broken down into smaller stages until finally one finds that there are really only two basic types of element in use, and these can be combined in some way and supplied with a clock to implement virtually any process. Figure 3.7 shows that the first type is a *logic* element. This produces an output which is a logical function of the input with minimal delay. The second type is a *storage* element which samples the state of the input(s) when clocked and holds or delays that state. The strength of binary logic is that the signal has only two states, and considerable noise and distortion of the binary waveform can be tolerated before the state becomes uncertain. At every logic

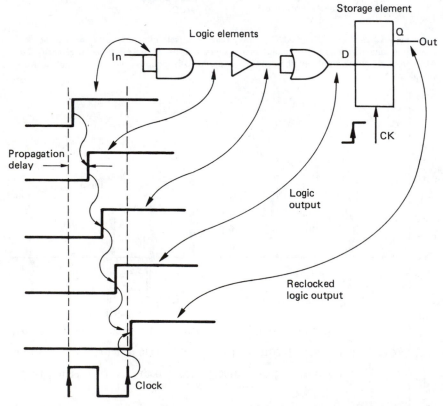

Figure 3.7 Logic elements have a finite propagation delay between input and output and cascading them delays the signal an arbitrary amount. Storage elements sample the input on a clock edge and can return a signal to near coincidence with the system clock. This is known as reclocking. Reclocking eliminates variations in propagation delay in logic elements.

element, the signal is compared with a threshold, and can thus can pass through any number of stages without being degraded.

In addition, the use of a storage element at regular locations throughout logic circuits eliminates time variations or jitter. Figure 3.7 also shows that if the inputs to a logic element change, the output will not change until the *propagation delay* of the element has elapsed. However, if the output of the logic element forms the input to a storage element, the output of that element will not change until the input is sampled *at the next clock edge*. In this way the signal edge is aligned to the system clock and the propagation delay of the logic becomes irrelevant. The process is known as reclocking.

The two states of the signal when measured with an oscilloscope are simply two voltages, usually referred to as high and low. As there are only two states, there can only be *true* or *false* meanings. The true state of the signal can be assigned by the designer to either voltage state. When a high voltage represents a true logic condition and a low voltage represents a false condition, the system is known as *positive logic*, or *high true logic*. This is the usual system, but sometimes the low voltage represents the true condition and the high voltage represents the false condition. This is known as *negative logic* or *low true logic*. Provided that everyone is aware of the logic convention in use, both work equally well.

In logic systems, all logical functions, however complex, can be configured from combinations of a few fundamental logic elements or *gates*. It is not profitable to spend too much time debating which are the truly fundamental ones, since most can be made from combinations of others. Figure 3.8 shows the important simple gates and their derivatives, and introduces the logical expressions to describe them, which can be compared with the truth-table notation. The figure also shows the important fact that when negative logic is used, the OR gate function interchanges with that of the AND gate.

If numerical quantities need to be conveyed down the two-state signal paths described here, then the only appropriate numbering system is binary, which has only two symbols, 0 and 1. Just as positive or negative logic could be used for the truth of a logical binary signal, it can also be used for a numerical binary signal. Normally, a high voltage level will represent a binary 1 and a low voltage will represent a binary 0, described as a 'high for a one' system. Clearly a 'low for a one' system is just as feasible. Decimal numbers have several columns, each of which represents a different power of ten; in binary the column position specifies the power of two.

Several binary digits or bits are needed to express the value of a binary audio sample. These bits can be conveyed at the same time by several signals to form a parallel system, which is most convenient inside equipment or for short distances because it is inexpensive, or one at a time down a single signal path, which is more complex, but convenient for cables between pieces of equipment because the connectors require fewer pins. When a binary system is used to convey numbers in this way, it can be called a digital system.

The basic memory element in logic circuits is the latch, which is constructed from two gates as shown in Figure 3.9(a), and which can be set or reset. A more useful variant is the D-type latch shown at (b) which remembers the state of the input at the time a separate clock either changes state for an edge-triggered device, or after it goes false for a level-triggered device. A shift register can be made from a series of latches by connecting the Q output of one latch to the D

Positive logic name	Boolean expression	Positive logic symbol	Positive logic truth table	Plain English
Inverter or NOT gate	$Q = \overline{A}$	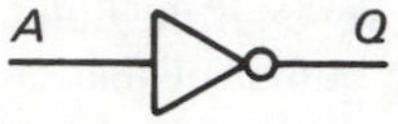	A | Q 0 | 1 1 | 0	Output is opposite of input
AND gate	$Q = A \cdot B$	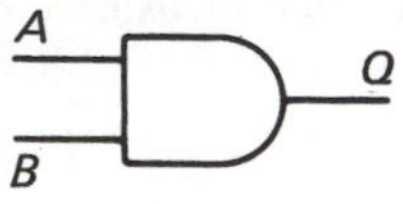	A B | Q 0 0 | 0 0 1 | 0 1 0 | 0 1 1 | 1	Output true when both inputs are true only
NAND (Not AND) gate	$Q = \overline{A \cdot B}$ $= \overline{A} + \overline{B}$	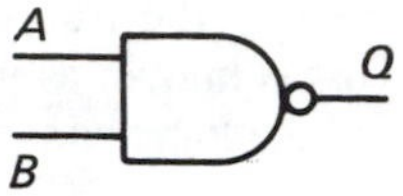	A B | Q 0 0 | 1 0 1 | 1 1 0 | 1 1 1 | 0	Output false when both inputs are true only
OR gate	$Q = A + B$	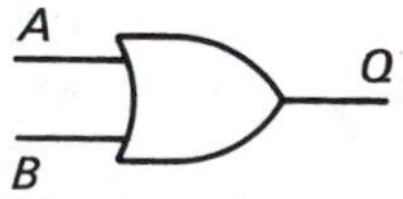	A B | Q 0 0 | 0 0 1 | 1 1 0 | 1 1 1 | 1	Output true if either or both inputs true
NOR (Not OR) gate	$Q = \overline{A + B}$ $= \overline{A} \cdot \overline{B}$	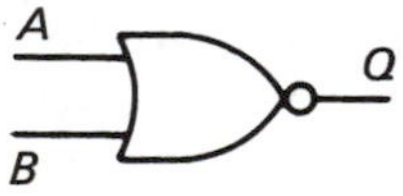	A B | Q 0 0 | 1 0 1 | 0 1 0 | 0 1 1 | 0	Output false if either or both inputs true
Exclusive OR (XOR) gate	$Q = A \oplus B$	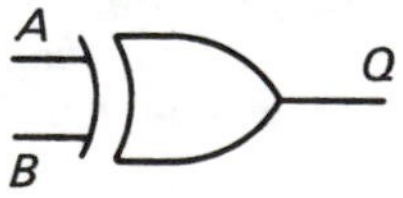	A B | Q 0 0 | 0 0 1 | 1 1 0 | 1 1 1 | 0	Output true if inputs are different

Figure 3.8 The basic logic gates compared.

input of the next and connecting all the clock inputs in parallel. Data are delayed by the number of stages in the register. Shift registers are also useful for converting between serial and parallel data transmissions. Where large numbers of bits are to be stored, cross-coupled latches are less suitable because they are more complicated to fabricate inside integrated circuits than dynamic memory, and consume more current.

In large random access memories (RAMs), the data bits are stored as the presence or absence of charge in a tiny capacitor as shown in Figure 3.9(c). The capacitor is formed by a metal electrode, insulated by a layer of silicon dioxide

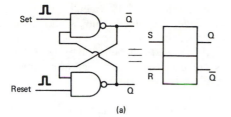

(a)

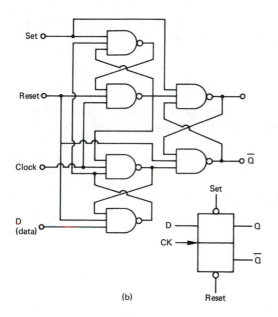

(b)

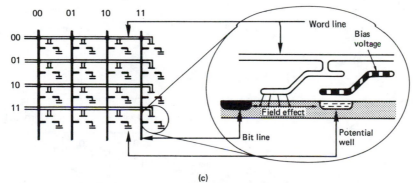

(c)

Figure 3.9 Digital semiconductor memory types. In (a), one data bit can be stored in a simple set-reset latch, which has little application because the D-type latch in (b) can store the state of the single data input when the clock occurs. These devices can be implemented with bipolar transistors of FETs, and are called static memories because they can store indefinitely. They consume a lot of power.

In (c), a bit is stored as the charge in a potential well in the substrate of a chip. It is accessed by connecting the bit line with the field effect from the word line. The single well where the two lines cross can then be written or read. These devices are called dynamic RAMs because the charge decays, and they must be read and rewritten (refreshed) periodically.

from a semiconductor substrate, hence the term MOS (metal oxide semi-conductor). The charge will suffer leakage, and the value would become indeterminate after a few milliseconds. Where the delay needed is less than this, decay is of no consequence, as data will be read out before they have had a chance to decay. Where longer delays are necessary, such memories must be refreshed periodically by reading the bit value and writing it back to the same place. Most modern MOS RAM chips have suitable circuitry built-in. In large RAMs it is clearly impractical to have a connection to each bit. Instead, the desired bit has to be addressed before it can be read or written. The size of the chip package restricts the number of pins available, so that large memories may use the same address pins more than once. The bits are arranged internally as rows and columns, and the row address and the column address are input sequentially on the same pins.

The circuitry necessary for adding pure binary or two's complement numbers is shown in Figure 3.10. Addition in binary requires two bits to be taken at a time from the same position in each word, starting at the least significant bit. Should both be ones, the output is zero, and there is a *carry-out* generated. Such a circuit is called a half adder, shown in Figure 3.10(a) and is suitable for the least significant bit of the calculation. All higher stages will require a circuit which can accept a carry input as well as two data inputs. This is known as a full adder (Figure 3.10(b)). When mixing by adding sample values, care has to be taken to ensure that if the sum of the two sample values exceeds the number range the result will be clipping rather than wraparound.

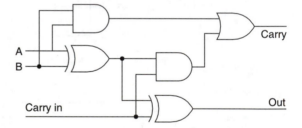
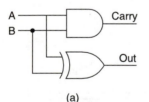

(a)

Data A	Bits B	Carry in	Out	Carry out
0	0	0	0	0
0	0	1	1	0
0	1	0	1	0
0	1	1	0	1
1	0	0	1	0
1	0	1	0	1
1	1	0	0	1
1	1	1	1	1

(b)

Figure 3.10 (a) Half adder; (b) full-adder circuit and truth table.

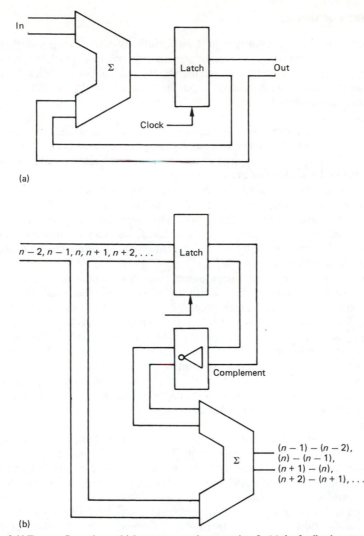

Figure 3.11 Two configurations which are common in processing. In (a) the feedback around the adder adds the previous sum to each input to perform accumulation or digital integration. In (b) an invertor allows the difference between successive inputs to be computed. This is differentiation.

A storage element can be combined with an adder to obtain a number of useful functional blocks which will frequently be found in audio equipment. Figure 3.11(a) shows that a latch is connected in a feedback loop around an adder. The latch contents are added to the input each time it is clocked. The configuration is known as an accumulator in computation because it adds up or accumulates values fed into it. In filtering, it is known as an discrete time integrator. If the input is held at some constant value, the output increases by that amount on each clock. The output is thus a sampled ramp. Figure 3.11(b) shows that the addition of an invertor allows the difference between successive inputs to be obtained. This is digital differentiation. The output is proportional to the slope of the input.

3.3 The computer

The computer is now a vital part of digital audio systems, being used both for control purposes and to process audio signals as data. In control, the computer finds applications in database management, automation, editing, and in electromechanical systems such as tape drives and robotic cassette handling. Once processing speeds advanced sufficiently, computers became able to manipulate digital audio in real time.

The computer is a programmable device in that its operation is not determined by its construction alone, but instead by a series of *instructions* forming a *program*. The program is supplied to the computer one instruction at a time so that the desired sequence of events takes place.

Programming of this kind has been used for over a century in electro-mechanical devices, including automated knitting machines and street organs which are programmed by punched cards. However, the computer differs from these devices in that the program is not fixed, but can be modified by the computer itself. This possibility led to the creation of the term *software* to suggest a contrast to the constancy of hardware.

Computer instructions are binary numbers each of which is interpreted in a specific way. As these instructions don't differ from any other kind of data, they can be stored in RAM. The computer can change its own instructions by accessing the RAM. Most types of RAM are volatile, in that they lose data when power is removed. Clearly if a program is entirely stored in this way, the computer will not be able to recover fom a power failure. The solution is that a very simple starting or *bootstrap* program is stored in non-volatile ROM which will contain instructions that will bring in the main program from a storage system such as a disk drive after power is applied. As programs in ROM cannot be altered, they are sometimes referred to as *firmware* to indicate that they are classified between hardware and software.

Making a computer do useful work requires more than simply a program which performs the required computation. There is also a lot of mundane activity which does not differ significantly from one program to the next. This includes deciding which part of the RAM will be occupied by the program and which by the data, producing commands to the storage disk drive to read the input data from a file and write back the results. It would be very inefficient if all programs had to handle these processes themselves. Consequently the concept of an *operating system* was developed. This manages all the mundane decisions and creates an environment in which useful programs or *applications* can execute.

The ability of the computer to change its own instructions makes it very powerful, but it also makes it vulnerable to abuse. Programs exist which are deliberately written to do damage. These *viruses* are generally attached to plausible messages or data files and enter computers through storage media or communications paths.

There is also the possibility that programs contain logical errors such that in certain combinations of circumstances the wrong result is obtained. If this results in the unwitting modification of an instruction, the next time that instruction is accessed the computer will crash. In consumer-grade software, written for the vast personal computer market, this kind of thing is unfortunately accepted.

For critical applications, software must be *verified*. This is a process which can prove that a program can recover from absolutely every combination of

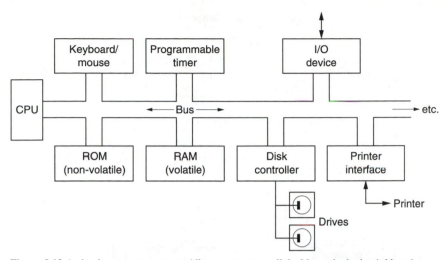

Figure 3.12 A simple computer system. All components are linked by a single data/address/
control bus. Although cheap and flexible, such a bus can only make one connection at a time, so
it is slow.

circumstances and keep running properly. This is a non-trivial process, because
the number of combinations of states a computer can get into is staggering. As
a result most software is unverified. It is of the utmost importance that networked
computers which can suffer virus infection or computers running unverified
software are never used in a life-support or critical application.

Figure 3.12 shows a simple computer system. The various parts are linked by
a bus which allows binary numbers to be transferred from one place to another.
This will generally use tri-state logic so that when one device is sending to
another, all other devices present a high impedance to the bus.

The ROM stores the startup program, the RAM stores the operating system,
applications programs and the data to be processed. The disk drive stores large
quantities of data in a non-volatile form. The RAM only needs to be able to hold
part of one program as other parts can be brought from the disk as required. A
program executes by *fetching* one instruction at a time from the RAM to the
processor along the bus.

The bus also allows keyboard/mouse inputs and outputs to the display and
printer. Inputs and outputs are generally abbreviated to I/O. Finally a
programmable timer will be present which acts as a kind of alarm clock for the
processor.

The processor or CPU (central processing unit) is the heart of the system.
Figure 3.13 shows the data path of a simple CPU. The CPU has a bus interface
which allows it to generate bus addresses and input or output data. Sequential
instructions are stored in RAM at contiguously increasing locations so that a
program can be executed by fetching instructions from a RAM address specified
by the program counter (PC) to the instruction register in the CPU. As each
instruction is completed, the PC is incremented so that it points to the next
instruction. In this way the time taken to execute the instruction can vary.

The processor is notionally divided into data paths and control paths. The CPU
contains a number of general-purpose registers or scratchpads which can be used

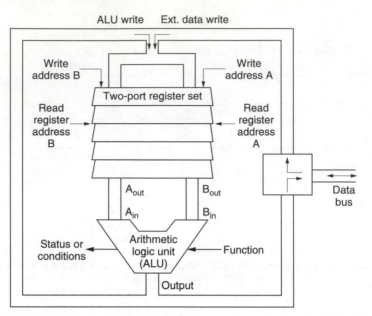

Figure 3.13 The data path of a simple CPU. Under control of an instruction, the ALU will perform some function on a pair of input values from the registers and store or output the result.

to store partial results in complex calculations. Pairs of these registers can be addressed so that their contents go to the ALU (arithmetic logic unit). This performs various arithmetic (add, subtract, etc.) or logical (and, or, etc.) functions on the input data. The output of the ALU may be routed back to a register or output. By reversing this process it is possible to get data into the registers from the RAM. The ALU also outputs the conditions resulting from the calculation, which can control conditional instructions. Which function the ALU performs and which registers are involved are determined by the instruction currently in the instruction register then is decoded in the control path. One pass through the ALU can be completed in one cycle of the processor's clock. Instructions vary in complexity as do the number of clock cycles needed to complete them. Incoming instructions are decoded and used to access a look-up table which converts them into *microinstructions*, one of which controls the CPU at each clock cycle.

3.4 Timebase correction

In Chapter 1 it was stated that a strength of digital technology is the ease with which delay can be provided. Accurate control of delay is the essence of timebase correction, necessary whenever the instantaneous time of arrival or rate from a data source does not match the destination. In digital audio, the destination will almost always require perfectly regular timing, namely the sampling rate clock of the final DAC. Timebase correction consists of aligning jittery signals from storage media or transmission channels with that stable reference.

A further function of timebase correction is to reverse the time compression applied prior to recording or transmission. As was shown in section 1.6, digital recorders compress data into blocks to facilitate editing and error correction as well as to permit head switching between blocks in rotary-head machines. Owing to the spaces between blocks, data arrive in bursts on replay, but must be fed to the output convertors in an unbroken stream at the sampling rate.

In computer hard-disk drives, which are used in digital audio workstations, time compression is also used, but a converse problem also arises. Data from the disk blocks arrive at a reasonably constant rate, but cannot necessarily be accepted at a steady rate by the logic because of contention for the use of buses and memory by the different parts of the system. In this case the data must be buffered by a relative of the timebase corrector which is usually referred to as a silo.

Although delay is easily implemented, it is not possible to advance a data stream. Most real machines cause instabilities balanced about the correct timing: the output jitters between too early and too late. Since the information cannot be advanced in the corrector, only delayed, the solution is to run the machine in advance of real time. In this case, correctly timed output signals will need a nominal delay to align them with reference timing. Early output signals will receive more delay, and late output signals will receive less delay.

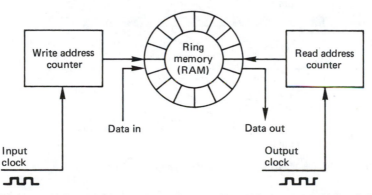

Data in   Data out

Input   Output
clock   clock

Figure 3.14 Most TBCs are implemented as a memory addressed by a counter which periodically overflows to give a ring structure. The memory allows the read and write sides to be asynchronous.

Section 2.2 showed the principles of digital storage elements which can be used for delay purposes. The shift-register approach and the RAM approach to delay are very similar, as a shift register can be thought of as a memory whose address increases automatically when clocked. The data rate and the maximum delay determine the capacity of the RAM required. Figure 3.14 shows that the addressing of the RAM is by a counter that overflows endlessly from the end of the memory back to the beginning, giving the memory a ring-like structure. The write address is determined by the incoming data, and the read address is determined by the outgoing data.

In hard disk systems, the data transfers to and from the disk itself must be at a rate determined by the rotation of the disk. If the data cannot be supplied or accepted at this rate, data will be lost. The solution is to use a relative of the timebase corrector, known as a silo, which is a kind of memory that can provide or accept data from the disk as required, and buffer the timing of that data from the timing of the rest of the system. With the use of a silo, a disk write would not be affected if the computer briefly suspended bus data flow to service an interrupt as the data would come from the silo.

3.5 Multiplexing

Multiplexing is used where several signals are to be transmitted down the same channel. The channel bit rate must be the same as or greater than the sum of the source bit rates. Figure 3.15 shows that when multiplexing is used, the data from each source have to be time compressed. This is done by buffering source data in a memory at the multiplexer. It is written into the memory in real time as it arrives, but will be read from the memory with a clock which has a much higher rate. This means that the readout occurs in a smaller timespan. If, for example, the clock frequency is raised by a factor of ten, the data for a given signal will be transmitted in a tenth of the normal time, leaving time in the multiplex for nine more such signals.

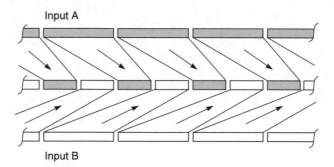

Figure 3.15 Multiplexing requires time compression on each input.

In the demultiplexer another buffer memory will be required. Only the data for the selected signal will be written into this memory at the bit rate of the multiplex. When the memory is read at the correct speed, the data will emerge with their original timebase.

In practice it is essential to have mechanisms to identify the separate signals to prevent them being mixed up and to convey the original signal clock frequency to the demultiplexer. In time-division multiplexing the timebase of the transmission is broken into equal slots, one for each signal. This makes it easy for the demultiplexer, but forces a rigid structure on all the signals such that they must all be locked to one another and have an unchanging bit rate. Packet multiplexing overcomes these limitations.

The multiplexer must switch between different time-compressed signals to create the bitstream and this is much easier to organize if each signal is in the form of data packets of constant size. Figure 3.16 shows a packet multiplexing system.

Each packet consists of two components: the header, which identifies the packet, and the payload, which is the data to be transmitted. The header will

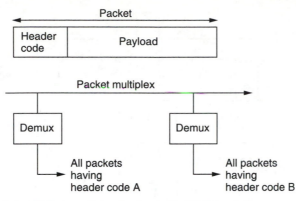

Figure 3.16 Packet multiplexing relles on headers to identify the packets.

contain at least an identification code (ID) which is unique for each signal in the multiplex. The demultiplexer checks the ID codes of all incoming packets and discards those which do not have the wanted ID.

Packet multiplexing has advantages over time-division multiplexing because it does not set the bit rate of each signal. A demultiplexer simply checks packet IDs and selects all packets with the wanted code. It will do this however frequently such packets arrive. Consequently it is practicable to have variable bit rate signals in a packet multiplex. The multiplexer has to ensure that the total bit rate does not exceed the rate of the channel, but that rate can be allocated arbitrarily between the various signals.

As a practical matter is is usually necessary to keep the bit rate of the multiplex constant. With variable rate inputs this is done by creating null packets which are generally called *stuffing* or *packing*. The headers of these packets contain an unique ID which the demultiplexer does not recognize and so these packets are discarded on arrival.

In an MPEG environment, statistical multiplexing can be extremely useful because it allows for the varying difficulty of real program material. In a multiplex of several digital radio stations channels, it is unlikely that all the programs will encounter difficult material simultaneously. When one program encounters complex program material, more data rate can be allocated at the allowable expense of the remaining programs which are handling easy material.

3.6 Gain control

When making a digital recording, the gain of the analog input will usually be adjusted so that the quantizing range is fully exercised in order to make a recording of maximum signal-to-noise ratio. During post-production, the recording may be played back and mixed with other signals, and the desired effect can only be achieved if the level of each can be controlled independently. Gain is controlled in the digital domain by multiplying each sample value by a coefficient. If that coefficient is less than one, attenuation will result; if it is greater than one, amplification can be obtained, provided that the resultant larger sample values can be handled without clipping.

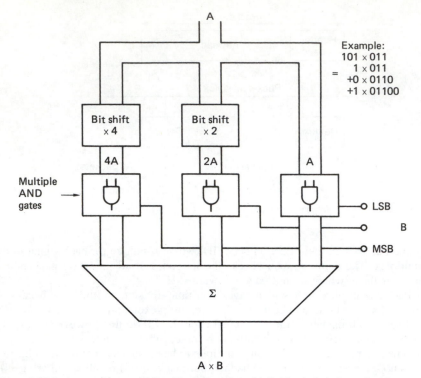

Figure 3.17 Structure of fast multiplier. The input A is multiplied by 1, 2, 4, 8, etc. by bit shifting. The digits of the B input then determine which multiples of A should be added together by enabling AND gates between the shifters and the adder. For long wordlengths, the number of gates required becomes enormous, and the device is best implemented in a chip.

Multiplication in binary circuits is difficult. It can be performed by repeated adding, but this is too slow to be of any use. In fast multiplication, one of the inputs will be simultaneously multiplied by one, two, four, etc., by hard-wired bit shifting. Figure 3.17 shows that the other input bits will determine which of these powers will be added to produce the final sum, and which will be neglected. If multiplying by five, the process is the same as multiplying by four, multiplying by one, and adding the two products. This is achieved by adding the input to itself shifted two places. As the wordlength of such a device increases, the complexity increases exponentially, so this is a natural application for an integrated circuit.

3.7 Digital faders and controls

In a digital mixer, the gain coefficients will originate in hand-operated faders, just as in analog. Analog mixers having automated mixdown employ a system similar to the one shown in Figure 3.18. Here, the faders produce a varying voltage and this is converted to a digital code or gain coefficient in an ADC and recorded alongside the audio tracks. On replay the coefficients are converted back to analog voltages which control VCAs (voltage-controlled amplifiers) in series with the analog audio channels. A digital mixer has a similar structure, and the

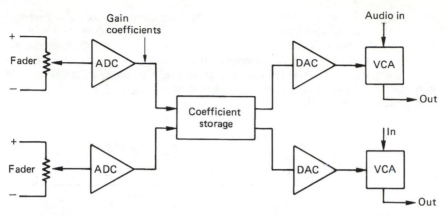

Figure 3.18 The automated mixdown system of an audio console digitizes fader positions for storage and uses the coefficients later to drive VCAs via convertors.

coefficients can be obtained in the same way. However, on replay, the coefficients are not converted back to analog, but remain in the digital domain and control multipliers in the digital audio channels directly. As the coefficients are digital, it is so easy to add automation to a digital mixer that there is not much point in building one without.

Gain coefficients can be obtained by digitizing the output of an analog fader, or directly in a digital fader. This is a form of displacement transducer in which the mechanical position of the knob is converted directly to a digital code. In practical equipment, the position of other controls, such as for equalizers or scrub wheels, will also need to be digitized. Controls can be linear or rotary, and absolute or relative. In an absolute control, the position of the knob determines the output directly. These are inconvenient in automated systems because unless the knob is motorized, the operator cannot see the setting the automation system

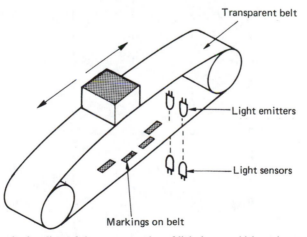

Figure 3.19 An absolute linear fader uses a number of light beams which are interrupted in various combinations according to the position of a grating. A Gray code shown in Figure 3.20 must be used to prevent false codes.

has selected. In a relative control, the knob can be moved to increase or decrease the output, but its absolute position is meaningless. The absolute setting is displayed on a bar LED nearby. In a rotary control, the bar LED may take the form of a ring of LEDs around the control. The automation system setting can be seen on the display and no motor is needed. In a relative linear fader, the control may take the form of an endless ridged belt like a caterpillar track. If this is transparent, the bar LED may be seen through it.

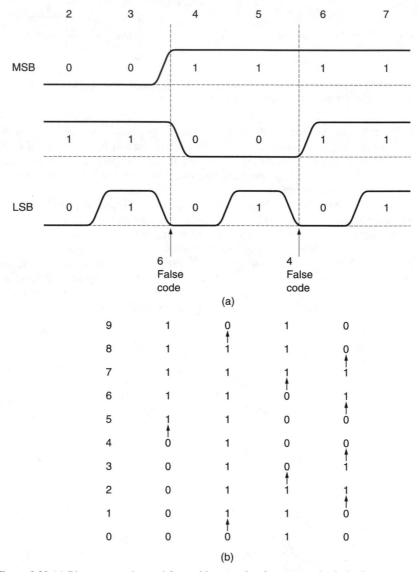

Figure 3.20 (a) Binary cannot be used for position encoders because mechanical tolerances cause false codes to be produced. (b) In Gray code, only one bit (arrowed) changes in between positions, so no false codes can be generated.

Figure 3.19 shows an absolute linear fader. A grating is moved with respect to several light beams, one for each bit of the coefficient required. The interruption of the beams by the grating determines which photocells are illuminated. It is not possible to use a pure binary pattern on the grating because this results in transient false codes due to mechanical tolerances. Figure 3.20 shows some examples of these false codes. For example, on moving the fader from 3 to 4, the MSB goes true slightly before the middle bit goes false. This results in a

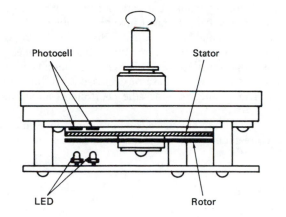

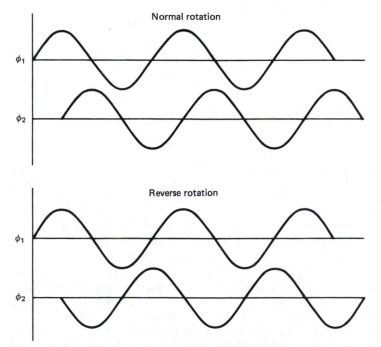

Figure 3.21 The fixed and rotating gratings produce moiré fringes which are detected by two light paths as quadrature sinusoids. The relative phase determines the direction, and the frequency is proportional to speed of rotation.

momentary value of $4 + 2 = 6$ between 3 and 4. The solution is to use a code in which only one bit ever changes in going from one value to the next. One such code is the Gray code, which was devised to overcome timing hazards in relay logic but is now used extensively in position encoders. Gray code can be converted to binary in a ROM, a gate array or by software.

Figure 3.21 shows a rotary incremental encoder. This produces a sequence of pulses whose number is proportional to the angle through which it has been turned. The rotor carries a radial grating over its entire perimeter. This turns over a second fixed radial grating whose bars are not parallel to those of the first grating. The resultant moiré fringes travel inward or outward depending on the direction of rotation. Two suitably positioned light beams falling on photocells will produce outputs in quadrature. The relative phase determines the direction and the frequency is proportional to speed. The encoder outputs can be connected to a counter whose contents will increase or decrease according to the direction the rotor is turned. The counter provides the coefficient output and drives the display.

For audio use, a logarithmic characteristic is required in gain control. Linear coefficients can conveniently be rendered logarithmic in a ROM or by software.

3.8 A digital mixer

The signal path of a simple digital mixer is shown in Figure 3.22. The two inputs are multiplied by their respective coefficients, and added together in two's complement to achieve the mix as was shown in Figure 3.6. Peak limiting will be required. The sampling rate of the inputs must be exactly the same, and in the same phase, or the circuit will not be able to add on a sample-by-sample basis. If the inputs have come from different sources, those sources must be synchronized by the same master clock, and/or timebase correction must be provided on the inputs. Synchronization of audio sources follows the principle long established in video in which a reference signal is fed to all devices which then slave or *genlock* to it.

Some thought must be given to the wordlength of the system. If a sample is attenuated, it will develop bits which are below the radix point. For example, if an eight-bit sample is attenuated by 24 dB, the sample value will be shifted four places down. Extra bits must be available within the mixer to accommodate this shift. Digital mixers can have an internal wordlength of up to 32 bits. When several attenuated sources are added together to produce the final mix, the result will be a stream of 32-bit or longer samples. As the output will generally need to be of the same format as the input, the wordlength must be shortened. This must be done very carefully, as it is a form of quantizing and will require dithering. The necessary techniques will be treated in Chapter 4.

In practice a digital mixer would not have one multiplier for every input. Figure 3.22 also shows that a more economical system results when a time-shared bus system is used with only one multiplier followed by an accumulator. In one sample period, each of the input samples is fed in turn to the lower input of the multiplier at the same time as the corresponding coefficient is fed to the upper input. The products from the multiplier are accumulated during the sample period, so that at the end of the sample period, the accumulator holds the sum of all the products, which is the digitally mixed sample. The process then repeats for

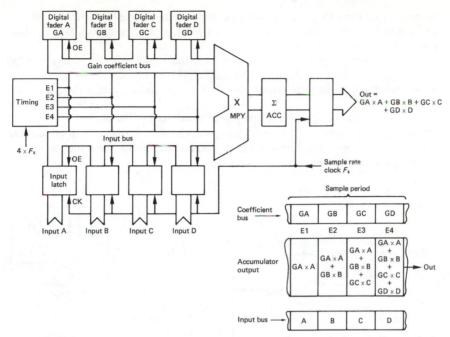

Figure 3.22 One multiplier/accumulator can be time shared between several signals by operating at a multiple of sampling rate. In this example, four multiplications are performed during one sample period.

the next sample period. To facilitate the sharing of common circuits by many signals, tri-state logic devices can be used. The outputs of such devices can be wired in parallel, and the state of the parallel connection will be the state of the device whose output is enabled. Clearly only one output can be enabled at a time, and this will be ensured by a sequencer circuit connected to all the device enables. In digital signal processing (DSP), the processes shown above can be simulated in software.

In analog audio mixers, the controls have to be positioned close to the circuitry for performance reasons; thus one control knob is needed for every variable, and the control panel is physically large. Remote control is difficult with such construction. The order in which the signal passes through the various stages of the mixer is determined at the time of design, and any changes are difficult.

In a digital mixer,[1,2] all the filters are controlled by simply changing the coefficients, and remote control is easy. Since control is by digital parameters, it is possible to use assignable controls, such that there need only be one set of filter and equalizer controls, whose setting is conveyed to any channel chosen by the operator.[3] The use of digital processing allows the console to include a video display of the settings.

Since the audio processing in a digital mixer is by program control, the configuration of the desk can be changed at will by running the programs for the various functions in a different order. The operator can configure the desk to his own requirements by entering symbols on a block diagram on the video display, for example. The configuration and the setting of all the controls can be stored in memory or for a longer term, on disk, and recalled instantly. Such a desk can be in almost constant use, because it can be put back exactly to a known state easily after someone else has used it.

A further advantage of working in the digital domain is that delay can be controlled individually in the audio channels.[4] This allows for the time of arrival of wavefronts at various microphones to be compensated despite their physical position.

Figure 3.23 shows a typical digital mixer installation.[3] The analog microphone inputs are from remote units containing ADCs so that the length of analog cabling can be kept short. The input units communicate with the signal processor using digital fibre-optic links.

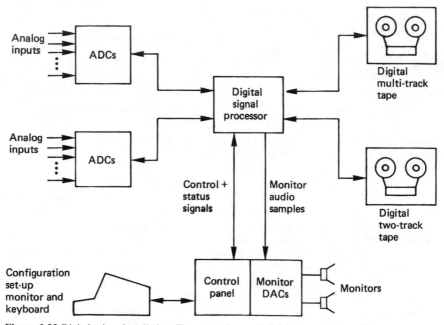

Figure 3.23 Digital mixer installation. The convenience of digital transmission without degradation allows the control panel to be physically remote from the processor.

The sampling rate of a typical digital audio signal is low compared to the speed at which typical logic gates can operate. It is sensible to minimize the quantity of hardware necessary by making each perform many functions in one sampling period. Although general-purpose computers can be programmed to process digital audio, they are not ideal for the purpose. This has resulted in the development of specialized digital audio signal processors, almost always called DSP.[5-7] These units are implemented with more internal registers than data processors to facilitate multi-point filter algorithms. The arithmetic unit will be designed to offer high-speed multiply/accumulate using techniques such as pipelining, which allows operations to overlap.[8] The functions of the register set and the arithmetic unit are controlled by a microsequencer.

External control of a DSP will generally be by a smaller processor, often in the operator's console, which passes coefficients to the DSP as the operator moves the controls. In large systems, it is possible for several different consoles to control different sections of the DSP.[9]

The steadily improving economics of digital logic means that the true cost to the user of digital mixing consoles continues to fall, despite the inclusion of an increasing number of features. The traditional analog console is therefore under threat, except possibly at the highest quality level where equipment economics are not so dominant.

3.9 Filters

One of the most important processes in digital audio is filtering, and its parallel topic of transforms. Filters and transforms are relevant to sampling, conversion, recording, transmission and compression systems. Filtering is unavoidable. Sometimes a process has a filtering effect which is undesirable, for example the limited frequency response of a microphone, and we try to minimize it. On other occasions a filtering effect is specifically required. Filters are required in ADCs, DACs, in the data channels of digital recorders and transmission systems, in compression systems and in DSP.

Figure 3.24 shows that impulse response testing tells a great deal about a filter. In a perfect filter, all frequencies should experience the same time delay. If some groups of frequencies experience a different delay from others, there is a group-delay error. As an impulse contains an infinite spectrum, a filter suffering from group-delay error will separate the different frequencies of an impulse along the time axis.

A pure delay will cause a phase shift proportional to frequency, and a filter with this characteristic is said to be phase-linear. The impulse response of a phase-linear filter is symmetrical. If a filter suffers from group-delay error it cannot be phase-linear. It is almost impossible to make a perfectly phase-linear analog filter, and many filters have a group-delay equalization stage following them which is often as complex as the filter itself. In the digital domain it is straightforward to make a phase-linear filter, and phase equalization becomes unnecessary.

Because of the sampled nature of the signal, whatever the response at low frequencies may be, all PCM channels act as low-pass filters because they cannot contain frequencies above the Nyquist limit of half the sampling frequency.

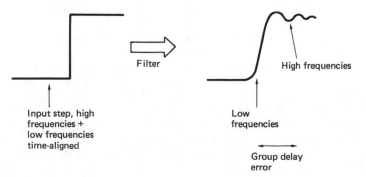

Figure 3.24 Group delay time-displaces signals as a function of frequency.

Transforms are a useful subject because they can help either to understand processes which cause undesirable filtering or to design filters. The information itself may be subject to a transform, especially in compression schemes. Transforming converts the information into another analog. The information is still there, but expressed with respect to temporal or spatial frequency rather than time or space. Instead of binary numbers representing the magnitude of samples, there are binary numbers representing the magnitude of frequency coefficients. What happens in the frequency domain must always be consistent with what happens in the time or space domains. Every combination of frequency and phase response has a corresponding impulse response in the time domain.

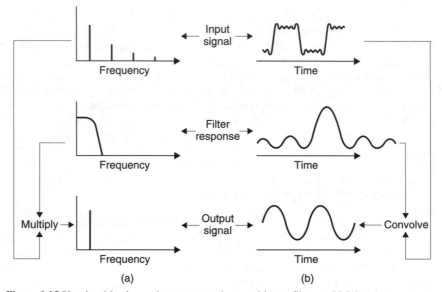

(a)   (b)

Figure 3.25 If a signal having a given spectrum is passed into a filter, multiplying the two spectra will give the output spectrum at (a). Equally transforming the filter frequency response will yield the impulse response of the filter. If this is convolved with the time domain waveform, the result will be the output waveform, whose transform is the output spectrum (b).

Figure 3.25 shows the relationship between the domains. On the left is the frequency domain. Here an input signal having a given spectrum is input to a filter having a given frequency response. The output spectrum will be the product of the two functions. If the functions are expressed logarithmically in deciBels, the product can be obtained by simple addition.

On the right, the time-domain output waveform represents the convolution of the impulse response with the input waveform. However, if the frequency transform of the output waveform is taken, it must be the same as the result obtained from the frequency response and the input spectrum. This is a useful result because it means that when audio sampling is considered, it will be possible to explain the process in both domains.

When a waveform is input to a system, the output waveform will be the convolution of the input waveform and the impulse response of the system. Convolution can be followed by reference to a graphic example shown in Figure 3.26. Where

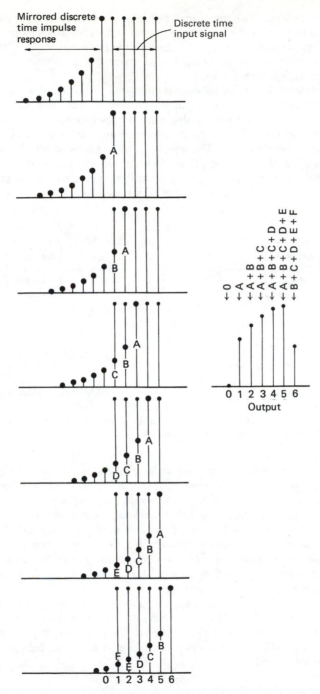

Figure 3.26 In time discrete convolution, the mirrored impulse response is stepped through the input one sample period at a time. At each step, the sum of the cross-products is used to form an output value. As the input in this example is a constant-height pulse, the output is simply proportional to the sum of the coincident impulse response samples.

the impulse response is asymmetrical, the decaying tail occurs *after* the input. To obtain the correct result it is necessary to reverse the impulse response in time so that it is mirrored prior to sweeping it through the input waveform. If the impulse response is symmetrical, as would be the case with a linear phase filter, the mirroring process is superfluous.

In the continuous domain, the output voltage is proportional to area where the two impulses overlap. However, in the sampled, or discrete time domain, both the impulse and the input are a set of discrete samples which clearly must have the same sample spacing. The impulse response only has value where impulses coincide. Elsewhere it is zero. The impulse response is therefore stepped through the input one sample period at a time. At each step, the area is still proportional to the output, but as the time steps are of uniform width, the area is proportional to the impulse height and so the output is obtained by adding up the lengths of overlap.

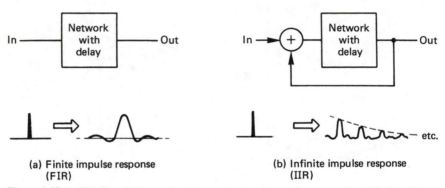

(a) Finite impulse response
    (FIR)

(b) Infinite impulse response
    (IIR)

Figure 3.27 An FIR filter (a) responds only to an input, whereas the output of an IIR filter (b) continues indefinitely rather like a decaying echo.

Filters can be described in two main classes, as shown in Figure 3.27, according to the nature of the impulse response. Finite-impulse response (FIR) filters are always stable and, as their name suggests, respond to an impulse once, as they have only a forward path. In the temporal domain, the time for which the filter responds to an input is finite, fixed and readily established. The same is therefore true about the distance over which a FIR filter responds in the spatial domain. FIR filters can be made perfectly phase-linear if a significant processing delay is accepted. Most filters used for sampling rate conversion and oversampling fall into this category.

Infinite-impulse response (IIR) filters respond to an impulse indefinitely and are not necessarily stable, as they have a return path from the output to the input. For this reason they are also called recursive filters. As the impulse response is not symmetrical, IIR filters are not phase-linear. Audio equalizers often employ recursive filters.

3.10 FIR filters

A FIR filter performs convolution of the input waveform with its own impulse response. It does this by graphically constructing the impulse response for every input sample and superimposing all these responses. It is first necessary to establish the correct impulse response. Figure 3.28(a) shows an example of a low-pass filter which cuts off at one quarter of the sampling rate. The impulse response of an ideal low-pass filter is a $\sin x/x$ curve where the time between the two central zero crossings is the reciprocal of the cut-off frequency. According to the mathematical model, the waveform has always existed and carries on for ever.

The peak value of the output coincides with the input impulse. This means that the filter cannot be causal, because the output has changed before the input is known. Thus in all practical applications it is necessary to truncate the extreme ends of the impulse response, which causes an aperture effect, and to introduce a time delay in the filter equal to half the duration of the truncated impulse in order to make the filter causal. As an input impulse is shifted through the series of registers in Figure 3.28(b), the impulse response is created, because at each point it is multiplied by a coefficient as in (c).

These coefficients are simply the result of sampling and quantizing the desired impulse response. Clearly the sampling rate used to sample the impulse must be the same as the sampling rate for which the filter is being designed. In practice the coefficients are calculated, rather than attempting to sample an actual impulse response. The coefficient wordlength will be a compromise between cost and performance. Because the input sample shifts across the system registers to create the shape of the impulse response, the configuration is also known as a transversal filter. In operation with real sample streams, there will be several consecutive sample values in the filter registers at any time in order to convolve the input with the impulse response.

Simply truncating the impulse response causes an abrupt transition from input samples which matter and those which do not. Truncating the filter superimposes a rectangular shape on the time-domain impulse response. In the frequency domain the rectangular shape transforms to a $\sin x/x$ characteristic which is superimposed on the desired frequency response as a ripple. One consequence of this is known as Gibb's phenomenon; a tendency for the response to peak just before the cut-off frequency.[10,11] As a result, the length of the impulse which must be considered will depend not only on the frequency response, but also on the amount of ripple which can be tolerated. If the relevant period of the impulse is measured in sample periods, the result will be the number of points or multiplications needed in the filter. Figure 3.29 compares the performance of filters with different numbers of points. A high-quality digital audio FIR filter may need in excess of 100 points. Rather than simply truncate the impulse response in time, it is better to make a smooth transition from samples which do not count to those that do. This can be done by multiplying the coefficients in the filter by a window function which peaks in the centre of the impulse.

If the coefficients are not quantized finely enough, it will be as if they had been calculated inaccurately, and the performance of the filter will be less than expected. Figure 3.30 shows an example of quantizing coefficients. Conversely, raising the wordlength of the coefficients increases cost.

The FIR structure is inherently phase-linear because it is easy to make the impulse response absolutely symmetrical. The individual samples in a digital system do not know in isolation what frequency they represent, and they can only pass through the filter at a rate determined by the clock. Because of this inherent phase-linearity, a FIR filter can be designed for a specific impulse response, and the frequency response will follow.

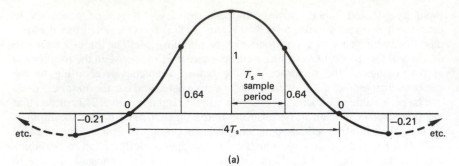

(a)

Figure 3.28 (a) The impulse response of an LPF is a sin$x$/x curve which stretches from $-\infty$ to $+\infty$ in time. The ends of the response must be neglected, and a delay introduced to make the filter causal.

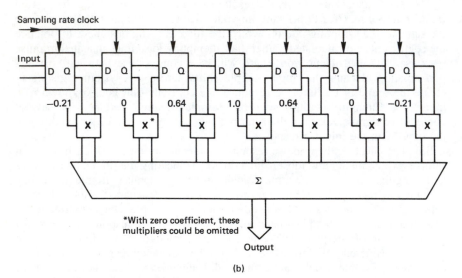

(b)

Figure 3.28 (b) The structure of an FIR LPF. Input samples shift across the register and at each point are multiplied by different coefficients.

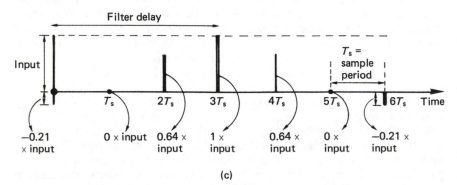

(c)

Figure 3.28 (c) When a single unit sample shifts across the circuit of Figure 3.28(b), the impulse response is created at the output as the impulse is multiplied by each coefficient in turn.

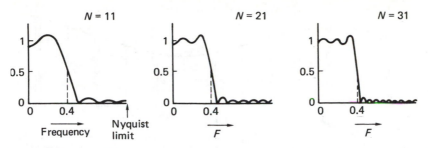

Figure 3.29 The truncation of the impulse in an FIR filter caused by the use of a finite number of points (*N*) results in ripple in the response. Shown here are three different numbers of points for the same impulse response. The filter is an LPF which rolls off at 0.4 of the fundamental interval. (Courtesy *Philips Technical Review*)

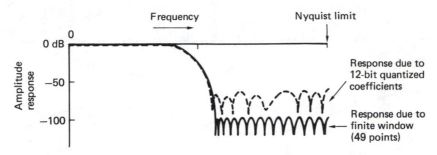

Figure 3.30 Frequency response of a 49-point transversal filter with infinite precision (solid line) shows ripple due to finite window size. Quantizing coefficients to twelve bits reduces attenuation in the stopband. (Responses courtesy *Philips Technical Review*)

The frequency response of the filter can be changed at will by changing the coefficients. A programmable filter only requires a series of ROMs to supply the coefficients; the address supplied to the ROMs will select the response. The frequency response of a digital filter will also change if the clock rate is changed, so it is often less ambiguous to specify a frequency of interest in a digital filter in terms of a fraction of the fundamental interval rather than in absolute terms.

3.11 Sampling-rate conversion

Sampling-rate conversion is an important enabling technology on which a large number of practical devices are based. There are numerous standard sampling rates for audio and it may be necessary to convert between them. In some low-bit rate applications such as Internet audio, the sampling rate may deliberately be reduced. To take advantage of oversampling convertors, an increase in sampling rate is necessary for DACs and a reduction in sampling rate is necessary for ADCs. In oversampling the factors by which the rates are changed are usually simpler than in other applications.

There are three basic but related categories of rate conversion, as shown in Figure 3.31. The most straightforward (a) changes the rate by an integer ratio, up or down. The timing of the system is thus simplified because all samples (input

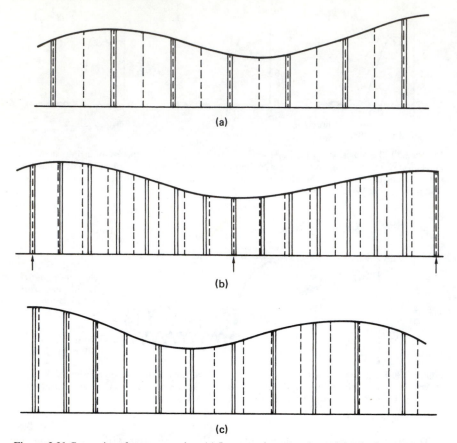

Figure 3.31 Categories of rate conversion. (a) Integer-ratio conversion, where the lower-rate samples are always coincident with those of the higher rate. There are a small number of phases needed. (b) Fractional-ratio conversion, where sample coincidence is periodic. A larger number of phases are required. Example here is conversion from 50.4 kHz to 44.1 kHz (8/7). (c) Variable-ratio conversion, where there is no fixed relationship, and a large number of phases are required.

and output) are present on edges of the higher-rate sampling clock. Such a system is generally adopted for oversampling convertors; the exact sampling rate immediately adjacent to the analog domain is not critical, and will be chosen to make the filters easier to implement.

Next in order of difficulty is the category shown at (b) where the rate is changed by the ratio of two small integers. Samples in the input periodically time-align with the output.

The most complex rate-conversion category is where there is no simple relationship between input and output sampling rates, and in fact they may vary. This situation, shown at (c), is known as variable-ratio conversion. The temporal or spatial relationship of input and output samples is arbitrary.

The technique of integer-ratio conversion is used in conjunction with oversampling convertors and in compression systems where sub-sampled versions of an audio waveform may be required.

Rate convertors incorporate two steps. The first is to control the system bandwidth. If the sampling rate is to be reduced, the bandwidth of the input signal must also be reduced to prevent aliasing. This stage is not required if the rate is to be increased. Next is the interpolation stage which represents the original waveform by samples at new locations. Performing the steps of rate increase separately is inefficient. The combination of the two processes into an interpolating filter minimizes the amount of computation.

As the purpose of the system is purely to change the sampling rate, the filter must be as transparent as possible, and this suggests the use of an FIR structure displaying linear phase. The theoretical impulse response of such a filter is a $\sin x/x$ curve which has zero value at the position of adjacent input samples. In practice this impulse cannot be implemented because it is infinite. The impulse response used will be truncated and windowed as described earlier. To simplify this discussion, assume that a $\sin x/x$ impulse is to be used. There is a strong parallel with the operation of a DAC where the analog voltage is returned to the time-continuous state by summing the analog impulses due to each sample. In a digital interpolating filter, this process is duplicated.[12]

If the sampling rate is to be doubled, new samples must be interpolated exactly half-way between existing samples. The necessary impulse response is shown in Figure 3.32; it can be sampled at the *output* sample period and quantized to form coefficients. If a single input sample is multiplied by each of these coefficients in turn, the impulse response of that sample at the new sampling rate will be obtained. Note that every other coefficient is zero, which confirms that no computation is necessary on the existing samples; they are just transferred to the output. The intermediate sample is then computed by adding together the impulse responses of every input sample in the window. The figure shows how this mechanism operates. If the sampling rate is to be increased by a factor of four, three sample values must be interpolated between existing input samples. It is then necessary to sample the impulse response at one-quarter the period of input samples to obtain three sets of coefficients which will be used in turn. In hardware-implemented filters, the input sample which is passed straight to the output is transferred by using a fourth filter phase where all coefficients are zero except the central one, which is unity.

Fractional ratio conversion allows interchange between different images having different pixel array sizes. Fractional ratios also occur in the vertical axis of standards convertors. Figure 3.31 showed that when the two sampling rates have a simple fractional relationship m/n, there is a periodicity in the relationship between samples in the two streams. It is possible to have a system clock running at the least-common multiple frequency which will divide by different integers to give each sampling rate.[13]

In a variable-ratio interpolator, values will exist for the points at which input samples were made, but it is necessary to compute what the sample values would have been at absolutely any point between available samples. The general concept of the interpolator is the same as for the fractional-ratio convertor, except that an infinite number of filter phases is ideally necessary. Since a realizable filter will have a finite number of phases, it is necessary to study the degradation this causes. The desired continuous temporal or spatial axis of the interpolator is quantized by the phase spacing, and a sample value needed at a particular point will be replaced by a value for the nearest available filter phase. The number of phases in the filter therefore determines the accuracy of the interpolation. The

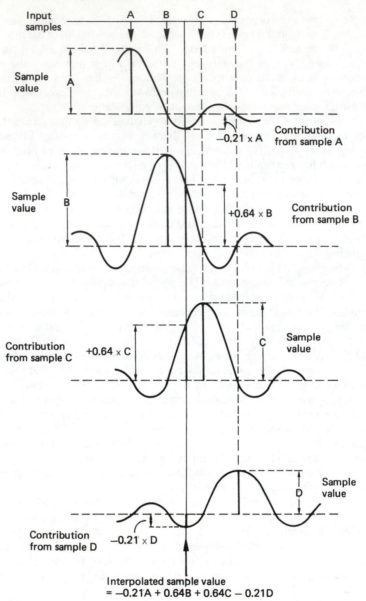

Figure 3.32 A two times oversampling interpolator. To compute an intermediate sample, the input samples are imagined to be sin$x$/x impulses, and the contributions from each at the point of interest can be calculated. In practice, rather more samples on either side need to be taken into account.

effects of calculating a value for the wrong point are identical to those of sampling with clock jitter, in that an error occurs proportional to the slope of the signal. The result is program-modulated noise. The higher the noise specification, the greater the desired time accuracy and the greater the number of phases required. The number of phases is equal to the number of sets of coefficients available, and should not be confused with the number of points in the filter, which is equal to the number of coefficients in a set (and the number of multiplications needed to calculate one output value).

3.12 Transforms and duality

The duality of transforms provides an interesting insight into what is happening in common processes. Fourier analysis holds that any periodic waveform can be reproduced by adding together an arbitrary number of harmonically related sinusoids of various amplitudes and phases. Figure 3.33 shows how a square wave can be built up of harmonics. The spectrum can be drawn by plotting the amplitude of the harmonics against frequency. It will be seen that this gives a spectrum which is a decaying wave. It passes through zero at all even multiples of the fundamental. The shape of the spectrum is a $\sin x/x$ curve. If a square wave has a $\sin x/x$ spectrum, it follows that a filter with a rectangular impulse response will have a $\sin x/x$ spectrum.

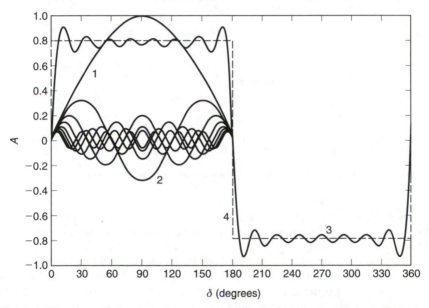

Figure 3.33 Fourier analysis of a square wave into fundamental and harmonics. A, amplitude, δ, phase of fundamental wave in degrees; 1, first harmonic (fundamental); 2, odd harmonics 3–15; 3, sum of harmonics 1–15; 4, ideal square wave.

A low-pass filter has a rectangular spectrum, and this has a $\sin x/x$ impulse response. These characteristics are known as a transform pair. In transform pairs, if one domain has one shape of the pair, the other domain will have the other shape. Figure 3.34 shows a number of transform pairs.

At (a) a square wave has a $\sin x/x$ spectrum and a $\sin x/x$ impulse has a square spectrum. In general the product of equivalent parameters on either side of a transform remains constant, so that if one increases, the other must fall. If (a)

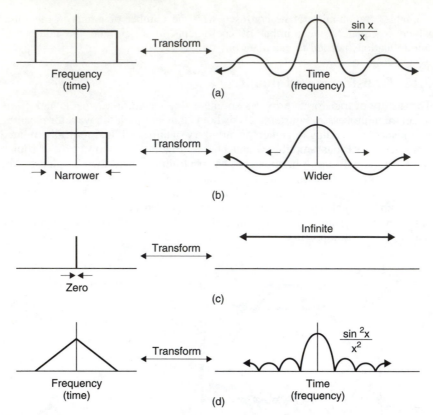

Figure 3.34 Transform pairs. At (a) the dual of a rectangle is a sin$x$/x function. If one is time domain, the other is frequency domain. At (b), narrowing one domain widens the other. The limiting case of this is (c). Transform of the sin$x$/x squared function is triangular (d).

shows a filter with a wider bandwidth, having a narrow impulse response, then (b) shows a filter of narrower bandwidth which has a wide impulse response. This is duality in action. The limiting case of this behaviour is where one parameter becomes zero, the other goes to infinity. At (c) a time-domain pulse of infinitely short duration has a flat spectrum. Thus a flat waveform, i.e. DC, has only zero in its spectrum. The impulse response of the optics of a laser disk (d) has a sin^2x/x^2 intensity function, and this is responsible for the triangular falling frequency response of the pickup. The lens is a rectangular aperture, but as there is no such thing as negative light, a sin$x$/x impulse response is impossible. The squaring process is consistent with a positive-only impulse response. Interestingly the transform of a Gaussian response in still Gaussian.

Duality also holds for sampled systems. A sampling process is periodic in the time domain. This results in a spectrum which is periodic in the frequency domain. If the time between the samples is reduced, the bandwidth of the system rises. Figure 3.35(a) shows that a continuous time signal has a continuous spectrum whereas at (b) the frequency transform of a sampled signal is also discrete. In other words sampled signals can only be analysed into a finite number of frequencies. The more accurate the frequency analysis has to be, the

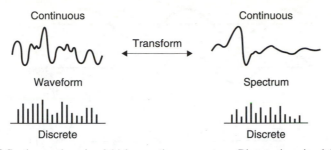

Figure 3.35 Continuous time signal (a) has continuous spectrum. Discrete time signal (b) has discrete spectrum.

more samples are needed in the block. Making the block longer reduces the ability to locate a transient in time. This is the Heisenberg inequality, which is the limiting case of duality, because when infinite accuracy is achieved in one domain, there is no accuracy at all in the other.

3.13 The Fourier transform

Figure 3.33 showed that if the amplitude and phase of each frequency component is known, linearly adding the resultant components in an inverse transform results in the original waveform. In digital systems the waveform is expressed as a number of discrete samples. As a result the Fourier transform analyses the signal into an equal number of discrete frequencies. This is known as a discrete Fourier transform or DFT in which the number of frequency coefficients is equal to the number of input samples. The fast Fourier transform is no more than an efficient way of computing the DFT.[14]

It will be evident from Figure 3.33 that the knowledge of the phase of the frequency component is vital, as changing the phase of any component will seriously alter the reconstructed waveform. Thus the DFT must accurately analyse the phase of the signal components.

Section 2.10 showed a point rotating about a fixed axis at constant speed. One way of defining the phase of a waveform is to specify the angle through which the point has rotated at time zero ($T = 0$). If a second point is made to revolve at 90° to the first, it would produce a cosine wave when translated. It is possible to produce a waveform having arbitrary phase by adding together the sine and cosine waves in various proportions and polarities. For example, adding the sine and cosine waves in equal proportion results in a waveform lagging the sine wave by 45°. The proportions necessary are respectively the sine and the cosine of the phase angle. Thus the two methods of describing phase can be readily interchanged.

The discrete Fourier transform spectrum-analyses a string of samples by searching separately for each discrete target frequency. It does this by multiplying the input waveform by a sine wave, known as the basis function, having the target frequency and adding up or integrating the products. Figure 3.36(a) shows that multiplying by basis functions gives a non-zero integral when the input frequency is the same, whereas (b) shows that with a different input frequency (in fact all other different frequencies) the integral is zero showing that no component of the target frequency exists. Thus from a real waveform

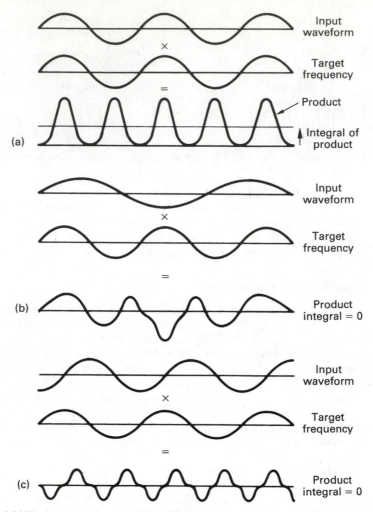

Figure 3.36 The input waveform is multiplied by the target frequency and the result is averaged or integrated. At (a) the target frequency is present and a large integral results. With another input frequency the integral is zero as at (b). The correct frequency will also result in a zero integral shown at (c) if it is at 90° to the phase of the search frequency. This is overcome by making two searches in quadrature.

containing many frequencies all frequencies except the target frequency are excluded. The magnitude of the integral is proportional to the amplitude of the target component.

Figure 3.36(c) shows that the target frequency will not be detected if it is phase shifted 90° as the product of quadrature waveforms is always zero. Thus the discrete Fourier transform must make a further search for the target frequency using a cosine basis function. It follows from the arguments above that the relative proportions of the sine and cosine integrals reveal the phase of the input component. Thus each discrete frequency in the spectrum must be the result of a pair of quadrature searches.

Searching for one frequency at a time as above will result in a DFT, but only after considerable computation. However, a lot of the calculations are repeated many times over in different searches. The fast Fourier transform gives the same result with less computation by logically gathering together all the places where the same calculation is needed and making the calculation once.

3.14 The discrete cosine transform (DCT)

The DCT is a special case of a discrete Fourier transform in which the sine components of the coefficients have been eliminated leaving a single number. This is actually quite easy. Figure 3.37(a) shows a block of input samples to a transform process. By repeating the samples in a time-reversed order and performing a discrete Fourier transform on the double-length sample set a DCT is obtained. The effect of mirroring the input waveform is to turn it into an even function whose sine coefficients are all zero. The result can be understood by considering the effect of individually transforming the input block and the reversed block.

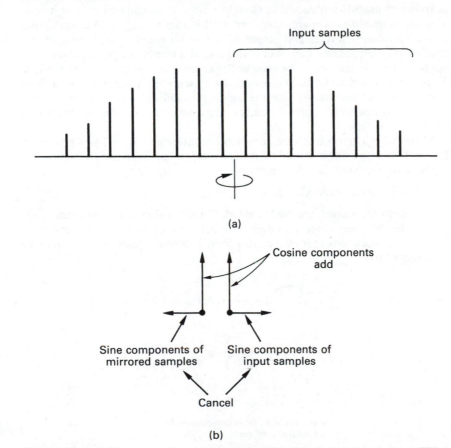

Figure 3.37 The DCT is obtained by mirroring the input block as shown at (a) prior to an FFT. The mirroring cancels out the sine components as at (b), leaving only cosine coefficients.

Figure 3.37(b) shows that the phase of all the components of one block are in the opposite sense to those in the other. This means that when the components are added to give the transform of the double length block all the sine components cancel out, leaving only the cosine coefficients, hence the name of the transform.[15] In practice the sine component calculation is eliminated. Another advantage is that doubling the block length by mirroring doubles the frequency resolution, so that twice as many useful coefficients are produced. In fact a DCT produces as many useful coefficients as input samples.

The DCT is primarily used in audio coding because it converts the input waveform into a form where redundancy can be easily detected and removed. More details of the DCT can be found in Chapter 5.

3.15 Modulo-*n* arithmetic

Conventional arithmetic which is in everyday use relates to the real world of counting actual objects, and to obtain correct answers the concepts of borrow and carry are necessary in the calculations.

There is an alternative type of arithmetic which has no borrow or carry which is known as modulo arithmetic. In modulo-*n* no number can exceed *n*. If it does, *n* or whole multiples of *n* are subtracted until it does not. Thus 25 modulo-16 is 9 and 12 modulo-5 is 2. The output of a four-bit counter overflows when it reaches 1111 because the carry-out is ignored. If a number of clock pulses *m* are applied from the zero state, the state of the counter will be given by *m* Mod.16. Thus modulo arithmetic is appropriate to systems in which there is a fixed wordlength and this means that the range of values the system can have is restricted by that wordlength. A number range which is restricted in this way is called a finite field.

Modulo-2 is a numbering scheme which is used frequently in digital processes. Figure 3.38 shows that in modulo-2 the conventional addition and subtraction are replaced by the XOR function such that:

A + B Mod.2 = A XOR B

When multi-bit values are added Mod.2, each column is computed quite independently of any other. This makes Mod.2 circuitry very fast in operation as it is not necessary to wait for the carries from lower-order bits to ripple up to the high-order bits.

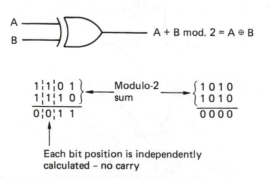

Each bit position is independently
calculated – no carry

Figure 3.38 In modulo-2 calculations, there can be no carry or borrow operations and conventional addition and subtraction become identical. The XOR gate is a modulo-2 adder.

Modulo-2 arithmetic is not the same as conventional arithmetic and takes some getting used to. For example, adding something to itself in Mod.2 always gives the answer zero.

3.16 The Galois field

Figure 3.39 shows a simple circuit consisting of three D-type latches which are clocked simultaneously. They are connected in series to form a shift register. At (a) a feedback connection has been taken from the output to the input and the result is a ring counter where the bits contained will recirculate endlessly. At (b) one XOR gate is added so that the output is fed back to more than one stage. The result is known as a twisted-ring counter and it has some interesting properties. Whenever the circuit is clocked, the left-hand bit moves to the right-hand latch, the centre bit moves to the left-hand latch and the centre latch becomes the XOR of the two outer latches. The figure shows that whatever the starting condition of the three bits in the latches, the same state will always be reached again after seven clocks, except if zero is used. The states of the latches form an endless ring of non-sequential numbers called a Galois field after the French mathematical prodigy Evariste Galois who discovered them. The states of the circuit form a maximum length sequence because there are as many states as are permitted by the wordlength. As the states of the sequence have many of the characteristics of random numbers, yet are repeatable, the result can also be called a pseudo-random sequence (prs). As the all-zeros case is disallowed, the length of a maximum length sequence generated by a register of m bits cannot exceed (2^m-1) states. The Galois field, however, includes the zero term. It is useful to explore the bizarre mathematics of Galois fields which use modulo-2 arithmetic.

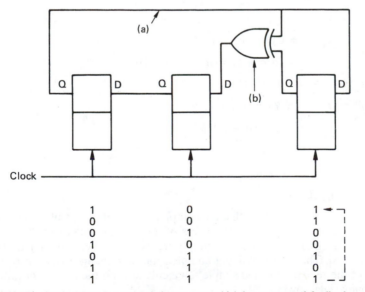

Figure 3.39 The circuit shown is a twisted-ring counter which has an unusual feedback arrangement. Clocking the counter causes it to pass through a series of non-sequential values. See text for details.

Familiarity with such manipulations is helpful when studying the error correction, particularly the Reed–Solomon codes used in recorders and treated in Chapter 6. They will also be found in processes which require pseudo-random numbers such as digital dither, considered in Chapter 3, and randomized channel codes used in, for example, DAB and discussed in Chapter 9.

The circuit of Figure 3.39 can be considered as a counter and the four points shown will then be representing different powers of 2 from the MSB on the left to the LSB on the right. The feedback connection from the MSB to the other stages means that whenever the MSB becomes 1, two other powers are also forced to one so that the code of 1011 is generated.

Each state of the circuit can be described by combinations of powers of x, such as

$$x^2 = 100$$

$$x = 010$$

$$x^2 + x = 110, \text{ etc.}$$

The fact that three bits have the same state because they are connected together is represented by the Mod.2 equation:

$$x^3 + x + 1 = 0$$

Let $x = a$, which is a primitive element. Now

$$a^3 + a + 1 = 0 \tag{3.1}$$

In modulo-2

$$a + a = a^2 + a^2 = 0$$

$$a = x = 010$$

$$a^2 = x^2 = 100$$

$$a^3 = a + 1 = 011 \text{ from (3.1)}$$

$$a^4 = a \times a^3 = a(a + 1) = a^2 + a = 110$$

$$a^5 = a^2 + a + 1 = 111$$

$$a^6 = a \times a^5 = a(a^2 + a + 1)$$

$$= a^3 + a^2 + a = a + 1 + a^2 + a$$

$$= a^2 + 1 = 101$$

$$a^7 = a(a^2 + 1) = a^3 + a$$

$$= a + 1 + a = 1 = 001$$

In this way it can be seen that the complete set of elements of the Galois field can be expressed by successive powers of the primitive element. Note that the twisted-ring circuit of Figure 3.39 simply raises a to higher and higher powers as it is clocked. Thus the seemingly complex multibit changes caused by a single clock of the register become simple to calculate using the correct primitive and the appropriate power.

The numbers produced by the twisted-ring counter are not random; they are completely predictable if the equation is known. However, the sequences produced are sufficiently similar to random numbers that in many cases they will be useful. They are thus referred to as pseudo-random sequences. The feedback connection is chosen such that the expression it implements will not factorize. Otherwise a maximum-length sequence could not be generated because the circuit might sequence around one or other of the factors depending on the initial condition. A useful analogy is to compare the operation of a pair of meshed gears. If the gears have a number of teeth which is relatively prime, many revolutions are necessary to make the same pair of teeth touch again. If the number of teeth have a common multiple, far fewer turns are needed.

3.17 The phase-locked loop

All digital audio systems need to be clocked at the appropriate rate in order to function properly. Whilst a clock may be obtained from a fixed-frequency oscillator such as a crystal, many operations in video require *genlocking* or synchronizing the clock to an external source. The phase-locked loop excels at this job, and many others, particularly in connection with recording and transmission.

In phase-locked loops, the oscillator can run at a range of frequencies according to the voltage applied to a control terminal. This is called a voltage-controlled oscillator or VCO. Figure 3.40 shows that the VCO is driven by a

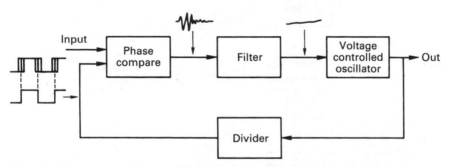

Figure 3.40 A phase-locked loop requires these components as a minimum. The filter in the control voltage serves to reduce clock jitter.

phase error measured between the output and some reference. The error changes the control voltage in such a way that the error is reduced, so that the output eventually has the same frequency as the reference. A low-pass filter is fitted in the control voltage path to prevent the loop becoming unstable. If a divider is placed between the VCO and the phase comparator, as in the figure, the VCO frequency can be made to be a multiple of the reference. This also has the effect of making the loop more heavily damped, so that it is less likely to change frequency if the input is irregular.

Figure 3.41 shows how the 48 kHz sampling rate clock of professional digital audio may be obtained from the sync pulses of an analog video signal by such a multiplication process.

Figure 3.42 shows the NLL or numerically locked loop. This is similar to a phase-locked loop, except that the two phases concerned are represented by the

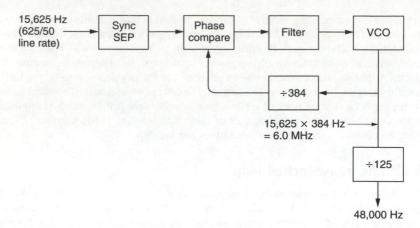

Figure 3.41 Obtaining a 48 kHz sampling clock from the line frequency of 625/50 video using a phase-locked loop.

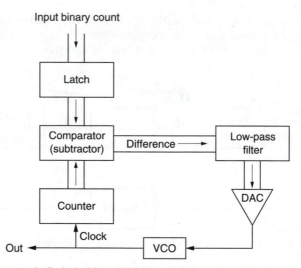

Figure 3.42 The numerically locked loop (NLL) is a digital version of the phase-locked loop.

state of a binary number. The NLL is useful to generate a remote clock from a master. The state of a clock count in the master is periodically transmitted to the NLL which will re-create the same clock frequency. The technique is used in MPEG transport streams.

References

1. Richards, J.W., Digital audio mixing. *Radio and Electron. Eng.*, **53**, 257–264 (1983)
2. Richards, J.W. and Craven, I., An experimental 'all digital' studio mixing desk. *J. Audio Eng. Soc.*, **30**, 117–126 (1982)
3. Jones, M.H., Processing systems for the digital audio studio. In *Digital Audio*, edited by B. Blesser, B. Locanthi and T.G. Stockham Jr, pp. 221–225, New York: Audio Engineering Society (1982)

4. Lidbetter, P.S., A digital delay processor and its applications. Presented at the 82nd Audio Engineering Society Convention (London, 1987), Preprint 2474(K-4)
5. McNally, G.J., COPAS: A high speed real time digital audio processor. *BBC Research Dept Report*, RD 1979/26
6. McNally, G.W., Digital audio: COPAS-2, a modular digital audio signal processor for use in a mixing desk. *BBC Research Dept Report*, RD 1982/13
7. Vandenbulcke, C. *et al.*, An integrated digital audio signal processor. Presented at the 77th Audio Engineering Society Convention (Hamburg, 1985), Preprint 2181(B-7)
8. Moorer, J.A., The audio signal processor: the next step in digital audio. In *Digital Audio*, edited by B. Blesser, B. Locanthi and T.G. Stockham Jr, pp. 205–215, New York: Audio Engineering Society (1982)
9. Gourlaoen, R. and Delacroix, P., The digital sound mixing desk: architecture and integration in the future all-digital studio. Presented at the 80th Audio Engineering Society Convention (Montreux, 1986), Preprint 2327(D-1)
10. van den Enden, A.W.M. and Verhoeckx, N.A.M., Digital signal processing: theoretical background. *Philips Tech. Rev.*, **42**, 110–144, (1985)
11. McClellan, J.H., Parks, T.W. and Rabiner, L.R., A computer program for designing optimum FIR linear-phase digital filters. *IEEE Trans. Audio and Electroacoustics*, **AU-21**, 506–526 (1973)
12. Crochiere, R.E. and Rabiner, L.R., Interpolation and decimation of digital signals – a tutorial review. *Proc. IEEE*, **69**, 300–331 (1981)
13. Rabiner, L.R., Digital techniques for changing the sampling rate of a signal. In *Digital Audio*, edited by B. Blesser, B. Locanthi and T.G. Stockham Jr, pp. 79–89, New York: Audio Engineering Society (1982)
14. Kraniauskas, P., *Transforms in Signals and Systems*, Wokingham: Addison-Wesley (1992)
15. Ahmed, N., Natarajan, T. and Rao, K., Discrete Cosine Transform, *IEEE Trans. Computers*, **C-23**, 90–93 (1974).

4

Conversion

Chapter 1 introduced the fundamental characteristic of digital audio which is that the quality is independent of the storage or transmission medium and is determined instead by the accuracy of conversion between the analog and digital domains. This chapter will examine in detail the theory and practice of this critical aspect of digital audio.

4.1 Introduction to conversion

Any analog audio source can be characterized by a given useful bandwidth and signal-to-noise ratio. If a well-engineered digital channel having a wider bandwidth and a greater signal-to-noise ratio is put in series with such a source, it is only necessary to set the levels correctly and the analog signal is then subject to no loss of information whatsoever.

The sound conveyed through a digital system travels as a stream of bits. Because the bits are discrete, it is easy to quantify the flow, just by counting the number per second. It is much harder to quantify the amount of information in an analog signal (from a microphone, for example) but if this were done using the same units, it would be possible to decide just what bit rate was necessary to convey that signal without loss of information. If a signal can be conveyed without loss of information, and without picking up any unwanted signals on the way, it will have been transmitted perfectly.

The connection between analog signals and information capacity was made by Shannon, in one of the most significant papers in the history of this technology,[1] and those parts which are important for this subject are repeated here. The principles are straightforward, and offer an immediate insight into the relative performances and potentials of different modulation methods, including digitizing.

Figure 4.1 shows an analog signal with a certain amount of superimposed noise, as is the case for all real audio signals. Noise is defined as a random superimposed signal which is not correlated with the wanted signal. To avoid pitfalls in digital audio, this definition must be applied with what initially seems like pedantry. The noise is random, and so the actual voltage of the wanted signal is uncertain; it could be anywhere in the range of the noise amplitude. If the signal amplitude is, for the sake of argument, sixteen times the noise amplitude, it would only be possible to convey sixteen different signal levels unambiguously, because the levels have to be sufficiently different that noise will not make

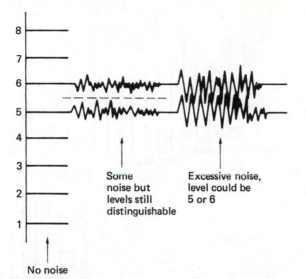

Figure 4.1 To receive eight different levels in a signal unambiguously, the peak-to-peak noise must be less than the difference in level. Signal-to-noise ratio must be at least 8:1 or 18 dB to convey eight levels. This can also be conveyed by three bits ($2^3 = 8$). For 16 levels, SNR would have to be 24 dB, which would be conveyed by four bits.

one look like another. It is possible to convey sixteen different levels in all combinations of four data bits, and so the connection between the analog and quantized domains is established.

The choice of sampling rate (the rate at which the signal voltage must be examined to convey the information in a changing signal) is important in any system; if it is too low, the signal will be degraded, and if it is too high, the number of samples to be recorded will rise unnecessarily, as will the cost of the system. Here it will be established just what sampling rate is necessary in a given situation, initially in theory, then taking into account practical restrictions. By multiplying the number of bits needed to express the signal voltage by the rate at which the process must be updated, the bit rate of the digital data stream resulting from a particular analog signal can be determined.

There are a number of ways in which an audio waveform can be digitally represented, but the most useful and therefore common is pulse code modulation or PCM which was introduced in Chapter 1. The input is a continuous-time, continuous-voltage waveform, and this is converted into a discrete-time, discrete-voltage format by a combination of sampling and quantizing. These two processes are independent and can be performed in either order. Figure 4.2(a) shows an analog sampler preceding a quantizer, whereas (b) shows an asynchronous quantizer preceding a digital sampler. Ideally, both will give the same results; in practice each has different advantages and suffers from different deficiencies. Both approaches will be found in real equipment.

The independence of sampling and quantizing allows each to be discussed quite separately in some detail, prior to combining the processes for a full understanding of conversion.

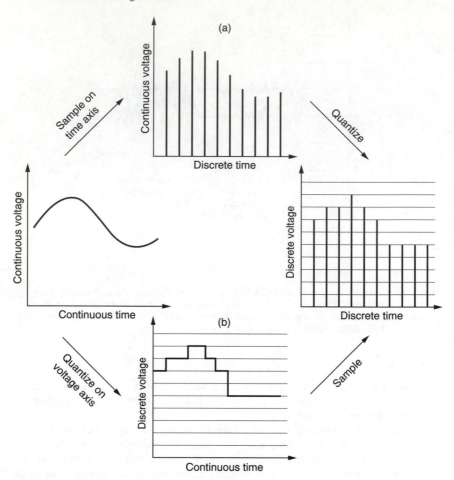

Figure 4.2 Since sampling and quantizing are orthogonal, the order in which they are performed is not important. In (a) sampling is performed first and the samples are quantized. This is common in audio convertors. In (b) the analog input is quantized into an asynchronous binary code. Sampling takes place when this code is latched on sampling clock edges. This approach is universal in video convertors.

4.2 Sampling and aliasing

Sampling is no more than periodic measurement, and it will be shown here that there is no theoretical need for sampling to be audible. Practical equipment may of course be less than ideal, but, given good engineering practice, the ideal may be approached quite closely.

Audio sampling must be regular, because the process of timebase correction prior to conversion back to analog assumes a regular original process as was shown in Chapter 1. The sampling process originates with a pulse train which is shown in Figure 4.3(a) to be of constant amplitude and period. The audio waveform amplitude-modulates the pulse train in much the same way as the carrier is modulated in an AM radio transmitter. One must be careful to avoid

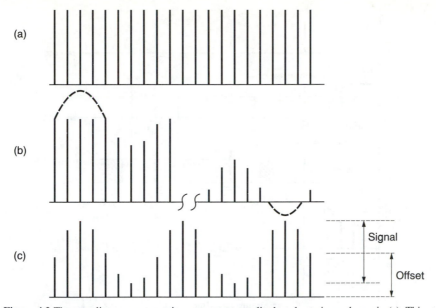

Figure 4.3 The sampling process requires a constant-amplitude pulse train as shown in (a). This is amplitude modulated by the waveform to be sampled. If the input waveform has excessive amplitude or incorrect level, the pulse train clips as shown in (b). For an audio waveform, the greatest signal level is possible when an offset of half the pulse amplitude is used to centre the waveform as shown in (c).

over-modulating the pulse train as shown in (b) and this is achieved by applying a DC offset to the analog waveform so that silence corresponds to a level half-way up the pulses as at (c). Clipping due to any excessive input level will then be symmetrical.

In the same way that AM radio produces sidebands or images above and below the carrier, sampling also produces sidebands although the carrier is now a pulse train and has an infinite series of harmonics as shown in Figure 4.4(a). The sidebands repeat above and below each harmonic of the sampling rate as shown in (b).

The sampled signal can be returned to the continuous-time domain simply by passing it into a low-pass filter. This filter has a frequency response which prevents the images from passing, and only the baseband signal emerges, completely unchanged. If considered in the frequency domain, this filter can be called an anti-image filter; if considered in the time domain it can be called a reconstruction filter.

If an input is supplied having an excessive bandwidth for the sampling rate in use, the sidebands will overlap (Figure 4.4(c)) and the result is aliasing, where certain output frequencies are not the same as their input frequencies but instead become difference frequencies (d). It will be seen from Figure 4.4 that aliasing does not occur when the input frequency is equal to or less than half the sampling rate, and this derives the most fundamental rule of sampling, which is that the sampling rate must be at least twice the highest input frequency. Sampling theory is usually attributed to Shannon[2] who applied it to information theory at around

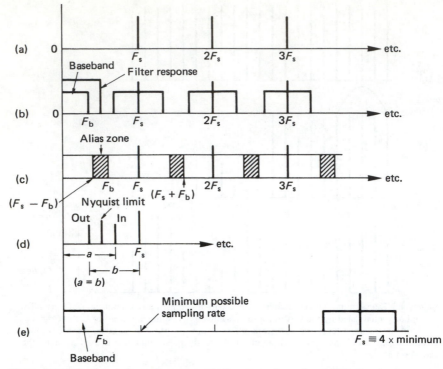

Figure 4.4 (a) Spectrum of sampling pulses. (b) Spectrum of samples. (c) Aliasing due to sideband overlap. (d) Beat-frequency production (e) 4× oversampling.

the same time as Kotelnikov in Russia. These applications were pre-dated by Whittaker. Despite that it is often referred to as Nyquist's theorem.

Whilst aliasing has been described above in the frequency domain, it can be described equally well in the time domain. In Figure 4.5(a) the sampling rate is obviously adequate to describe the waveform, but at (b) it is inadequate and aliasing has occurred.

Aliasing is commonly seen on television and in the cinema, owing to the relatively low frame rates used. With a frame rate of 24 Hz, a film camera will alias on any object changing at more than 12 Hz. Such objects include the spokes

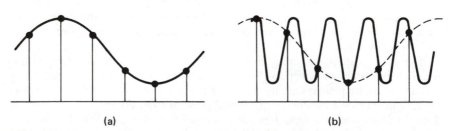

Figure 4.5 In (a) the sampling is adequate to reconstruct the original signal. In (b) the sampling rate is inadequate, and reconstruction produces the wrong waveform (dashed). Aliasing has taken place.

of stagecoach wheels. When the spoke-passing frequency reaches 24 Hz the wheels appear to stop. Aliasing does, however, have useful applications, including the stroboscope, which makes rotating machinery appear stationary, the sampling oscilloscope, which can display periodic waveforms of much greater frequency than the sweep speed of the tube normally allows, and the spectrum analyser.

One often has no control over the spectrum of input signals and in practice it is necessary also to have a low-pass filter at the input to prevent aliasing. This anti-aliasing filter prevents frequencies of more than half the sampling rate from reaching the sampling stage.

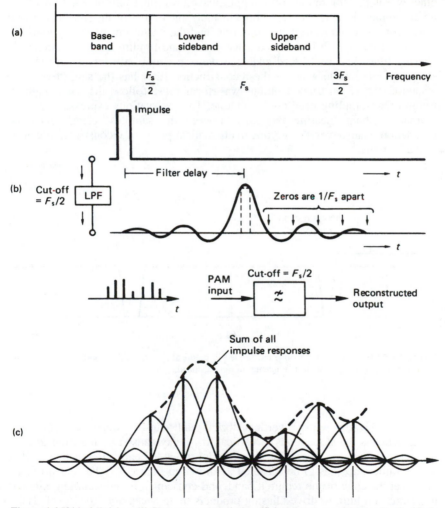

Figure 4.6 If ideal 'brick wall' filters are assumed, the efficient spectrum of (a) results. An ideal low-pass filter has an impulse response shown in (b). The impulse passes through zero at intervals equal to the sampling period. When convolved with a pulse train at the sampling rate, as shown in (c), the voltage at each sample instant is due to that sample alone as the impulses from all other samples pass through zero there.

4.3 Reconstruction

If ideal low-pass anti-aliasing and anti-image filters are assumed, having a vertical cut-off slope at half the sampling rate, an ideal spectrum shown in Figure 4.6(a) is obtained. It was shown in Chapter 3 that the impulse response of a phase linear ideal low-pass filter is a $\sin x/x$ waveform in the time domain, and this is repeated in (b). Such a waveform passes through zero volts periodically. If the cut-off frequency of the filter is one-half of the sampling rate, the impulse passes through zero *at the sites of all other samples*. It can be seen from Figure 4.6(c) that at the output of such a filter, the voltage at the centre of a sample is due to that sample alone, since the value of *all* other samples is zero at that instant. In other words the continuous time output waveform must join up the tops of the input samples. In between the sample instants, the output of the filter is the sum of the contributions from many impulses, and the waveform smoothly joins the tops of the samples. It is a consequence of the band-limiting of the original anti-aliasing filter that the filtered analog waveform could only travel between the sample points in one way. As the reconstruction filter has the same frequency response, the reconstructed output waveform must follow the same path. It follows that sampling need not be audible. The reservations expressed by some journalists about 'hearing the gaps between the samples' clearly have no foundation whatsoever. A rigorous mathematical proof of reconstruction can be found in Betts.[3]

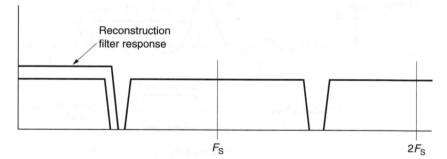

Figure 4.7 As filters with finite slope are needed in practical systems, the sampling rate is raised slightly beyond twice the highest frequency in the baseband.

The ideal filter with a vertical 'brick-wall' cut-off slope is difficult to implement. As the slope tends to vertical, the delay caused by the filter goes to infinity: the quality is marvellous but you don't live to hear it. In practice, a filter with a finite slope has to be accepted as shown in Figure 4.7. The cut-off slope begins at the edge of the required band, and consequently the sampling rate has to be raised a little to drive aliasing products to an acceptably low level. There is no absolute factor by which the sampling rate must be raised; it depends upon the filters which are available and the level of aliasing products which are acceptable. The latter will depend upon the wordlength to which the signal will be quantized.

4.4 Filter design

The discussion so far has assumed that perfect anti-aliasing and reconstruction filters are used. Perfect filters are not available, of course, and because designers must use devices with finite slope and rejection, aliasing can still occur. It is not easy to specify anti-aliasing filters, particularly the amount of stopband rejection needed. The amount of aliasing resulting would depend on, among other things, the amount of out-of-band energy in the input signal.

It could be argued that the reconstruction filter is unnecessary, since all the images are outside the range of human hearing. However, the slightest non-linearity in subsequent stages would result in gross intermodulation distortion. Most transistorized audio power amplifiers become grossly non-linear when fed with signals far beyond the audio band. It is this non-linearity which enables audio equipment to demodulate strong radio transmissions. The simple solution is to curtail the response of equipment somewhat beyond the audio band so that they become immune to passing taxis and refrigerator thermostats. This is seldom done in Hi-Fi amplifiers because of the mistaken belief that response far beyond the audio band is needed for high fidelity. The truth of the belief is academic as all known recorded or broadcast music sources, whether analog or digital, are band-limited. As a result there is nothing to which a power amplifier of excess bandwidth can respond except RF interference and inadequately suppressed images from digital sources.

Every signal which has been through the digital domain has passed through both an anti-aliasing filter and a reconstruction filter. These filters must be carefully designed in order to prevent artifacts, particularly those due to lack of phase linearity as they may be audible.[4-6] The nature of the filters used has a great bearing on the subjective quality of the system. Entire books have been written about analog filters, so they will only be treated briefly here.

Figures 4.8 and 4.9 show the terminology used to describe the common elliptic low-pass filter. These filters are popular because they can be realized with fewer components than other filters of similar response. It is a characteristic of these elliptic filters that there are ripples in the passband and stopband. Lagadec and

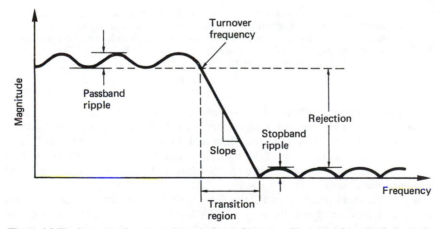

Figure 4.8 The important features and terminology of low-pass filters used for anti-aliasing and reconstruction.

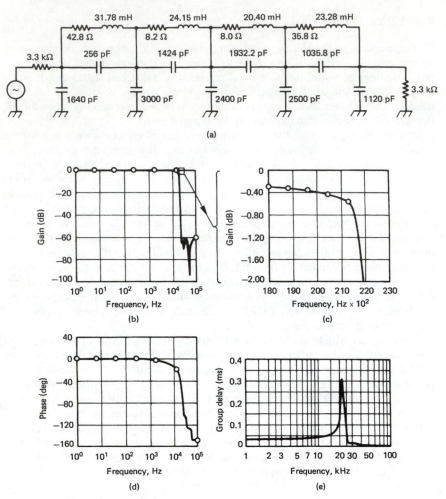

Figure 4.9 (a) Circuit of typical nine-pole elliptic passive filter with frequency response in (b) shown magnified in the region of cut-off in (c). Note phase response in (d) beginning to change at only 1 kHz, and group delay in (e), which require compensation for quality applications. Note that in the presence of out-of-band signals, aliasing might only be 60 dB down. A 13-pole filter manages in excess of 80 dB, but phase response is worse.

Stockham[7] found that filters with passband ripple cause dispersion: the output signal is smeared in time and, on toneburst signals, pre-echoes can be detected. In much equipment the anti-aliasing filter and the reconstruction filter will have the same specification, so that the passband ripple is doubled with a corresponding increase in dispersion. Sometimes slightly different filters are used to reduce the effect.

It is difficult to produce an analog filter with low distortion. Passive filters using inductors suffer non-linearity at high levels due to the *B/H* curve of the cores. It seems a shame to go to such great lengths to remove the non-linearity of magnetic tape from a recording using digital techniques only to pass the signal through magnetic inductors in the filters. Active filters can simulate inductors which are linear using opamp techniques, but they tend to suffer non-linearity at high frequencies where the falling open-loop gain reduces the effect of feedback. Active filters can also contribute noise, but this is not necessarily a bad thing in controlled amounts, since it can act as a dither source.

It is instructive to examine the phase response of such filters. Since a sharp cut-off is generally achieved by cascading many filter sections which cut at a similar frequency, the phase responses of these sections will accumulate. The phase may start to leave linearity at only a few kiloHertz, and near the cut-off frequency the phase may have completed several revolutions. As stated, these phase errors can be audible and phase equalization is necessary. An advantage of linear phase filters is that ringing is minimized, and there is less possibility of clipping on transients.

It is possible to construct a ripple-free phase-linear filter with the required stopband rejection,[8,9] but it is expensive and the money may be better spent in avoiding the need for such a filter. Much effort can be saved in analog filter design by using oversampling. Strictly, oversampling means no more than that a higher sampling rate is used than is required by sampling theory. In the loose sense an 'oversampling convertor' generally implies that some combination of high sampling rate and various other techniques has been applied.

Oversampling is treated in depth in a later section of this chapter. The audible superiority and economy of oversampling convertors has led them to be almost universal. Accordingly the treatment of oversampling in this volume is more prominent than that of filter design.

4.5 Choice of sampling rate

Sampling theory is only the beginning of the process which must be followed to arrive at a suitable sampling rate. The finite slope of realizable filters will compel designers to raise the sampling rate. For consumer products, the lower the sampling rate, the better, since the cost of the medium is directly proportional to the sampling rate: thus sampling rates near to twice 20 kHz are to be expected. For professional products, there is a need to operate at variable speed for pitch correction. When the speed of a digital recorder is reduced, the offtape sampling rate falls, and Figure 4.10 shows that with a minimal sampling rate the first image frequency can become low enough to pass the reconstruction filter. If the sampling frequency is raised without changing the response of the filters, the speed can be reduced without this problem. It follows that variable-speed recorders, generally those with stationary heads, must use a higher sampling rate.

In the early days of digital audio research, the necessary bandwidth of about 1 megabit per second per audio channel was difficult to store. Disk drives had the bandwidth but not the capacity for long recording time, so attention turned to video recorders. In Chapter 8 it will be seen that these were adapted to store audio samples by creating a pseudo-video waveform which could convey binary as black and white levels. The sampling rate of such a system is constrained to relate simply to the field rate and field structure of the television standard used, so that an integer number of samples can be stored on each usable TV line in the field. Such a recording can be made on a monochrome recorder, and these recordings are made in two standards, 525 lines at 60 Hz and 625 lines at 50 Hz.

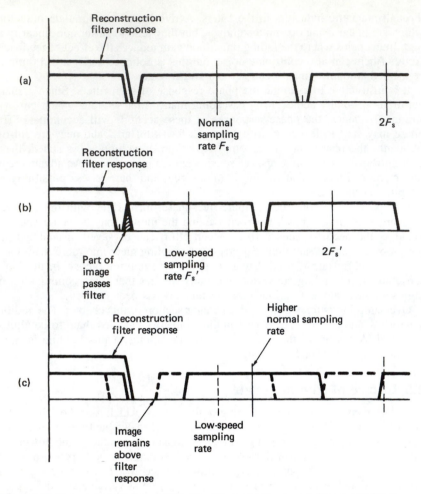

Figure 4.10 At normal speed, the reconstruction filter correctly prevents images entering the baseband, as at (a). When speed is reduced, the sampling rate falls, and a fixed filter will allow part of the lower sideband of the sampling frequency to pass. If the sampling rate of the machine is raised, but the filter characteristic remains the same, the problem can be avoided, as at (c).

Thus it is possible to find a frequency which is a common multiple of the two and also suitable for use as a sampling rate.

The allowable sampling rates in a pseudo-video system can be deduced by multiplying the field rate by the number of active lines in a field (blanked lines cannot be used) and again by the number of samples in a line. By careful choice of parameters it is possible to use either 525/60 or 625/50 video with a sampling rate of 44.1 kHz.

In 60 Hz video, there are 35 blanked lines, leaving 490 lines per frame, or 245 lines per field for samples. If three samples are stored per line, the sampling rate becomes

$$60 \times 245 \times 3 = 44.1 \, \text{kHz}$$

In 50 Hz video, there are 37 lines of blanking, leaving 588 active lines per frame, or 294 per field, so the same sampling rate is given by

$$50.00 \times 294 \times 3 = 44.1\,kHz.$$

The sampling rate of 44.1 kHz came to be that of the Compact Disc. Even though CD has no video circuitry, the equipment originally used to make CD masters was video based and determines the sampling rate.

For landlines to FM stereo broadcast transmitters having a 15 kHz audio bandwidth, the sampling rate of 32 kHz is more than adequate, and has been in use for some time in the United Kingdom and Japan. This frequency is also in use in the NICAM 728 stereo TV sound system and in DAB. It is also used for the Sony NT format mini-cassette. The professional sampling rate of 48 kHz was proposed as having a simple relationship to 32 kHz, being far enough above 40 kHz for variable-speed operation.

Although in a perfect world the adoption of a single sampling rate might have had virtues, for practical and economic reasons digital audio now has essentially three rates to support: 32 kHz for broadcast, 44.1 kHz for CD and its mastering equipment, and 48 kHz for 'professional' use.[10] 48 kHz is extensively used in television where it can be synchronized to both line standards relatively easily. The currently available DVTR formats offer only 48 kHz audio sampling. A number of formats can operate at more than one sampling rate. Both DAT and DASH formats are specified for all three rates, although not all available hardware implements every possibility. Most hard disk recorders will operate at a range of rates.

Recently there have been proposals calling for dramatically increased audio sampling rates. There is no scientific basis for this. Sub-optimal equipment may show an improvement in quality if the sampling rate is significantly raised, but properly engineered equipment should not need excess bandwidth.

4.6 Sample and hold

In practice many analog to digital convertors require a finite time to operate, and instantaneous samples must be extended by a device called a sample-and-hold or, more accurately, a track-hold circuit.

The simplest possible track-hold circuit is shown in Figure 4.11(a). When the switch is closed, the output will follow the input. When the switch is opened, the capacitor holds the signal voltage which existed at the instant of opening. This simple arrangement has a number of shortcomings, particularly the time constant of the on-resistance of the switch with the capacitor, which extends the settling time. The effect can be alleviated by putting the switch in a feedback loop as shown in Figure 4.11(b). The buffer amplifiers must meet a stringent specification, because they need bandwidth well in excess of audio frequencies to ensure that operation is always feedback controlled between holding periods. When the switch is opened, the slightest change in input voltage causes the input buffer to saturate, and it must be able to rapidly recover from this condition when the switch next closes. The feedback minimizes the effect of the on-resistance of the switch, but the off-resistance must be high to prevent the input signal affecting the held voltage. The leakage current of the integrator must be low to prevent droop which is the term given to an unwanted slow change in the held voltage.

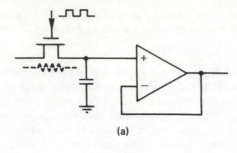

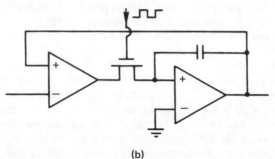

(a)

(b)

Figure 4.11 (a) The simple track-hold circuit shown has poor frequency response as the resistance of the FET causes a rolloff in conjunction with the capacitor. In (b) the resistance of the FET is now inside a feedback loop and will be eliminated, provided the left-hand op-amp never runs out of gain or swing.

Figure 4.12 shows the various events during a track-hold sequence and catalogs the various potential sources of inaccuracy. A further phenomenon which is not shown in Figure 4.12 is that of dielectric relaxation. When a capacitor is discharged rapidly by connecting a low resistance path across its terminals, not all the charge is removed. After the discharge circuit is

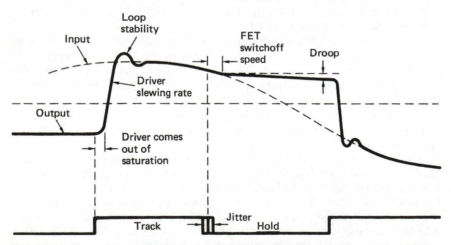

Figure 4.12 Characteristics of the feedback track-hold circuit of Figure 4.11(b) showing major sources of error.

disconnected, the capacitor voltage may rise again slightly as charge which was trapped in the high-resistivity dielectric slowly leaks back to the electrodes. In track-hold circuits dielectric relaxation can cause the value of one sample to be affected by the previous one. Some dielectrics display less relaxation than others. Mica capacitors, traditionally regarded as being of high quality, actually display substantially worse relaxation characteristics than many other types. Polypropylene and teflon are significantly better.

The track-hold circuit is extremely difficult to design because of the accuracy demanded by audio applications. In particular it is very difficult to meet the droop specification for much more than sixteen-bit applications. Greater accuracy has been reported by modelling the effect of dielectric relaxation and applying an inverse correction signal.[11]

When a performance limitation such as the track-hold stage is found, it is better to find an alternative approach. It will be seen later in this chapter that more advanced conversion techniques allow the track-hold circuit and its shortcomings to be eliminated.

4.7 Sampling clock jitter

The instants at which samples are taken in an ADC and the instants at which DACs make conversions must be evenly spaced, otherwise unwanted signals can be added to the audio. Figure 4.13 shows the effect of sampling clock jitter on a sloping waveform. Samples are taken at the wrong times. When these samples have passed through a system, the timebase correction stage prior to the DAC will remove the jitter, and the result is shown at (b). The magnitude of the unwanted signal is proportional to the slope of the audio waveform and so the amount of jitter which can be tolerated falls at 6 dB per octave. As the resolution of the system is increased by the use of longer sample wordlength, tolerance to jitter is further reduced. The nature of the unwanted signal depends on the spectrum of the jitter. If the jitter is random, the effect is noise-like and relatively benign unless the amplitude is excessive. Figure 4.14 shows the effect of differing amounts of random jitter with respect to the noise floor of various wordlengths. Note that even small amounts of jitter can degrade a twenty-bit convertor to the performance of a good sixteen-bit unit. There is thus no point in upgrading to higher-resolution convertors if the clock stability of the system is insufficient to allow their performance to be realized.

Clock jitter is not necessarily random. Figure 4.15 shows that one source of clock jitter is crosstalk or interference on the clock signal. A balanced clock line will be more immune to such crosstalk, but the consumer electrical digital audio interface is unbalanced and prone to external interference. The unwanted additional signal changes the time at which the sloping clock signal appears to cross the threshold voltage of the clock receiver. This is simply the same phenomenon as that of Figure 4.13 but in reverse. The threshold itself may be changed by ripple on the clock receiver power supply. There is no reason why these effects should be random; they may be periodic and potentially audible.[12,13]

The allowable jitter is measured in picoseconds, as shown in Figure 4.13 and clearly steps must be taken to eliminate it by design. Convertor clocks must be generated from clean power supplies which are well decoupled from the power used by the logic because a convertor clock must have a signal-to-noise ratio of

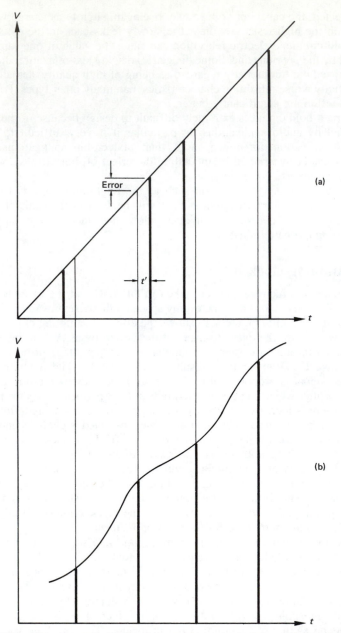

Figure 4.13 The effect of sampling timing jitter on noise, and calculation of the required accuracy for a sixteen-bit system. (a) Ramp sampled with jitter has error proportional to slope. (b) When jitter is removed by later circuits, error appears as noise added to samples. For a sixteen-bit system there are $2^{16}Q$, and the maximum slope at 20 kHz will be $20\,000\,\pi \times 2^{16}Q$ per second. If jitter is to be neglected, the noise must be less than $\frac{1}{2}Q$, thus timing accuracy t' multiplied by maximum slope $= \frac{1}{2}Q$ or $20\,000\,\pi \times 2^{16}Qt' = \frac{1}{2}Q$

$$\therefore 2' = \frac{1}{2 \times 20\,000 \times \pi \times 2^{16}} = 121\,\text{ps}$$

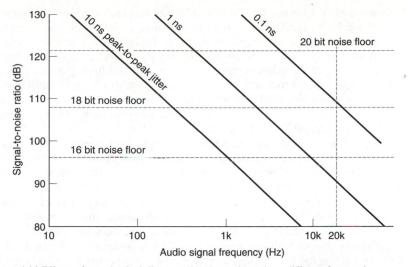

Figure 4.14 Effects of sample clock jitter on signal-to-noise ratio at different frequencies, compared with theoretical noise floors of systems with different resolutions. (After W.T. Shelton, with permission)

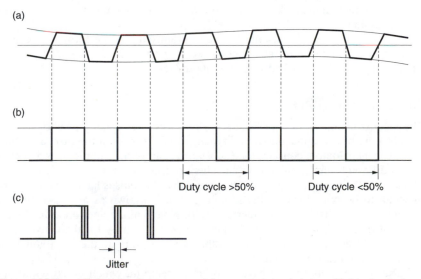

Figure 4.15 Crosstalk in transmission can result in unwanted signals being added to the clock waveform. It can be seen here that a low-frequency interference signal affects the slicing of the clock and causes a periodic jitter.

the same order as that of the audio. Otherwise noise on the clock causes jitter which in turn causes noise in the audio.

If an external clock source is provided, it cannot be used directly, but must be fed through a well-designed, well-damped phase-locked loop which will filter out the jitter. The operation of a phase-locked loop was described in Chapter 3.

The phase-locked loop must be built to a higher accuracy standard than in most applications. Noise reaching the frequency control element will cause the very jitter the device is meant to eliminate. Some designs use a crystal oscillator whose natural frequency can be shifted slightly by a varicap diode. The high Q of the crystal produces a cleaner clock. Unfortunately this high Q also means that the frequency swing which can be achieved is quite small. It is sufficient for locking to a single standard sampling rate reference, but not for locking to a range of sampling rates or for variable-speed operation. In this case a conventional varicap VCO is required.

Although it has been documented for many years, attention to control of clock jitter is not as great in actual hardware as it might be. It accounts for much of the slight audible differences between convertors reproducing the same data. A well-engineered convertor should substantially reject jitter on an external clock and should sound the same when reproducing the same data irrespective of the source of the data. A remote convertor which sounds different when reproducing, for example, the same Compact Disc via the digital outputs of a variety of CD players is simply not well engineered and should be rejected. Similarly if the effect of changing the type of digital cable feeding the convertor can be heard, the unit is deficient. Unfortunately many consumer external DACs fall into this category, as the steps outlined above have not been taken. Some consumer external DACs, however, have RAM timebase correction which has a large enough correction range that the convertor can run from a local fixed frequency crystal. The incoming clock does no more than control the memory write cycles. Any incoming jitter is rejected totally.

Many portable digital machines have compromised jitter performance because their small size and weight constraints make the provision of adequate screening, decoupling and phase-locked loop circuits difficult.

4.8 Aperture effect

The reconstruction process of Figure 4.6 only operates exactly as shown if the impulses are of negligible duration. In many DACs this is not the case, and many keep the analog output constant for a substantial part of the sample period or even until a different sample value is input. This produces a waveform which is more like a staircase than a pulse train. The case where the pulses have been extended in width to become equal to the sample period is known as a zero-order-hold system and has a 100 per cent aperture ratio. Note that the aperture effect is not apparent in a track-hold system; the holding period is only for the convenience of the quantizer which then outputs a value corresponding to the input voltage at the instant hold mode was entered.

Pulses of negligible width have a uniform spectrum, which is flat within the audio band, whereas pulses of 100 per cent aperture ratio have a $\sin x/x$ spectrum which is shown in Figure 4.16. The frequency response falls to a null at the sampling rate, and as a result is about 4 dB down at the edge of the audio band. If the pulse width is stable, the reduction of high frequencies is constant and predictable, and an appropriate equalization circuit can render the overall response flat once more. An alternative is to use resampling which is shown in Figure 4.17. Resampling passes the zero-order-hold waveform through a further synchronous sampling stage which consists of an analog switch which closes briefly in the centre of each sample period. The output of the switch will be

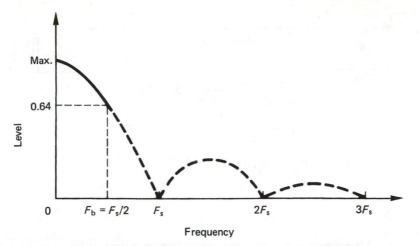

Figure 4.16 Frequency response with 100 per cent aperture has nulls at multiples of sampling rate. Area of interest is up to half sampling rate.

pulses which are narrower than the original. If, for example, the aperture ratio is reduced to 50 per cent of the sample period, the first frequency response null is now at twice the sampling rate, and the loss at the edge of the audio band is reduced. As the figure shows, the frequency response becomes flatter as the aperture ratio falls. The process should not be carried too far, as with very small aperture ratios there is little energy in the pulses and noise can be a problem. A practical limit is around 12.5 per cent where the frequency response is virtually ideal. The term resampling will also be found in descriptions of sampling rate convertors, where it refers to the process of finding samples at new locations to describe the original waveform. The context usually makes it clear which meaning is intended.

4.9 Quantizing

Quantizing is the process of expressing some infinitely variable quantity by discrete or stepped values. Quantizing turns up in a remarkable number of everyday guises. Figure 4.18 shows that an inclined ramp enables infinitely variable height to be achieved, whereas a step-ladder allows only discrete heights to be had. A step-ladder quantizes height. When accountants round off sums of money to the nearest pound or dollar they are quantizing. Time passes continuously, but the display on a digital clock changes suddenly every minute because the clock is quantizing time.

In audio the values to be quantized are infinitely variable voltages from an analog source. Strict quantizing is a process which operates in the voltage domain only. For the purpose of studying the quantizing of a single sample, time is assumed to stand still. This is achieved in practice either by the use of a track-hold circuit or the adoption of a quantizer technology which operates before the sampling stage.

Figure 4.19(a) shows that the process of quantizing divides the voltage range up into quantizing intervals Q, also referred to as steps S. In applications such as

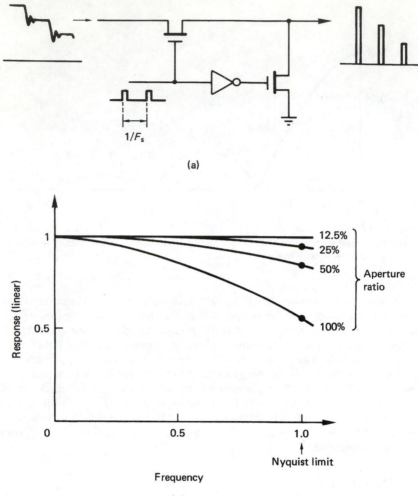

(a)

(b)

Figure 4.17 (a) Resampling circuit eliminates transients and reduces aperture ratio. (b) Response of various aperture ratios.

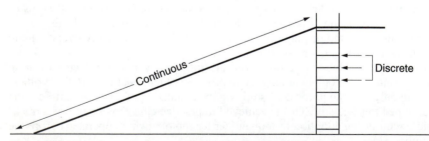

Figure 4.18 An analog parameter is continuous whereas a quantized parameter is restricted to certain values. Here the sloping side of a ramp can be used to obtain any height whereas a ladder only allows discrete heights.

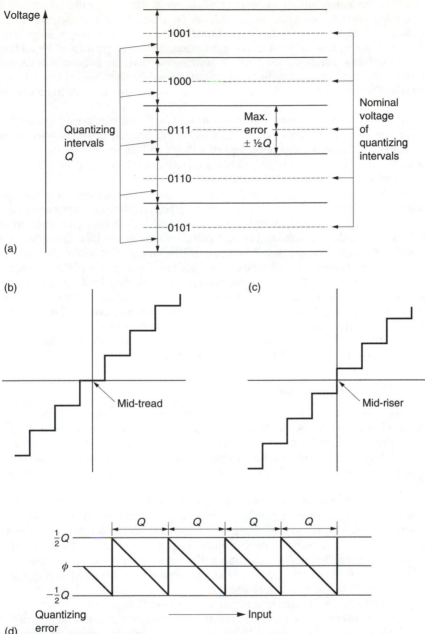

Figure 4.19 Quantizing assigns discrete numbers to variable voltages. All voltages within the same quantizing interval are assigned the same number which causes a DAC to produce the voltage at the centre of the intervals shown by the dashed lines in (a). This is the characteristic of the mid-tread quantizer shown in (b). An alternative system is the mid-riser system shown in (c). Here 0 volts analog falls between two codes and there is no code for zero. Such quantizing cannot be used prior to signal processing because the number is no longer proportional to the voltage. Quantizing error cannot exceed $\pm\frac{1}{2}Q$ as shown in (d).

telephony these may advantageously be of differing size, but for digital audio the quantizing intervals are made as identical as possible. If this is done, the binary numbers which result are truly proportional to the original analog voltage, and the digital equivalents of mixing and gain changing can be performed by adding and multiplying sample values. If the quantizing intervals are unequal this cannot be done. When all quantizing intervals are the same, the term uniform quantizing is used. The term linear quantizing will be found, but this is a contradiction in terms.

The term LSB (least significant bit) will also be found in place of quantizing interval in some treatments, but this is a poor term because quantizing works in the voltage domain. A bit is not a unit of voltage and can have only two values. In studying quantizing, voltages within a quantizing interval will be discussed, but there is no such thing as a fraction of a bit.

Whatever the exact voltage of the input signal, the quantizer will locate the quantizing interval in which it lies. In what may be considered a separate step, the quantizing interval is then allocated a code value which is typically some form of binary number. The information sent is the number of the quantizing interval in which the input voltage lies. Whereabouts that voltage lies within the interval is not conveyed, and this mechanism puts a limit on the accuracy of the quantizer. When the number of the quantizing interval is converted back to the analog domain, it will result in a voltage at the centre of the quantizing interval as this minimizes the magnitude of the error between input and output. The number range is limited by the wordlength of the binary numbers used. In a sixteen-bit system, 65 536 different quantizing intervals exist, although the ones at the extreme ends of the range have no outer boundary.

4.10 Quantizing error

It is possible to draw a transfer function for such an ideal quantizer followed by an ideal DAC, and this is also shown in Figure 4.19. A transfer function is simply a graph of the output with respect to the input. In audio, when the term linearity is used, this generally means the straightness of the transfer function. Linearity is a goal in audio, yet it will be seen that an ideal quantizer is anything but linear.

Figure 4.19(b) shows the transfer function is somewhat like a staircase, and zero volts analog, corresponding to all zeros digital or muting, is half-way up a quantizing interval, or on the centre of a tread. This is the so-called mid-tread quantizer which is universally used in audio. Figure 4.19(c) shows the alternative mid-riser transfer function which causes difficulty in audio because it does not have a code value at muting level and as a result the numerical code value is not proportional to the analog signal voltage.

Quantizing causes a voltage error in the audio sample which is given by the difference between the actual staircase transfer function and the ideal straight line. This is shown in Figure 4.19(d) to be a sawtooth-like function which is periodic in Q. The amplitude cannot exceed $\pm\frac{1}{2}Q$ peak-to-peak unless the input is so large that clipping occurs.

Quantizing error can also be studied in the time domain where it is better to avoid complicating matters with the aperture effect of the DAC. For this reason it is assumed here that output samples are of negligible duration. Then impulses from the DAC can be compared with the original analog waveform and the

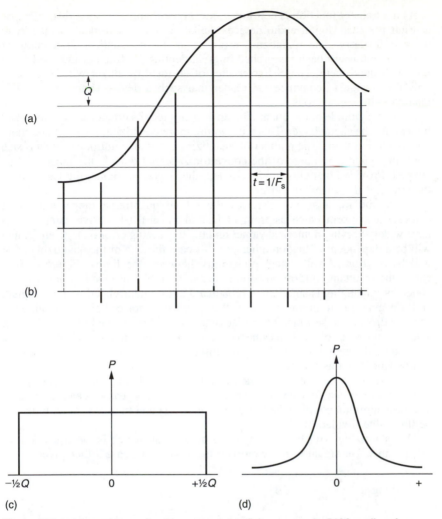

Figure 4.20 At (a) an arbitrary signal is represented to finite accuracy by PAM needles whose peaks are at the centre of the quantizing intervals. The errors caused can be thought of as an unwanted signal (b) added to the original. In (c) the amplitude of a quantizing error needle will be from $-\tfrac{1}{2}Q$ to $+\tfrac{1}{2}Q$ with equal probability. Note, however, that white noise in analog circuits generally has Gaussian amplitude distribution, shown in (d).

difference will be impulses representing the quantizing error waveform. This has been done in Figure 4.20. The horizontal lines in the drawing are the boundaries between the quantizing intervals, and the curve is the input waveform. The vertical bars are the quantized samples which reach to the centre of the quantizing interval. The quantizing error waveform shown at (b) can be thought of as an unwanted signal which the quantizing process adds to the perfect original. If a very small input signal remains within one quantizing interval, the quantizing error *is* the signal.

As the transfer function is non-linear, ideal quantizing can cause distortion. As a result practical digital audio devices deliberately use non-ideal quantizers to achieve linearity. The quantizing error of an ideal quantizer is a complex function, and it has been researched in great depth.[14–16] It is not intended to go into such depth here. The characteristics of an ideal quantizer will be pursued only far enough to convince the reader that such a device cannot be used in quality audio applications.

As the magnitude of the quantizing error is limited, its effect can be minimized by making the signal larger. This will require more quantizing intervals and more bits to express them. The number of quantizing intervals multiplied by their size gives the quantizing range of the convertor. A signal outside the range will be clipped. Provided that clipping is avoided, the larger the signal, the less will be the effect of the quantizing error.

Where the input signal exercises the whole quantizing range and has a complex waveform (such as from orchestral music), successive samples will have widely varying numerical values and the quantizing error on a given sample will be independent of that on others. In this case the size of the quantizing error will be distributed with equal probability between the limits. Figure 4.20(c) shows the resultant uniform probability density. In this case the unwanted signal added by quantizing is an additive broadband noise uncorrelated with the signal, and it is appropriate in this case to call it quantizing noise. This is not quite the same as thermal noise which has a Gaussian probability shown in Figure 4.20(d). The difference is of no consequence as in the large signal case the noise is masked by the signal. Under these conditions, a meaningful signal-to-noise ratio can be calculated as follows.

In a system using n-bit words. there will be 2^n quantizing intervals. The largest sinusoid which can fit without clipping will have this peak-to-peak amplitude. The peak amplitude will be half as great, i.e. $2^{n-1}Q$ and the r.m.s. amplitude will be this value divided by $\sqrt{2}$.

The quantizing error has an amplitude of $\frac{1}{2}Q$ peak which is the equivalent of $Q/\sqrt{12}$ r.m.s. The signal-to-noise ratio for the large signal case is then given by:

$$20 \log_{10} \frac{\sqrt{12} \times 2^{n-1}}{\sqrt{2}} \text{ dB}$$

$$= 20 \log_{10} (\sqrt{6} \times 2^{n-1}) \text{ dB}$$

$$= 20 \log (2^n \times \frac{\sqrt{6}}{2} \text{ dB}$$

$$= 20n \log 2 + 20 \log \frac{\sqrt{6}}{2} \text{ dB}$$

$$= 6.02n + 1.76 \text{ dB} \tag{4.1}$$

By way of example, a sixteen-bit system will offer around 98.1 dB SNR.

Whilst the above result is true for a large complex input waveform, treatments which then assume that quantizing error is *always* noise give results which are at variance with reality. The expression above is only valid if the probability density of the quantizing error is uniform. Unfortunately at low levels, and particularly with pure or simple waveforms, this is simply not the case.

At low audio levels, quantizing error ceases to be random, and becomes a function of the input waveform and the quantizing structure as Figure 4.20 showed. Once an unwanted signal becomes a deterministic function of the wanted signal, it has to be classed as a distortion rather than a noise. Distortion can also be predicted from the non-linearity, or staircase nature, of the transfer function. With a large signal, there are so many steps involved that we must stand well back, and a staircase with 65 000 steps appears to be a slope. With a small signal there are few steps and they can no longer be ignored.

The non-linearity of the transfer function results in distortion, which produces harmonics. Unfortunately these harmonics are generated *after* the anti-aliasing filter, and so any which exceed half the sampling rate will alias. Figure 4.21

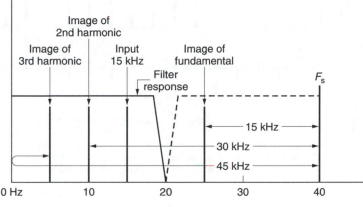

Figure 4.21 Quantizing produces distortion *after* the anti-aliasing filter, thus the distortion products will fold back to produce anharmonics in the audio band. Here the fundamental of 15 kHz produces second and third harmonic distortion at 30 and 45 kHz. This results in aliased products at $40 - 30 = 10$ kHz and $40 - 45 = (-)5$ kHz.

shows how this results in anharmonic distortion within the audio band. These anharmonics result in spurious tones known as birdsinging. When the sampling rate is a multiple of the input frequency the result is harmonic distortion. This is shown in Figure 4.22. Where more than one frequency is present in the input, intermodulation distortion occurs, which is known as granulation.

As the input signal is further reduced in level, it may remain within one quantizing interval. The output will be silent because the signal is now the quantizing error. In this condition, low-frequency signals such as air-conditioning rumble can shift the input in and out of a quantizing interval so that the quantizing distortion comes and goes, resulting in noise modulation.

Needless to say, any one of the above effects would preclude the use of an ideal quantizer for high-quality work. There is little point in studying the adverse effects further as they should be and can be eliminated completely in practical equipment by the use of dither. The importance of correctly dithering a quantizer cannot be emphasized enough, since failure to dither irrevocably distorts the converted signal: there can be no process which will subsequently remove that distortion.

The signal-to-noise ratio derived above has no relevance to practical audio applications as it will be modified by the dither and by any noise shaping used.

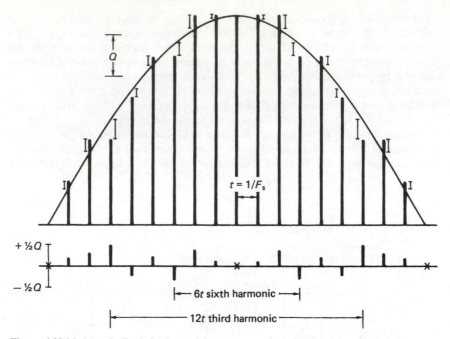

Figure 4.22 Mathematically derived quantizing error waveform for sine wave sampled at a multiple of itself. The numerous autocorrelations between quantizing errors show that there are harmonics of the signal in the error, and that the error is not random, but deterministic.

At high signal levels, quantizing error is effectively noise. As the audio level falls, the quantizing error of an ideal quantizer becomes more strongly correlated with the signal and the result is distortion. If the quantizing error can be decorrelated from the input in some way, the system can remain linear but noisy. Dither performs the job of decorrelation by making the action of the quantizer unpredictable and gives the system a noise floor like an analog system.

The first documented use of dither was by Roberts[17] in picture coding. In this system, pseudo-random noise (see Chapter 3) was added to the input signal prior to quantizing, but was subtracted after reconversion to analog. This is known as subtractive dither and has the advantages that the dither amplitude is non-critical and that the noise has full statistical independence from the signal. Unfortunately, it suffers from practical drawbacks, since the original noise waveform must accompany the samples or must be synchronously recreated at the DAC. This is virtually impossible in a system where the audio may have been edited or where its level has been changed by processing, as the noise needs to remain synchronous and be processed in the same way. Almost all practical digital audio systems use non-subtractive dither where the dither signal is added prior to quantization and no attempt is made to remove it at the DAC.[18] The introduction of dither prior to a conventional quantizer inevitably causes a slight reduction in the signal-to-noise ratio attainable, but this reduction is a small price to pay for the elimination of non-linearities. The technique of noise shaping in conjunction with dither will be seen to overcome this restriction and produce performance in excess of the subtractive dither example above.

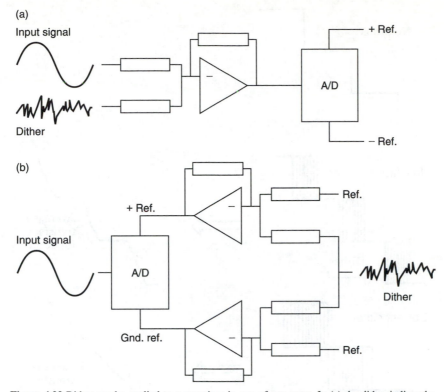

Figure 4.23 Dither can be applied to a quantizer in one of two ways. In (a) the dither is linearly added to the analog input signal, whereas in (b) it is added to the reference voltages of the quantizer.

The ideal (noiseless) quantizer of Figure 4.19 has fixed quantizing intervals and must always produce the same quantizing error from the same signal. In Figure 4.23 it can be seen that an ideal quantizer can be dithered by linearly adding a controlled level of noise either to the input signal or to the reference voltage which is used to derive the quantizing intervals. There are several ways of considering how dither works, all of which are equally valid.

The addition of dither means that successive samples effectively find the quantizing intervals in different places on the voltage scale. The quantizing error becomes a function of the dither, rather than a predictable function of the input signal. The quantizing error is not eliminated, but the subjectively unacceptable distortion is converted into a broadband noise which is more benign to the ear.

Some alternative ways of looking at dither are shown in Figure 4.24. Consider the situation where a low-level input signal is changing slowly within a quantizing interval. Without dither, the same numerical code is output for a number of sample periods, and the variations within the interval are lost. Dither has the effect of forcing the quantizer to switch between two or more states. The higher the voltage of the input signal within a given interval, the more probable it becomes that the output code will take on the next higher value. The lower the

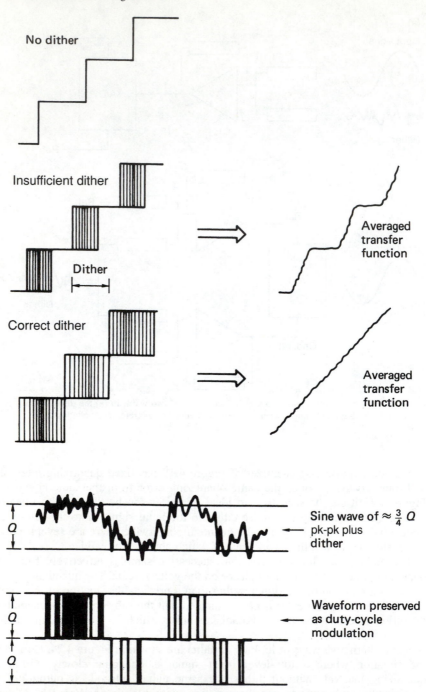

No dither

Insufficient dither

Dither

Correct dither

Averaged
transfer
function

Averaged
transfer
function

Q Sine wave of $\approx \frac{3}{4} Q$
pk-pk plus
dither

Q
Q Waveform preserved
as duty-cycle
modulation

Figure 4.24 Wideband dither of the appropriate level linearizes the transfer function to produce noise instead of distortion. This can be confirmed by spectral analysis. In the voltage domain, dither causes frequent switching between codes and preserves resolution in the duty cycle of the switching.

input voltage within the interval, the more probable it is that the output code will take the next lower value. The dither has resulted in a form of duty cycle modulation, and the resolution of the system has been extended indefinitely instead of being limited by the size of the steps.

Dither can also be understood by considering what it does to the transfer function of the quantizer. This is normally a perfect staircase, but in the presence of dither it is smeared horizontally until with a certain amplitude the average transfer function becomes straight.

In an extension of the application of dither, Blesser[19] has suggested digitally generated dither which is converted to the analog domain and added to the input signal prior to quantizing. That same digital dither is then subtracted from the digital quantizer output. The effect is that the transfer function of the quantizer is smeared diagonally (Figure 4.25). The significance of this diagonal smearing

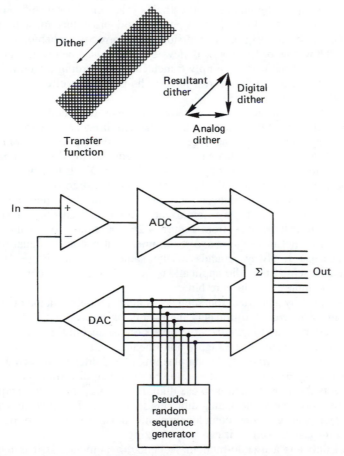

Figure 4.25 In this dither system, the dither added in the analog domain shifts the transfer function horizontally, but the same dither is subtracted in the digital domain, which shifts the transfer function vertically. The result is that the quantizer staircase is smeared diagonally as shown top left. There is thus no limit to dither amplitude, and excess dither can be used to improve differential linearity of the convertor.

is that the amplitude of the dither is not critical. However much dither is employed, the noise amplitude will remain the same. If dither of several quantizing intervals is used, it has the effect of making all the quantizing intervals in an imperfect convertor appear to have the same size.

The advanced ADC technology which is detailed later in this chapter allows as much as 24-bit resolution to be obtained, with perhaps more in the future. The situation then arises that an existing sixteen-bit device such as a digital recorder needs to be connected to the output of an ADC with greater wordlength. The words need to be shortened in some way.

Chapter 3 showed that when a sample value is attenuated, the extra low-order bits which come into existence below the radix point preserve the resolution of the signal and the dither in the least significant bit(s) which linearizes the system. The same word extension will occur in any process involving multiplication, such as digital filtering. It will subsequently be necessary to shorten the wordlength. Clearly the high-order bits cannot be discarded in two's complement as this would cause clipping of positive half-cycles and a level shift on negative half-cycles due to the loss of the sign bit. Low-order bits must be removed instead. Even if the original conversion was correctly dithered, the random element in the low-order bits will now be some way below the end of the intended word. If the word is simply truncated by discarding the unwanted low-order bits or rounded to the nearest integer the linearizing effect of the original dither will be lost.

Shortening the wordlength of a sample reduces the number of quantizing intervals available without changing the signal amplitude. As Figure 4.26 shows, the quantizing intervals become larger and the original signal is *requantized* with the new interval structure. This will introduce requantizing distortion having the same characteristics as quantizing distortion in an ADC. It then is obvious that when shortening the wordlength of a twenty-bit convertor to sixteen bits, the four low-order bits must be removed in a way that displays the same overall quantizing structure as if the original convertor had been only of sixteen-bit wordlength. It will be seen from Figure 4.26 that truncation cannot be used because it does not meet the above requirement but results in signal-dependent offsets because it always rounds in the same direction. Proper numerical rounding is essential in audio applications. Rounding in two's complement is a little more complex than in pure binary.

Requantizing by numerical rounding accurately simulates analog quantizing to the new interval size. Unfortunately the twenty-bit convertor will have a dither amplitude appropriate to quantizing intervals one sixteenth the size of a sixteen-bit unit and the result will be highly non-linear.

In practice, the wordlength of samples must be shortened in such a way that the requantizing error is converted to noise rather than distortion. One technique which meets this requirement is to use digital dithering[20] prior to rounding. This is directly equivalent to the analog dithering in an ADC. It will be shown later in this chapter that in more complex systems noise shaping can be used in requantizing just as well as it can in quantizing.

Digital dither is a pseudo-random sequence of numbers. If it is required to simulate the analog dither signal of Figures 4.23 and 4.24, then it is obvious that the noise must be bipolar so that it can have an average voltage of zero. Two's complement coding must be used for the dither values as it is for the audio samples.

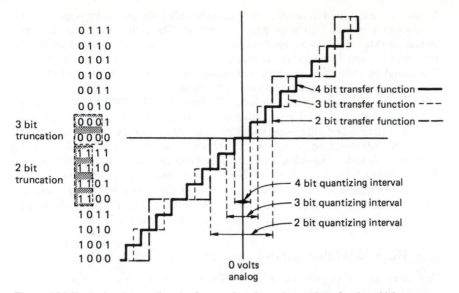

```
        0 1 1 1
        0 1 1 0
        0 1 0 1
        0 1 0 0
        0 0 1 1
        0 0 1 0
3 bit    0 0 0 1                        4 bit transfer function ——
truncation 0 0 0 0                       3 bit transfer function - - -
         1 1 1 1                         2 bit transfer function —— ——
2 bit    1 1 1 0
truncation 1 1 0 1                       4 bit quantizing interval
         1 1 0 0                         3 bit quantizing interval
         1 0 1 1                         2 bit quantizing interval
         1 0 1 0
         1 0 0 1
         1 0 0 0
                        0 volts
                        analog
```

Figure 4.26 Shortening the wordlength of a sample reduces the number of codes which can describe the voltage of the waveform. This makes the quantizing steps bigger, hence the term requantizing. It can be seen that simple truncation or omission of the bits does not give analogous behaviour. Rounding is necessary to give the same result as if the larger steps had been used in the original conversion.

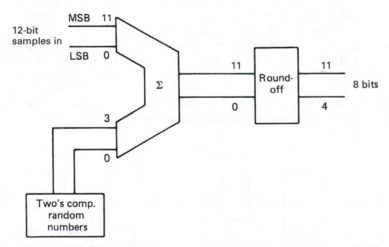

Figure 4.27 In a simple digital dithering system, two's complement values from a random number generator are added to low-order bits of the input. The dithered values are then rounded up or down according to the value of the bits to be removed. The dither linearizes the requantizing.

Figure 4.27 shows a simple digital dithering system (i.e. one without noise shaping) for shortening sample wordlength. The output of a two's complement pseudo-random sequence generator (see Chapter 3) of appropriate wordlength is added to input samples prior to rounding. The most significant of the bits to be discarded is examined in order to determine whether the bits to be removed sum to more or less than half a quantizing interval. The dithered sample is either rounded down, i.e. the unwanted bits are simply discarded, or rounded up, i.e. the unwanted bits are discarded but one is added to the value of the new short word. The rounding process is no longer deterministic because of the added dither which provides a linearizing random component.

If this process is compared with that of Figure 4.23 it will be seen that the principles of analog and digital dither are identical; the processes simply take place in different domains using two's complement numbers which are rounded or voltages which are quantized as appropriate. In fact quantization of an analog dithered waveform is identical to the hypothetical case of rounding after bipolar digital dither where the number of bits to be removed is infinite, and remains identical for practical purposes when as few as eight bits are to be removed. Analog dither may actually be generated from bipolar digital dither (which is no more than random numbers with certain properties) using a DAC.

4.11 Basic digital-to-analog conversion

This direction of conversion will be discussed first, since ADCs often use embedded DACs in feedback loops.

The purpose of a digital-to-analog convertor is to take numerical values and reproduce the continuous waveform that they represent. Figure 4.28 shows the major elements of a conventional conversion subsystem, i.e. one in which oversampling is not employed. The jitter in the clock needs to be removed with a VCO or VCXO. Sample values are buffered in a latch and fed to the convertor element which operates on each cycle of the clean clock. The output is then a voltage proportional to the number for at least a part of the sample period. A resampling stage may be found next, in order to remove switching transients, reduce the aperture ratio or allow the use of a convertor which takes a substantial part of the sample period to operate. The resampled waveform is then presented to a reconstruction filter which rejects frequencies above the audio band. This section is primarily concerned with the implementation of the convertor element. There are two main ways of obtaining an analog signal from PCM data. One is

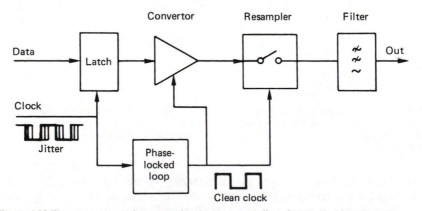

Figure 4.28 The components of a conventional convertor. A jitter-free clock drives the voltage conversion, whose output may be resampled prior to reconstruction.

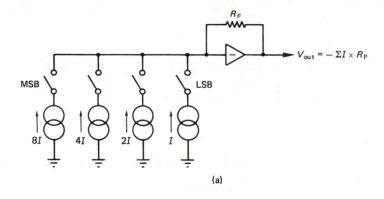

(a)

(b)

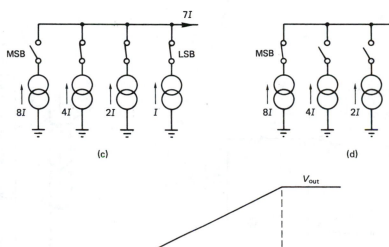

(c)   (d)

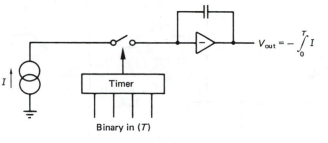

(e)

Figure 4.29 Elementary conversion: (a) weighted current DAC; (b) timed integrator DAC; (c) current flow with 0111 input; (d) current flow with 1000 input; (e) integrator ramps up for 15 cycles of clock for input 1111.

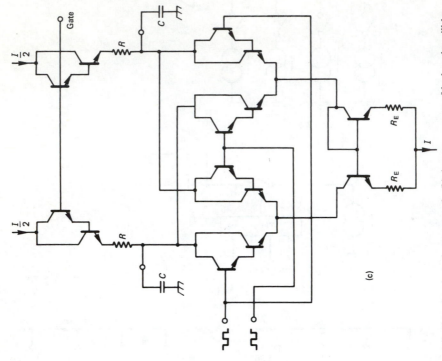

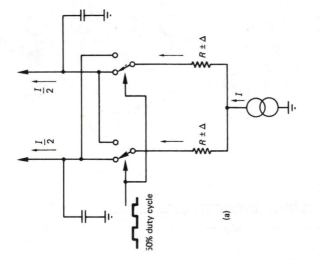

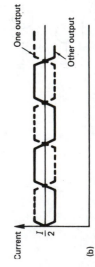

Figure 4.30 Dynamic element matching. (a) Each resistor spends half its time in each current path. (b) Average current of both paths will be identical if duty cycle is accurately 50 per cent. (c) Typical monolithic implementation. Note clock frequency is arbitrary.

to control binary-weighted currents and sum them; the other is to control the length of time a fixed current flows into an integrator. The two methods are contrasted in Figure 4.29. They appear simple, but are of no use for audio in these forms because of practical limitations. In Figure 4.29(c), the binary code is about to have a major overflow, and all the low-order currents are flowing. In Figure 4.29(d), the binary input has increased by one, and only the most significant current flows. This current must equal the sum of all the others plus one. The accuracy must be such that the step size is within the required limits. In this simple four-bit example, if the step size needs to be a rather casual 10 per cent accurate, the necessary accuracy is only one part in 160, but for a sixteen-bit system it would become one part in 655 360, or about 2 ppm. This degree of accuracy is almost impossible to achieve, let alone maintain in the presence of ageing and temperature change.

The integrator-type convertor in this four-bit example is shown in Figure 4.29(e); it requires a clock for the counter which allows it to count up to the maximum in less than one sample period. This will be more than sixteen times the sampling rate. However, in a sixteen-bit system, the clock rate would need to be 65 536 times the sampling rate, or about 3 GHz. Whilst there may be a market for a CD player which can defrost a chicken, clearly some refinements are necessary to allow either of these convertor types to be used in audio applications.

One method of producing currents of high relative accuracy is *dynamic element matching*.[21,22] Figure 4.30 shows a current source feeding a pair of nominally equal resistors. The two will not be the same owing to manufacturing tolerances and drift, and thus the current is only approximately divided between them. A pair of change-over switches places each resistor in series with each output. The average current in each output will then be identical, provided that the duty cycle of the switches is exactly 50 per cent. This is readily achieved in

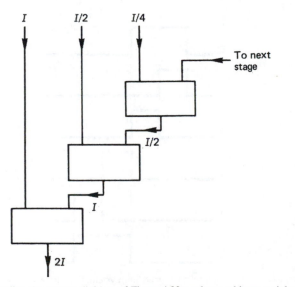

Figure 4.31 Cascading the current dividers of Figure 4.30 produces a binary-weighted series of currents.

a divide-by-two circuit. The accuracy criterion has been transferred from the resistors to the time domain in which accuracy is more readily achieved. Current averaging is performed by a pair of capacitors which do not need to be of any special quality. By cascading these divide-by-two stages, a binary-weighted series of currents can be obtained, as in Figure 4.31. In practice, a reduction in the number of stages can be obtained by using a more complex switching arrangement. This generates currents of ratio 1:1:2 by dividing the current into

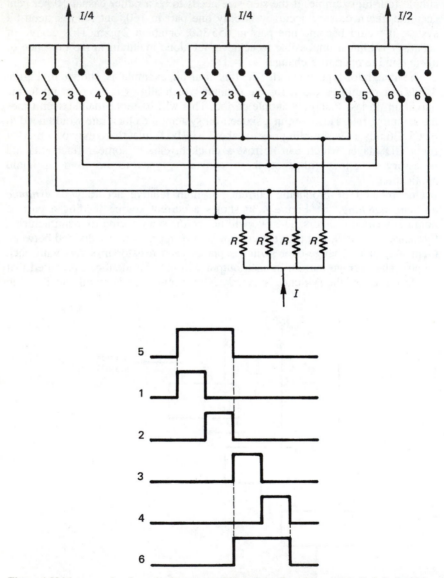

Figure 4.32 More complex dynamic element-matching system. Four drive signals (1, 2, 3, 4) of 25 per cent duty cycle close switches of corresponding number. Two signals (5, 6) have 50 per cent duty cycle, resulting in two current shares going to right-hand output. Division is thus into 1:1:2.

four paths and feeding two of them to one output, as shown in Figure 4.32. A major advantage of this approach is that no trimming is needed in manufacture, making it attractive for mass production. Freedom from drift is a further advantage.

To prevent interaction between the stages in weighted-current convertors, the currents must be switched to ground or into the virtual earth by change-over switches. The on-resistance of these switches is a source of error, particularly the MSB, which passes most current. A solution in monolithic convertors is to fabricate switches whose area is proportional to the weighted current, so that the voltage drops of all the switches are the same. The error can then be removed with a suitable offset. The layout of such a device is dominated by the MSB switch since, by definition, it is as big as all the others put together.

The practical approach to the integrator convertor is shown in Figures 4.33 and 4.34 where two current sources whose ratio is 256:1 are used; the larger is timed by the high byte of the sample and the smaller is timed by the low byte. The necessary clock frequency is reduced by a factor of 256.

Any inaccuracy in the current ratio will cause one quantizing step in every 256 to be of the wrong size as shown in Figure 4.35, but current tracking is easier to achieve in a monolithic device. The integrator capacitor must have low dielectric

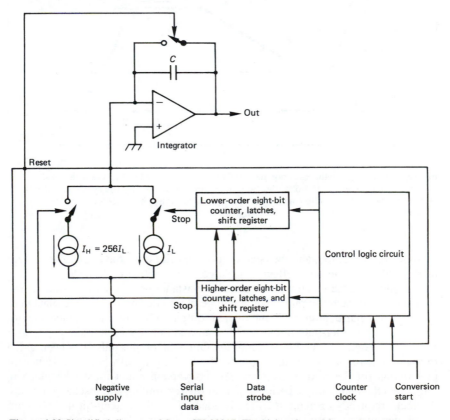

Figure 4.33 Simplified diagram of Sony CX-20017. The high-order and low-order current sources (I_H and I_L) and associated timing circuits can be seen. The necessary integrator is external.

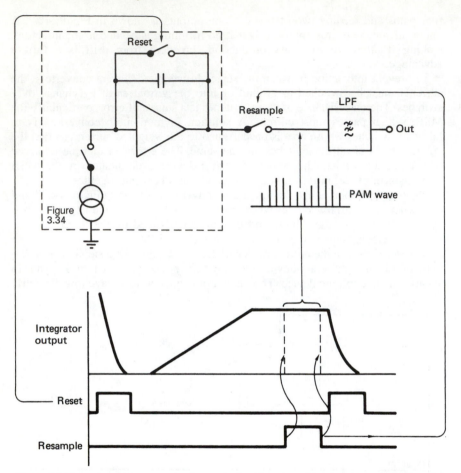

Figure 4.34 In an integrator convertor, the output level is only stable when the ramp finishes. An analog switch is necessary to isolate the ramp from subsequent circuits. The switch can also be used to produce a PAM (pulse amplitude modulated) signal which has a flatter frequency response than a zero-order-hold (staircase) signal.

leakage and relaxation, and the operational amplifier must have low bias current as this will have the same effect as leakage.

The output of the integrator will remain constant once the current sources are turned off, and the resampling switch will be closed during the voltage plateau to produce the pulse amplitude modulated output. Clearly this device cannot produce a zero-order-hold output without an additional sample-hold stage, so it is naturally complemented by resampling. Once the output pulse has been gated to the reconstruction filter, the capacitor is discharged with a further switch in preparation for the next conversion. The conversion count must take place in rather less than one sample period to permit the resampling and discharge phases. A clock frequency of about 20 MHz is adequate for a sixteen-bit 48 kHz unit, which permits the ramp to complete in 12.8 μs, leaving 8 μs for resampling and reset.

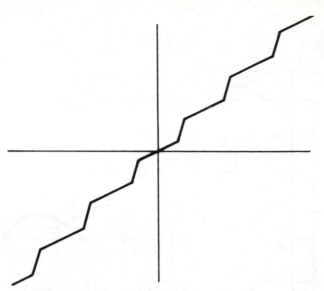

Figure 4.35 Imprecise tracking in a dual-slope convertor results in the transfer function shown here.

4.12 Basic analog-to-digital conversion

A conventional analog-to-digital subsystem is shown in Figure 4.36. Following the anti-aliasing filter there will be a sampling process. Many of the ADCs described here will need a finite time to operate, whereas an instantaneous sample must be taken from the input. The solution is to use a track-hold circuit, which was described in section 4.7. Following sampling the sample voltage is

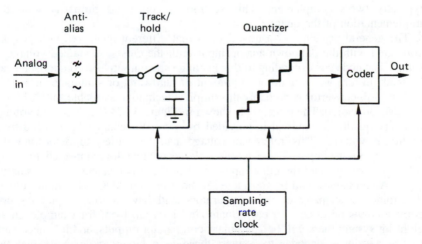

Figure 4.36 A conventional analog-to-digital subsystem. Following the anti-aliasing filter there will be a sampling process, which may include a track-hold circuit. Following quantizing, the number of the quantized level is then converted to a binary code, typically two's complement.

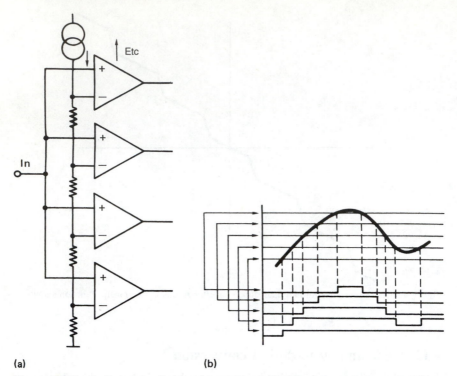

(a)   **(b)**

Figure 4.37 The flash convertor. In (a) each quantizing interval has its own comparator, resulting in waveforms of (b). A priority encoder is necessary to convert the comparator outputs to a binary code. Shown in (c) is a typical eight-bit flash convertor primarily intended for video applications. (Courtesy TRW)

quantized. The number of the quantized level is then converted to a binary code, typically two's complement. This section is concerned primarily with the implementation of the quantizing step.

The general principle of a quantizer is that different quantized voltages are compared with the unknown analog input until the closest quantized voltage is found. The code corresponding to this becomes the output. The comparisons can be made in turn with the minimal amount of hardware, or simultaneously.

The flash convertor is probably the simplest technique available for PCM and DPCM conversion. The principle is shown in Figure 4.37. The threshold voltage of every quantizing interval is provided by a resistor chain which is fed by a reference voltage. This reference voltage can be varied to determine the sensitivity of the input. There is one voltage comparator connected to every reference voltage, and the other input of all of these is connected to the analog input. A comparator can be considered to be a one-bit ADC. The input voltage determines how many of the comparators will have a true output. As one comparator is necessary for each quantizing interval, then, for example, in an eight-bit system there will be 255 binary comparator outputs, and it is necessary to use a priority encoder to convert these to a binary code. Note that the quantizing stage is asynchronous; comparators change state as and when the variations in the input waveform result in a reference voltage being crossed.

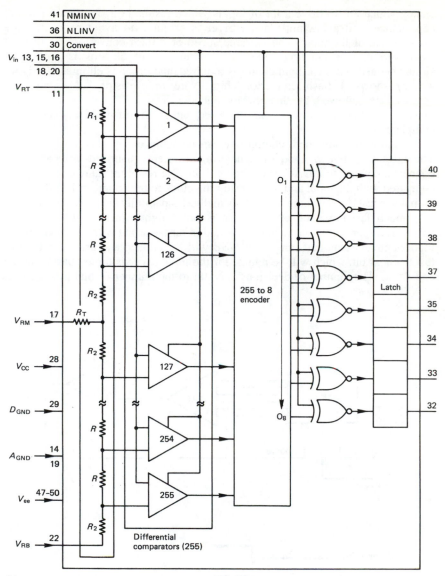

(c)

Figure 4.37 (c)

Sampling takes place when the comparator outputs are clocked into a subsequent latch. This is an example of quantizing before sampling as was illustrated in Figure 4.2. Although the device is simple in principle, it contains a lot of circuitry and can only be practicably implemented on a chip. A sixteen-bit device would need a ridiculous 65 535 comparators, and thus these convertors are not practicable for direct audio conversion, although they will be used to advantage in the DPCM and oversampling convertors described later in this chapter. The analog signal has to drive a lot of inputs which results in a significant parallel capacitance, and a low-impedance driver is essential to avoid restricting the slewing rate of the input. The extreme speed of a flash convertor is a distinct advantage in oversampling. Because computation of all bits is performed simultaneously, no track-hold circuit is required, and droop is eliminated. Figure 4.37(c) shows a flash convertor chip. Note the resistor ladder and the comparators followed by the priority encoder. The MSB can be selectively inverted so that the device can be used either in offset binary or two's complement mode.

Reduction in component complexity can be achieved by quantizing serially. The most primitive method of generating different quantized voltages is to connect a counter to a DAC as in Figure 4.38. The resulting staircase voltage is compared with the input and used to stop the clock to the counter when the DAC output has just exceeded the input. This method is painfully slow, and is not used, as a much faster method exists which is only slightly more complex. Using successive approximation, each bit is tested in turn, starting with the MSB. If the input is greater than half-range, the MSB will be retained and used as a base to test the next bit, which will be retained if the input exceeds three-quarters range and so on. The number of decisions is equal to the number of bits in the word,

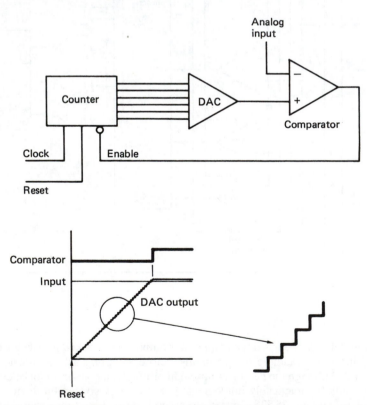

Figure 4.38 Simple ramp ADC compares output of DAC with input. Count is stopped when DAC output just exceeds input. This method, although potentially accurate, is much too slow for digital audio.

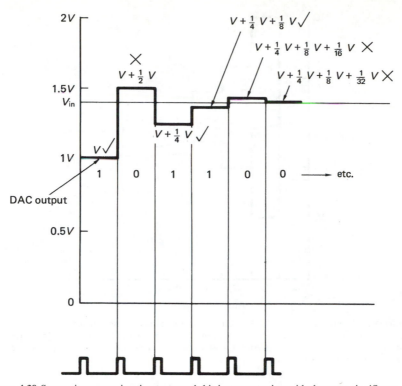

Figure 4.39 Successive approximation tests each bit in turn, starting with the most significant. The DAC output is compared with the input. If the DAC output is below the input (✓) the bit is made 1; if the DAC output is above the input (✓) the bit is made zero.

in contrast to the number of quantizing intervals which was the case in the previous example. A drawback of the successive approximation convertor is that the least significant bits are computed last, when droop is at its worst. Figures 4.39 and 4.40 show that droop can cause a successive approximation convertor to make a significant error under certain circumstances.

Analog-to-digital conversion can also be performed using the dual-current-source type DAC principle in a feedback system; the major difference is that the two current sources must work sequentially rather than concurrently. Figure 4.41 shows a sixteen-bit application in which the capacitor of the track-hold circuit is also used as the ramp integrator. The system operates as follows. When the track-hold FET switches off, the capacitor C will be holding the sample voltage. Two currents of ratio 128:1 are capable of discharging the capacitor. Owing to this ratio, the smaller current will be used to determine the seven least significant bits, and the larger current will determine the nine most significant bits. The currents are provided by current sources of ratio 127:1. When both run together, the current produced is 128 times that from the smaller source alone. This approach means that the current can be changed simply by turning off the larger source, rather than by attempting a change-over.

With both current sources enabled, the high-order counter counts up until the capacitor voltage has fallen below the reference of $-128Q$ supplied to comparator

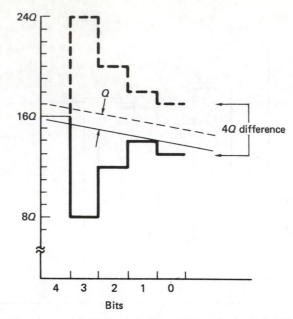

Figure 4.40 Two drooping track-hold signals (solid and dashed lines) which differ by one quantizing interval Q are shown here to result in conversions which are $4Q$ apart. Thus droop can destroy the monotonicity of a convertor. Low-level signals (near the midrange of the number system) are especially vulnerable.

1. At the next clock edge, the larger current source is turned off. Waiting for the next clock edge is important, because it ensures that the larger source can only run for entire clock periods, which will discharge the integrator by integer multiples of $128Q$. The integrator voltage will overshoot the $128Q$ reference, and the remaining voltage on the integrator will be less than $128Q$ and will be measured by counting the number of clocks for which the smaller current source runs before the integrator voltage reaches zero. This process is termed residual expansion. The break in the slope of the integrator voltage gives rise to the alternative title of gear-change convertor. Following ramping to ground in the conversion process, the track-hold circuit must settle in time for the next conversion. In this sixteen-bit example, the high-order conversion needs a maximum count of 512, and the low order needs 128: a total of 640. Allowing 25 per cent of the sample period for the track-hold circuit to operate, a 48 kHz convertor would need to be clocked at some 40 MHz. This is rather faster than the clock needed for the DAC using the same technology.

4.13 Alternative convertors

Although PCM audio is universal because of the ease with which it can be recorded and processed numerically, there are several alternative related methods of converting an analog waveform to a bitstream. The output of these convertor types is not Nyquist rate PCM, but this can be obtained from them by appropriate digital processing. In advanced conversion systems it is possible to adopt an alternative convertor technique specifically to take advantage of a particular

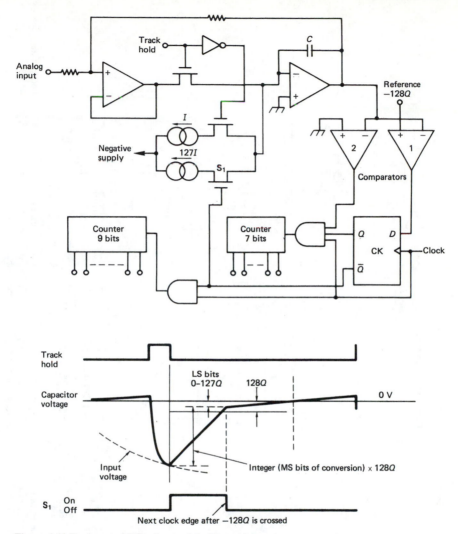

Figure 4.41 Dual-ramp ADC using track-hold capacitor as integrator.

characteristic. The output is then digitally converted to Nyquist rate PCM in order to obtain the advantages of both.

Conventional PCM has already been introduced. In PCM, the amplitude of the signal only depends on the number range of the quantizer, and is independent of the frequency of the input. Similarly, the amplitude of the unwanted signals introduced by the quantizing process is also largely independent of input frequency.

Figure 4.42 introduces the alternative convertor structures. The top half of the diagram shows convertors which are differential. In differential coding the value of the output code represents the difference between the current sample voltage and that of the previous sample. The lower half of the diagram shows convertors

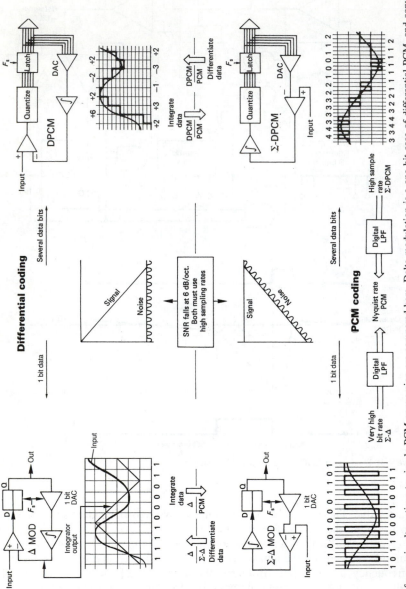

Figure 4.42 The four main alternatives to simple PCM conversion are compared here. Delta modulation is a one-bit case of differential PCM, and conveys the slope of the signal. The digital output of both can be integrated to give PCM. Σ–Δ (sigma–delta) is a one-bit case of Σ-DPCM. The application of integrator before differentiator makes the output true PCM, but tilts the noise floor; hence these can be referred to as 'noise-shaping' convertors.

which are PCM. In addition, the left side of the diagram shows single-bit convertors, whereas the right side shows multi-bit convertors.

In differential pulse code modulation (DPCM), shown at top right, the difference between the previous absolute sample value and the current one is quantized into a multi-bit binary code. It is possible to produce a DPCM signal from a PCM signal simply by subtracting successive samples; this is digital differentiation. Similarly the reverse process is possible by using an accumulator or digital integrator (see Chapter 3) to compute sample values from the differences received. The problem with this approach is that it is very easy to lose the baseline of the signal if it commences at some arbitrary time. A digital high-pass filter can be used to prevent unwanted offsets.

Differential convertors do not have an absolute amplitude limit. Instead there is a limit to the maximum rate at which the input signal voltage can change. They are said to be slew rate limited, and thus the permissible signal amplitude falls at 6 dB per octave. As the quantizing steps are still uniform, the quantizing error amplitude has the same limits as PCM. As input frequency rises, ultimately the signal amplitude available will fall down to it.

If DPCM is taken to the extreme case where only a binary output signal is available then the process is described as delta modulation (top-left in Figure 4.42). The meaning of the binary output signal is that the current analog input is above or below the accumulation of all previous bits. The characteristics of the system show the same trends as DPCM, except that there is severe limiting of the rate of change of the input signal. A DPCM decoder must accumulate all the difference bits to provide a PCM output for conversion to analog, but with a one-bit signal the function of the accumulator can be performed by an analog integrator.

If an integrator is placed in the input to a delta modulator, the integrator's amplitude response loss of 6 dB per octave parallels the convertor's amplitude limit of 6 dB per octave; thus the system amplitude limit becomes independent of frequency. This integration is responsible for the term sigma–delta modulation, since in mathematics sigma is used to denote summation. The input integrator can be combined with the integrator already present in a delta-modulator by a slight rearrangement of the components (bottom-left in Figure 4.42). The transmitted signal is now the amplitude of the input, not the slope; thus the receiving integrator can be dispensed with, and all that is necessary to after the DAC is an LPF to smooth the bits. The removal of the integration stage at the decoder now means that the quantizing error amplitude rises at 6 dB per octave, ultimately meeting the level of the wanted signal.

The principle of using an input integrator can also be applied to a true DPCM system and the result should perhaps be called sigma DPCM (bottom-right in Figure 4.42). The dynamic range improvement over delta–sigma modulation is 6 dB for every extra bit in the code. Because the level of the quantizing error signal rises at 6 dB per octave in both delta–sigma modulation and sigma DPCM, these systems are sometimes referred to as 'noise-shaping' convertors, although the word 'noise' must be used with some caution. The output of a sigma DPCM system is again PCM, and a DAC will be needed to receive it, because it is a binary code.

As the differential group of systems suffer from a wanted signal that converges with the unwanted signal as frequency rises, they must all use very high sampling rates.[23] It is possible to convert from sigma DPCM to conventional PCM by reducing the sampling rate digitally. When the sampling rate is reduced in this way, the reduction of bandwidth excludes a disproportionate amount of noise because the noise shaping concentrated it at frequencies beyond the audio band. The use of noise shaping and oversampling is the key to the high resolution obtained in advanced convertors.

4.14 Oversampling

Oversampling means using a sampling rate which is greater (generally substantially greater) than the Nyquist rate. Neither sampling theory nor quantizing theory *require* oversampling to be used to obtain a given signal quality, but Nyquist rate conversion places extremely high demands on component accuracy when a convertor is implemented. Oversampling allows a given signal quality to be reached without requiring very close tolerance, and therefore expensive, components. Although it can be used alone, the advantages of oversampling are better realized when it is used in conjunction with noise shaping. Thus in practice the two processes are generally used together and the terms are often seen used in the loose sense as if they were synonymous. For a detailed and quantitative analysis of oversampling having exhaustive references the serious reader is referred to Hauser.[24]

Figure 4.43 shows the main advantages of oversampling. At (a) it will be seen that the use of a sampling rate considerably above the Nyquist rate allows the anti-aliasing and reconstruction filters to be realized with a much more gentle cut-off slope. There is then less likelihood of phase linearity and ripple problems in the audio passband.

Figure 4.43(b) shows that information in an analog signal is two-dimensional and can be depicted as an area which is the product of bandwidth and the linearly expressed signal-to-noise ratio. The figure also shows that the same amount of information can be conveyed down a channel with a SNR of half as much (6 dB less) if the bandwidth used is doubled, with 12 dB less SNR if bandwidth is quadrupled, and so on, provided that the modulation scheme used is perfect.

The information in an analog signal can be conveyed using some analog modulation scheme in any combination of bandwidth and SNR which yields the appropriate channel capacity. If bandwidth is replaced by sampling rate and SNR is replaced by a function of wordlength, the same must be true for a digital signal as it is no more than a numerical analog. Thus raising the sampling rate potentially allows the wordlength of each sample to be reduced without information loss.

Oversampling permits the use of a convertor element of shorter wordlength, making it possible to use a flash convertor. The flash convertor is capable of working at very high frequency and so large oversampling factors are easily realized. The flash convertor needs no track-hold system as it works instantaneously. The drawbacks of track-hold set out in section 4.6 are thus eliminated. If the sigma-DPCM convertor structure of Figure 4.42 is realized with a flash convertor element, it can be used with a high oversampling factor. Figure 4.43(c) shows that this class of convertor has a rising noise floor. If the highly oversampled output is fed to a digital low-pass filter which has the same frequency response as an analog anti-aliasing filter used for Nyquist rate sampling, the result is a disproportionate reduction in noise because the majority of the noise was outside the audio band. A high-resolution convertor can be

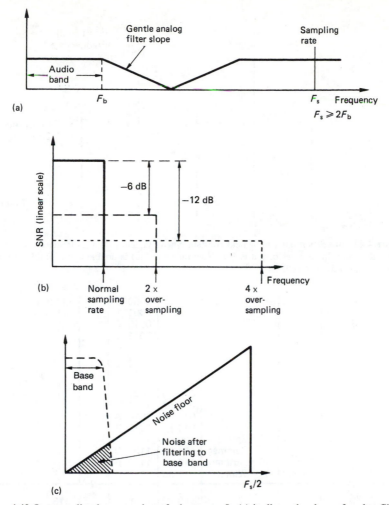

Figure 4.43 Oversampling has a number of advantages. In (a) it allows the slope of analog filters to be relaxed. In (b) it allows the resolution of convertors to be extended. In (c) *a noise-shaped* convertor allows a disproportionate improvement in resolution.

obtained using this technology without requiring unattainable component tolerances.

Information theory predicts that if an audio signal is spread over a much wider bandwidth by, for example, the use of an FM broadcast transmitter, the SNR of the demodulated signal can be higher than that of the channel it passes through, and this is also the case in digital systems. The concept is illustrated in Figure 4.44. At (a) four-bit samples are delivered at sampling rate F. As four bits have sixteen combinations, the information rate is $16F$. At (b) the same information rate is obtained with three-bit samples by raising the sampling rate to $2F$ and at (c) two-bit samples having four combinations require to be delivered at a rate of $4F$. Whilst the information rate has been maintained, it will be noticed that the bit-rate of (c) is twice that of (a). The reason for this is shown in Figure 4.45.

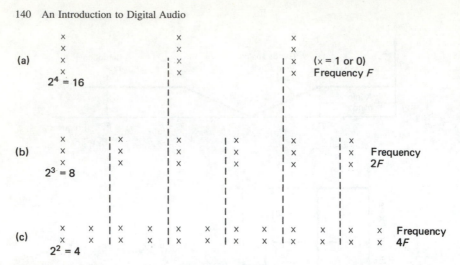

Figure 4.44 Information rate can be held constant when frequency doubles by removing one bit from each word. In all cases here it is 16F. Note bit rate of (c) is double that of (a). Data storage in oversampled form is inefficient.

	0 = No 1 = Yes	00 = Spring 01 = Summer 10 = Autumn 11 = Winter	000 do 001 re 010 mi 011 fa 100 so 101 la 110 te 111 do	0000 0 0001 1 0010 2 0011 3 0100 4 0101 5 0110 6 0111 7 1000 8 1001 9 1010 A 1011 B 1100 C 1101 D 1110 E 1111 F		0000 ↑ FFFF ↓ Digital audio sample values
No of bits	1	2	3	4		16
Information per word	2	4	8	16		65 536
Information per bit	2	2	≈3	4		4096

Figure 4.45 The amount of information per bit increases disproportionately as wordlength increases. It is always more efficient to use the longest words possible at the lowest word rate. It will be evident that sixteen-bit PCM is 2048 times as efficient as delta modulation. Oversampled data are also inefficient for storage.

A single binary digit can only have two states; thus it can only convey two pieces of information, perhaps 'yes' or 'no'. Two binary digits together can have four states, and can thus convey four pieces of information, perhaps 'spring summer autumn or winter', which is two pieces of information per bit. Three binary digits grouped together can have eight combinations, and convey eight pieces of information, perhaps 'doh re mi fah so lah te or doh', which is nearly three pieces of information per digit. Clearly the further this principle is taken, the greater the benefit. In a sixteen-bit system, each bit is worth 4K pieces of information. It is always more efficient, in information-capacity terms, to use the combinations of long binary words than to send single bits for every piece of information. The greatest efficiency is reached when the longest words are sent at the slowest rate which must be the Nyquist rate. This is one reason why PCM recording is more common than single-bit modulation schemes, despite the simplicity of implementation of the latter type of convertor. PCM simply makes more efficient use of the capacity of the binary channel.

As a result, oversampling is confined to convertor technology where it gives specific advantages in implementation. The storage or transmission system will usually employ PCM, where the sampling rate is a little more than twice the audio bandwidth. Figure 4.46 shows a digital audio tape recorder such as DAT using oversampling convertors. The ADC runs at n times the Nyquist rate, but once in the digital domain the rate needs to be reduced in a type of digital filter called a *decimator*. The output of this is conventional Nyquist rate PCM, according to the tape format, which is then recorded. On replay the sampling rate is raised once more in a further type of digital filter called an *interpolator*. The system now has the best of both worlds: using oversampling in the convertors overcomes the shortcomings of analog anti-aliasing and reconstruction filters and the wordlength of the convertor elements is reduced making them easier to construct; the recording is made with Nyquist rate PCM which minimizes tape consumption. Digital filters have the characteristic that their frequency response is proportional to the sampling rate. If a digital recorder is played at a reduced speed, the response of the digital filter will reduce automatically and prevent images passing the reconstruction process.

Oversampling is a method of overcoming practical implementation problems by replacing a single critical element or bottleneck by a number of elements whose overall performance is what counts. As Hauser[24] properly observed, oversampling tends to overlap the operations which are quite distinct in a

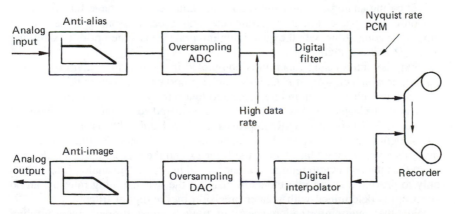

Figure 4.46 A recorder using oversampling in the convertors overcomes the shortcomings of analog anti-aliasing and reconstruction filters and the convertor elements are easier to construct; the recording is made with Nyquist rate PCM which minimizes tape consumption.

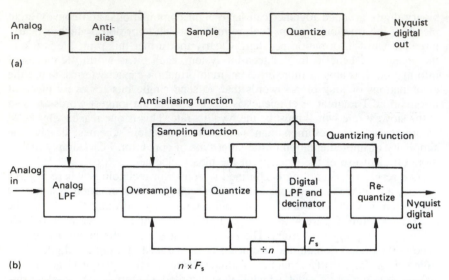

Figure 4.47 A conventional ADC performs each step in an identifiable location as in (a). With oversampling, many of the steps are distributed as shown in (b).

conventional convertor. In earlier sections of this chapter, the vital subjects of filtering, sampling, quantizing and dither have been treated almost independently. Figure 4.47(a) shows that it is possible to construct an ADC of predictable performance by taking a suitable anti-aliasing filter, a sampler, a dither source and a quantizer and assembling them like building bricks. The bricks are effectively in series and so the performance of each stage can only limit the overall performance. In contrast, Figure 4.47(b) shows that with oversampling the overlap of operations allows different processes to augment one another allowing a synergy which is absent in the conventional approach.

If the oversampling factor is n, the analog input must be bandwidth limited to $n.F_s/2$ by the analog anti-aliasing filter. This unit need only have flat frequency response and phase linearity within the audio band. Analog dither of an amplitude compatible with the quantizing interval size is added prior to sampling at $n.F_s/2$ and quantizing.

Next, the anti-aliasing function is completed in the digital domain by a low-pass filter which cuts off at $F_s/2$. Using an appropriate architecture this filter can be absolutely phase linear and implemented to arbitrary accuracy. Such filters are discussed in Chapter 3. The filter can be considered to be the demodulator of Figure 4.43 where the SNR improves as the bandwidth is reduced. The wordlength can be expected to increase. As Chapter 3 illustrated, the multiplications taking place within the filter extend the wordlength considerably more than the bandwidth reduction alone would indicate. The analog filter serves only to prevent aliasing into the audio band at the oversampling rate; the audio spectrum is determined with greater precision by the digital filter.

With the audio information spectrum now Nyquist limited, the sampling process is completed when the rate is reduced in the decimator. One sample in n is retained.

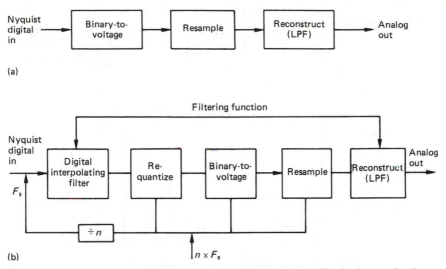

Figure 4.48 A conventional DAC in (a) is compared with the oversampling implementation in (b).

The excess wordlength extension due to the anti-aliasing filter arithmetic must then be removed. Digital dither is added, completing the dither process, and the quantizing process is completed by requantizing the dithered samples to the appropriate wordlength which will be greater than the wordlength of the first quantizer. Alternatively noise shaping may be employed.

Figure 4.48(a) shows the building-brick approach of a conventional DAC. The Nyquist rate samples are converted to analog voltages and then a steep-cut analog low-pass filter is needed to reject the sidebands of the sampled spectrum.

Figure 4.48(b) shows the oversampling approach. The sampling rate is raised in an interpolator which contains a low-pass filter which restricts the baseband spectrum to the audio bandwidth shown. A large frequency gap now exists between the baseband and the lower sideband. The multiplications in the interpolator extend the wordlength considerably and this must be reduced within the capacity of the DAC element by the addition of digital dither prior to requantizing. Again noise shaping may be used as an alternative.

4.15 Oversampling without noise shaping

If an oversampling convertor is considered which makes no attempt to shape the noise spectrum, it will be clear that if it contains a perfect quantizer, no amount of oversampling will increase the resolution of the system, since a perfect quantizer is blind to all changes of input within one quantizing interval, and looking more often is of no help. It was shown earlier that the use of dither would linearize a quantizer, so that input changes much smaller than the quantizing interval would be reflected in the output and this remains true for this class of convertor.

Figure 4.49 shows the example of a white-noise-dithered quantizer, oversampled by a factor of four. Since dither is correctly employed, it is valid to speak of the unwanted signal as noise. The noise power extends over the whole

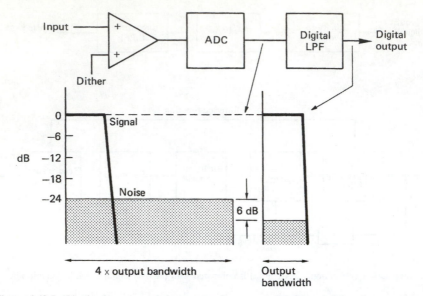

Figure 4.49 In this simple oversampled convertor, 4× oversampling is used. When the convertor output is low-pass filtered, the noise power is reduced to one-quarter, which in voltage terms is 6 dB. This is a suboptimal method and is not used.

baseband up to the Nyquist limit. If the basebandwidth is reduced by the oversampling factor of four back to the bandwidth of the original analog input, the noise bandwidth will also be reduced by a factor of four, and the noise power will be one-quarter of that produced at the quantizer. One-quarter noise power implies one-half the noise voltage, so the SNR of this example has been increased by 6 dB, the equivalent of one extra bit in the quantizer. Information theory predicts that an oversampling factor of four would allow an extension by two bits. This method is suboptimal in that very large oversampling factors would be needed to obtain useful resolution extension, but it would still realize some advantages, particularly the elimination of the steep-cut analog filter.

The division of the noise by a larger factor is the only route left open, since all the other parameters are fixed by the signal bandwidth required.

The reduction of noise power resulting from a reduction in bandwidth is only proportional if the noise is white, i.e. it has uniform power spectral density (PSD). If the noise from the quantizer is made spectrally non-uniform, the oversampling factor will no longer be the factor by which the noise power is reduced. The goal is to concentrate noise power at high frequencies, so that after low-pass filtering in the digital domain down to the audio input bandwidth, the noise power will be reduced by more than the oversampling factor.

4.16 Noise shaping

Noise shaping dates from the work of Cutler[25] in the 1950s. It is a feedback technique applicable to quantizers and requantizers in which the quantizing process of the current sample is modified in some way by the quantizing error of the previous sample.

When used with requantizing, noise shaping is an entirely digital process which is used, for example, following word extension due to the arithmetic in digital mixers or filters in order to return to the required wordlength. It will be found in this form in oversampling DACs. When used with quantizing, part of the noise-shaping circuitry will be analog. As the feedback loop is placed around an ADC it must contain a DAC. When used in convertors, noise shaping is primarily an implementation technology. It allows processes which are conveniently available in integrated circuits to be put to use in audio conversion. Once integrated circuits can be employed, complexity ceases to be a drawback and low-cost mass production is possible.

It has been stressed throughout this chapter that a series of numerical values or samples is just another analog of an audio waveform. Chapter 3 showed that all analog processes such as mixing, attenuation or integration all have exact numerical parallels. It has been demonstrated that digitally dithered requantizing is no more than a digital simulation of analog quantizing. It should be no surprise that in this section noise shaping will be treated in the same way. Noise shaping can be performed by manipulating analog voltages or numbers representing them or both. If the reader is content to make a conceptual switch between the two, many obstacles to understanding fall, not just in this topic, but in digital audio in general.

The term noise shaping is idiomatic and in some respects unsatisfactory because not all devices which are called noise shapers produce true noise. The caution which was given when treating quantizing error as noise is also relevant in this context. Whilst 'quantizing-error-spectrum shaping' is a bit of a mouthful, it is useful to keep in mind that noise shaping means just that in order to avoid some pitfalls. Some noise-shaper architectures do not produce a signal decorrelated quantizing error and need to be dithered.

Figure 4.50(a) shows a requantizer using a simple form of noise shaping. The low-order bits which are lost in requantizing are the quantizing error. If the value of these bits is added to the next sample before it is requantized, the quantizing error will be reduced. The process is somewhat like the use of negative feedback in an operational amplifier except that it is not instantaneous, but encounters a one-sample delay. With a constant input, the mean or average quantizing error will be brought to zero over a number of samples, achieving one of the goals of additive dither. The more rapidly the input changes, the greater the effect of the delay and the less effective the error feedback will be. Figure 4.50(b) shows the equivalent circuit seen by the quantizing error, which is created at the requantizer and subtracted from itself one sample period later. As a result the quantizing error spectrum is not uniform, but has the shape of a raised sine wave shown at (c), hence the term noise shaping. The noise is very small at DC and rises with frequency, peaking at the Nyquist frequency at a level determined by the size of the quantizing step. If used with oversampling, the noise peak can be moved outside the audio band.

Figure 4.51 shows a simple example in which two low-order bits need to be removed from each sample. The accumulated error is controlled by using the bits which were neglected in the truncation, and adding them to the next sample. In this example, with a steady input, the roundoff mechanism will produce an output of 01110111 . . . If this is low-pass filtered, the three ones and one zero result in a level of three-quarters of a quantizing interval, which is precisely the level which would have been obtained by direct conversion of the full digital input. Thus the resolution is maintained even though two bits have been removed.

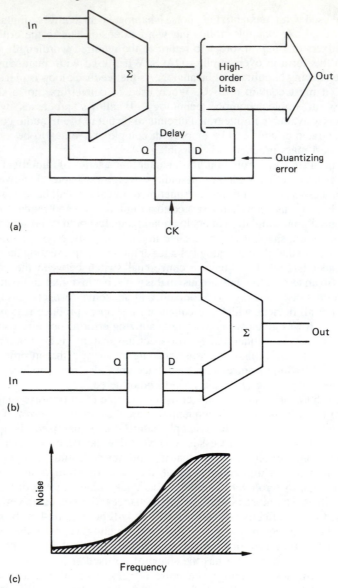

Figure 4.50 (a) A simple requantizer which feeds back the quantizing error to reduce the error of subsequent samples. The one-sample delay causes the quantizing error to see the equivalent circuit shown in (b) which results in a sinusoidal quantizing error spectrum shown in (c).

The noise-shaping technique was used in the first-generation Philips CD players which oversampled by a factor of four. Starting with sixteen-bit PCM from the disc, the 4× oversampling will in theory permit the use of an ideal fourteen-bit convertor, but only if the wordlength is reduced optimally. The oversampling DAC system used is shown in Figure 4.52.[26] The interpolator arithmetic extends the wordlength to 28 bits, and this is reduced to 14 bits using

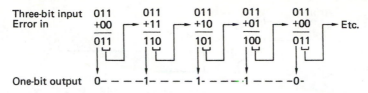

Three-bit input 011   011   011   011   011
Error in   +00 ⌐ +11 ⌐ +10 ⌐ +01 ⌐ +00 ⌐ Etc.
               ‾‾‾‾ ‾‾‾‾ ‾‾‾‾ ‾‾‾‾ ‾‾‾‾
               011   110   101   100   011

One-bit output 0- — — —1- — — — 1- — — —·1 — — — -0-

Figure 4.51 By adding the error caused by truncation to the next value, the resolution of the lost bits is maintained in the duty cycle of the output. Here, truncation of 011 by 2 bits would give continuous zeros, but the system repeats 0111, 0111, which, after filtering, will produce a level of three-quarters of a bit.

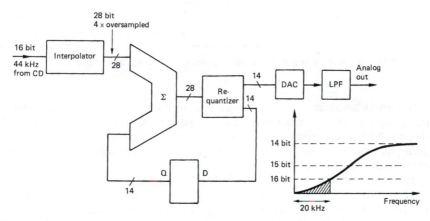

Figure 4.52 The noise-shaping system of the first generation of Phillips CD players.

the error feedback loop of Figure 4.50. The noise floor rises slightly towards the edge of the audio band, but remains below the noise level of a conventional sixteen-bit DAC which is shown for comparison.

The fourteen-bit samples then drive a DAC using dynamic element matching. The aperture effect in the DAC is used as part of the reconstruction filter response, in conjunction with a third-order Bessel filter which has a response 3 dB down at 30 kHz. Equalization of the aperture effect within the audio passband is achieved by giving the digital filter which produces the oversampled data a rising response. The use of a digital interpolator as part of the reconstruction filter results in extremely good phase linearity.

Noise shaping can also be used without oversampling. In this case the noise cannot be pushed outside the audio band. Instead the noise floor is shaped or weighted to complement the unequal spectral sensitivity of the ear to noise.[20,27,28] Unless we wish to violate Shannon's theory, this psychoacoustically optimal noise shaping can only reduce the noise power at certain frequencies by increasing it at others. Thus the average log PSD over the audio band remains the same, although it may be raised slightly by noise induced by imperfect processing.

Figure 4.53 shows noise shaping applied to a digitally dithered requantizer. Such a device might be used when, for example, making a CD master from a twenty-bit recording format. The input to the dithered requantizer is subtracted from the output to give the error due to requantizing. This error is filtered (and

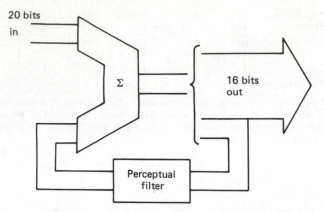

Figure 4.53 Perceptual filtering in a requantizer gives a subjectively improved SNR.

inevitably delayed) before being subtracted from the system input. The filter is not designed to be the exact inverse of the perceptual weighting curve because this would cause extreme noise levels at the ends of the band. Instead the perceptual curve is levelled off[29] such that it cannot fall more than e.g. 40 dB below the peak.

Psychoacoustically optimal noise shaping can offer nearly three bits of increased dynamic range when compared with optimal spectrally flat dither. Enhanced Compact Discs recorded using these techniques are now available.

4.17 Noise-shaping ADCs

The sigma DPCM convertor introduced in Figure 4.42 has a natural application here and is shown in more detail in Figure 4.54. The current digital sample from

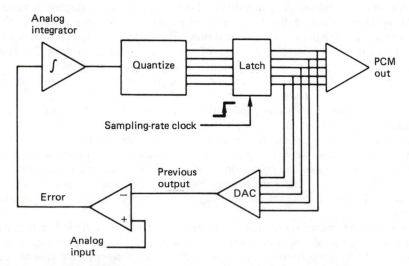

Figure 4.54 The sigma DPCM convertor of Figure 4.42 is shown here in more detail.

the quantizer is converted back to analog in the embedded DAC. The DAC output differs from the ADC input by the quantizing error. The DAC output is subtracted from the analog input to produce an error which is integrated to drive the quantizer in such a way that the error is reduced. With a constant input voltage the average error will be zero because the loop gain is infinite at DC. If the average error is zero, the mean or average of the DAC outputs must be equal to the analog input. The instantaneous output will deviate from the average in what is called an idling pattern. The presence of the integrator in the error feedback loop makes the loop gain fall with rising frequency. With the feedback falling at 6 dB per octave, the noise floor will rise at the same rate.

Figure 4.55 shows a simple oversampling system using a sigma-DPCM convertor and an oversampling factor of only four. The sampling spectrum shows that the noise is concentrated at frequencies outside the audio part of the oversampling baseband. Since the scale used here means that noise power is represented by the area under the graph, the area left under the graph after the filter shows the noise-power reduction. Using the relative areas of similar triangles shows that the reduction has been by a factor of sixteen. The corresponding noise-voltage reduction would be a factor of four, or 12 dB, which corresponds to an additional two bits in wordlength. These bits will be available in the wordlength extension which takes place in the decimating filter. Owing to the rise of 6 dB per octave in the PSD of the noise, the SNR will be 3 dB worse at the edge of the audio band.

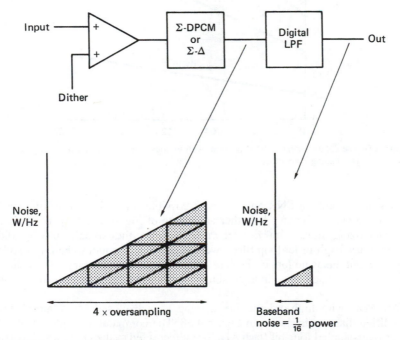

Figure 4.55 In a sigma-DPCM or Σ–Δ convertor, noise amplitude increases by 6 dB/octave, noise power by 12 dB/octave. In this 4× oversampling convertor, the digital filter reduces bandwidth by four, but noise power is reduced by a factor of 16. Noise voltage falls by a factor of four or 12 dB.

One way in which the operation of the system can be understood is to consider that the coarse DAC in the loop defines fixed points in the audio transfer function. The time averaging which takes place in the decimator then allows the transfer function to be interpolated between the fixed points. True signal-independent noise of sufficient amplitude will allow this to be done to infinite resolution, but by making the noise primarily outside the audio band the resolution is maintained but the audio band signal-to-noise ratio can be extended. A first-order noise shaping ADC of the kind shown can produce signal-dependent quantizing error and requires analog dither. However, this can be outside the audio band and so need not reduce the SNR achieved.

A greater improvement in dynamic range can be obtained if the integrator is supplanted to realize a higher-order filter.[30] The filter is in the feedback loop and so the noise will have the opposite response to the filter and will therefore rise more steeply to allow a greater SNR enhancement after decimation. Figure 4.56

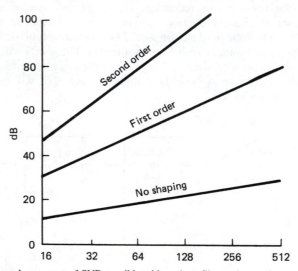

Figure 4.56 The enhancement of SNR possible with various filter orders and oversampling factors in noise-shaping convertors.

shows the theoretical SNR enhancement possible for various loop filter orders and oversampling factors. A further advantage of high-order loop filters is that the quantizing noise can be decorrelated from the signal, making dither unnecessary. High-order loop filters were at one time thought to be impossible to stabilize, but this is no longer the case, although care is necessary. One technique which may be used is to include some feedforward paths as shown in Figure 4.57.

An ADC with high-order noise shaping was disclosed by Adams[31] and a simplified diagram is shown in Figure 4.58. The comparator outputs of the 128 times oversampled four-bit flash ADC are directly fed to the DAC which consists of fifteen equal resistors fed by CMOS switches. As with all feedback loops, the transfer characteristic cannot be more accurate than the feedback, and in this case the feedback accuracy is determined by the precision of the DAC.[32] Driving the

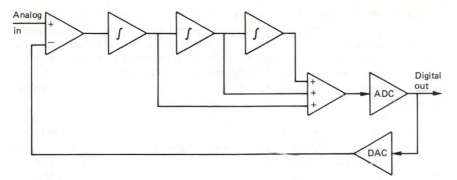

Figure 4.57 Stabilizing the loop filter in a noise-shaping convertor can be assisted by the incorporation of feedforward paths as shown here.

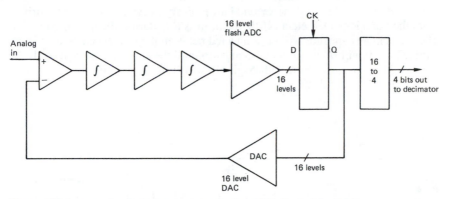

Figure 4.58 An example of a high-order noise-shaping ADC. See text for details.

DAC directly from the ADC comparators is more accurate because each input has equal weighting. The stringent MSB tolerance of the conventional binary weighted DAC is then avoided. The comparators also drive a 16 to 4 priority encoder to provide the four-bit PCM output to the decimator. The DAC output is subtracted from the analog input at the integrator. The integrator is followed by a pair of conventional analog operational amplifiers having frequency-dependent feedback and a passive network which gives the loop a fourth-order response overall. The noise floor is thus shaped to rise at 24 dB per octave beyond the audio band. The time constants of the loop filter are optimized to minimize the amplitude of the idling pattern as this is an indicator of the loop stability. The four-bit PCM output is low-pass filtered and decimated to the Nyquist frequency. The high oversampling factor and high-order noise shaping extend the dynamic range of the four-bit flash ADC to 108 dB at the output.

4.18 A one-bit DAC

It might be thought that the waveform from a one-bit DAC is simply the same as the digital input waveform. In practice this is not the case. The input signal is a logic signal which need only be above or below a threshold for its binary value to be correctly received. It may have a variety of waveform distortions and a duty cycle offset. The area under the pulses can vary enormously. In the DAC output the amplitude needs to be extremely accurate. A one-bit DAC uses only the binary information from the input, but reclocks to produce accurate timing and uses a reference voltage to produce accurate levels. The area of pulses produced is then constant. One-bit DACs will be found in noise-shaping ADCs as well as in the more obvious application of producing analog audio.

Figure 4.59(a) shows a one-bit DAC which is implemented with MOS field-effect switches and a pair of capacitors. Quanta of charge are driven into or out of a virtual earth amplifier configured as an integrator by the switched capacitor action. Figure 4.59(b) shows the associated waveforms. Each data bit period is divided into two equal portions; that for which the clock is high, and that for which it is low. During the first half of the bit period, pulse P+ is generated if the data bit is a 1, or pulse P– is generated if the data bit is a 0. The reference input is a clean voltage corresponding to the gain required.

C1 is *discharged* during the second half of every cycle by the switches driven from the complemented clock. If the next bit is a 1, during the next high period of the clock the capacitor will be connected between the reference and the virtual earth. Current will flow into the virtual earth until the capacitor is charged. If the next bit is not a 1, the current through C1 will flow to ground.

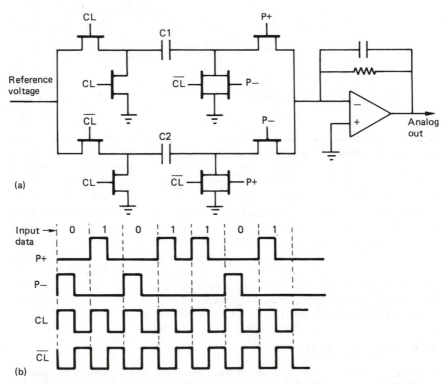

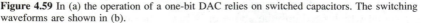

Figure 4.59 In (a) the operation of a one-bit DAC relies on switched capacitors. The switching waveforms are shown in (b).

C2 is *charged* to reference voltage during the second half of every cycle by the switches driven from the complemented clock. On the next high period of the clock, the reference end of C2 will be grounded, and so the op-amp end wil assume a negative reference voltage. If the next bit is a 0, this negative reference will be switched into the virtual earth, if not the capacitor will be discharged.

Thus on every cycle of the clock, a quantum of charge is either pumped into the integrator by C1 or pumped out by C2. The analog output therefore precisely reflects the ratio of ones to zeros.

4.19 One-bit noise-shaping ADCs

In order to overcome the DAC accuracy constraint of the sigma DPCM convertor, the sigma–delta convertor can be used as it has only one-bit internal resolution. A one-bit DAC cannot be non-linear by definition as it defines only two points on a transfer function. It can, however, suffer from other deficiencies such as DC offset and gain error although these are less offensive in audio. The one-bit ADC is a comparator.

As the sigma–delta convertor is only a one-bit device, clearly it must use a high oversampling factor and high-order noise shaping in order to have sufficiently good SNR for audio.[33] In practice the oversampling factor is limited not so much by the convertor technology as by the difficulty of computation in the decimator. A sigma–delta convertor has the advantage that the filter input 'words' are one bit long and this simplifies the filter design as multiplications can be replaced by selection of constants.

Conventional analysis of loops falls down heavily in the one-bit case. In particular the gain of a comparator is difficult to quantify, and the loop is highly non-linear so that considering the quantizing error as additive white noise in order to use a linear loop model gives rather optimistic results. In the absence of an accurate mathematical model, progress has been made empirically, with listening tests and by using simulation.

Single-bit sigma–delta convertors are prone to long idling patterns because the low resolution in the voltage domain requires more bits in the time domain to be integrated to cancel the error. Clearly the longer the period of an idling pattern, the more likely it is to enter the audio band as an objectionable whistle or 'birdie'. They also exhibit threshold effects or deadbands where the output fails to react to an input change at certain levels. The problem is reduced by the order of the filter and the wordlength of the embedded DAC. Second- and third-order feedback loops are still prone to audible idling patterns and threshold effect.[34] The traditional approach to linearizing sigma–delta convertors is to use dither. Unlike conventional quantizers, the dither used was of a frequency outside the audio band and of considerable level. Square-wave dither has been used and it is advantageous to choose a frequency which is a multiple of the final output sampling rate as then the harmonics will coincide with the troughs in the stopband ripple of the decimator. Unfortunately the level of dither needed to linearize the convertor is high enough to cause premature clipping of high-level signals, reducing the dynamic range. This problem is overcome by using in-band white noise dither at low level.[35]

An advantage of the one-bit approach is that in the one-bit DAC, precision components are replaced by precise timing in switched capacitor networks. The same approach can be used to implement the loop filter in an ADC. Figure 4.60

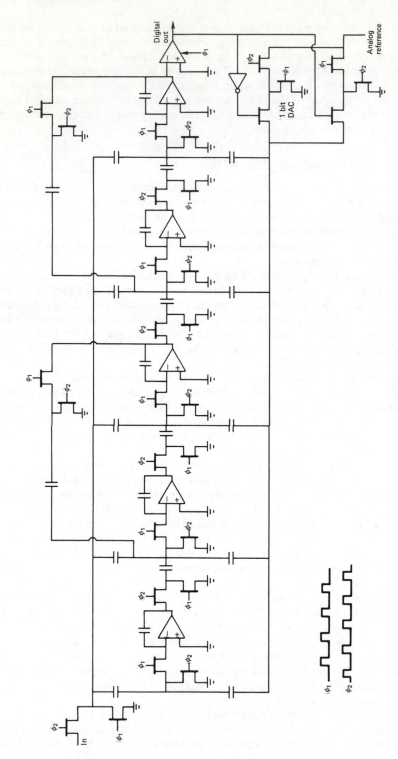

Figure 4.60 A third-order sigma–delta modulator using a switched capacitor loop filter.

shows a third-order sigma–delta modulator incorporating a DAC based on the principle of Figure 4.59. The loop filter is also implemented with switched capacitors.

References

1. Shannon, C.E., A mathematical theory of communication. *Bell Syst. Tech. J.*, **27**, 379 (1948)
2. Jerri, A.J., The Shannon sampling theorem – its various extensions and applications: a tutorial review. *Proc. IEEE*, **65**, 1565–1596 (1977)
3. Betts, J.A., *Signal Processing Modulation and Noise*, Sevenoaks: Hodder and Stoughton (1970)
4. Meyer, J., Time correction of anti-aliasing filters used in digital audio systems. *J. Audio Eng. Soc.*, **32**, 132–137 (1984)
5. Lipshitz, S.P., Pockock, M. and Vanderkooy, J., On the audibility of midrange phase distortion in audio systems. *J. Audio Eng. Soc.*, **30**, 580–595 (1982)
6. Preis, D. and Bloom, P.J., Perception of phase distortion in anti-alias filters. *J. Audio Eng. Soc.*, **32**, 842–848 (1984)
7. Lagadec, R. and Stockham, T.G., Jr, Dispersive models for A-to-D and D-to-A conversion systems. Presented at the 75th Audio Engineering Society Convention (Paris, 1984), Preprint 2097(H-8)
8. Blesser, B., Advanced A/D conversion and filtering: data conversion. In *Digital Audio*, edited by B.A. Blesser, B. Locanthl and T.G. Stockham Jr, pp. 37–53, New York: Audio Engineering Society (1983)
9. Lagadec, R., Weiss, D. and Greutmann, R., High-quality analog filters for digital audio. Presented at the 67th Audio Engineering Society Convention (New York, 1980), Preprint 1707(B-4)
10. Anon., AES recommended practice for professional digital audio applications employing pulse code modulation: preferred sampling frequencies. AES5–1984 (ANSI S4.28–1984), *J. Audio Eng. Soc.*, **32**, 781–785 (1984)
11. Pease, R., Understand capacitor soakage to optimise analog systems. *Electronics and Wireless World*, 832–835 (1992)
12. Harris, S., The effects of sampling clock jitter on Nyquist sampling analog to digital convertors and on oversampling delta-sigma ADCs. *J. Audio Eng. Soc.*, **38**, 537–542 (1990)
13. Nunn, J., Jitter specification and assessment in digital audio equipment. Presented at the 93rd Audio Engineering Society Convention. (San Francisco, 1992), Preprint No. 3361 (C-2)
14. Widrow, B., Statistical analysis of amplitude quantized sampled-data systems. *Trans. AIEE*, Part II, **79**, 555–568 (1961)
15. Lipshitz, S.P., Wannamaker, R.A. and Vanderkooy, J., Quantization and dither: a theoretical survey. *J. Audio Eng. Soc.*, **40**, 355–375 (1992)
16. Maher, R.C., On the nature of granulation noise in uniform quantization systems. *J. Audio Eng. Soc.*, **40**, 12–20 (1992)
17. Roberts, L.G., Picture coding using pseudo random noise. *IRE Trans. Inform. Theory*, **IT-8**, 145–154 (1962)
18. Vanderkooy, J. and Lipshitz, S.P., Resolution below the least significant bit in digital systems with dither. *J. Audio Eng. Soc.*, **32**, 106–113 (1984)
19. Blesser, B., Advanced A-D conversion and filtering: data conversion. In *Digital Audio*, edited by B.A. Blesser, B. Locanthl, and T.G. Stockham Jr., pp. 37–53. New York: Audio Engineering Society (1983)
20. Vanderkooy, J. and Lipshitz, S.P., Digital dither. Presented at the 81st Audio Engineering Society Convention (Los Angeles, 1986), Preprint 2412 (C-8)
21. v.d. Plassche, R.J., Dynamic element matching puts trimless convertors on chip. *Electronics*, 16 June 1983
22. v.d. Plassche, R.J. and Goedhart, D., A monolithic 14 bit D/A convertor. *IEEE J. Solid-State Circuits*, **SC-14**, 552–556 (1979)
23. Adams, R.W., Companded predictive delta modulation: a low-cost technique for digital recording. *J. Audio Eng. Soc.*, **32**, 659–672 (1984)
24. Hauser, M.W., Principles of oversampling A/D conversion. *J. Audio Eng. Soc.*, **39**, 3–26 (1991)
25. Cutler, C.C., Transmission systems employing quantization. US Pat. No. 2,927,962 (1960)

26. Gerzon, M. and Craven, P.G., Optimal noise shaping and dither of digital signals. Presented at the 87th Audio Engineers Society Convention (New York, 1989), Preprint No. 2822 (J-1)
27. Fielder, L.D., Human Auditory capabilities and their consequences in digital audio convertor design. In *Audio in Digital Times*, New York: Audio Engineering Society (1989)
28. Wannamaker, R.A., Psychoacoustically optimal noise shaping. *J. Audio Eng. Soc.*, **40**, 611–620 (1992)
29. Lipshitz, S.P., Wannamaker, R.A. and Vanderkooy, J., Minimally audible noise shaping. *J. Audio Eng. Soc.*, **39**, 836–852 (1991)
30. Adams, R.W., Design and implementation of an audio 18-bit A/D convertor using oversampling techniques. Presented at the 77th Audio Engineering Society Convention (Hamburg, 1985), Preprint 2182
31. Adams, R.W., An IC chip set for 20 bit A/D conversion. In *Audio in Digital Times*, New York: Audio Engineering Society (1989)
32. Richards, M., Improvements in oversampling analogue to digital convertors. Presented at the 84th Audio Engineering Society Convention (Paris, 1988), Preprint 2588 (D-8)
33. Inose, H. and Yasuda, Y., A unity bit coding method by negative feedback. *Proc. IEEE*, **51**, 1524–1535 (1963)
34. Naus, P.J. *et al.*, Low signal level distortion in sigma–delta modulators. Presented at the 84th Audio Engineering Society Convention (Paris, 1988), Preprint 2584
35. Stikvoort, E., High order one bit coder for audio applications. Presented at the 84th Audio Engineering Society Convention (Paris, 1988), Preprint 2583(D-3)

5

Compression

5.1 Introduction

Compression, bit rate reduction and data reduction are all terms which mean basically the same thing in this context. In essence the same (or nearly the same) audio information is carried using a smaller quantity and/or rate of data. It should be pointed out that in audio *compression* traditionally means a process in which the dynamic range of the sound is reduced, typically by broadcasters wishing their station to sound louder. However, when bit rate reduction is employed, the dynamics of the decoded signal are unchanged. Provided the context is clear, the two meanings can co-exist without a great deal of confusion.

There are several reasons why compression techniques are popular:

(a) Compression extends the playing time of a given storage device.
(b) Compression allows miniaturization. With fewer data to store, the same playing time is obtained with smaller hardware. This is useful in portable and consumer devices.
(c) Tolerances can be relaxed. With fewer data to record, storage density can be reduced, making equipment which is more resistant to adverse environments and which requires less maintenance.
(d) In transmission systems, compression allows a reduction in bandwidth which will generally result in a reduction in cost. This may make possible some process which would be uneconomic without it.
(e) If a given bandwidth is available to an uncompressed signal, compression allows faster than real-time transmission within that bandwidth.
(f) If a given bandwidth is available, compression allows a better-quality signal within that bandwidth.

Compression is summarized in Figure 5.1. It will be seen in (a) that the PCM audio data rate is reduced at source by the *compressor*. The compressed data are then passed through a communication channel and returned to the original audio rate by the *expander*. The ratio between the source data rate and the channel data rate is called the *compression factor*. The term *coding gain* is also used. Sometimes a compressor and expander in series are referred to as a *compander*. The compressor may equally well be referred to as a *coder* and the expander a *decoder* in which case the tandem pair may be called a *codec*.

Where the encoder is more complex than the decoder the system is said to be asymmetrical. Figure 5.1(b) shows that MPEG[1,2] audio coders work in this way, as

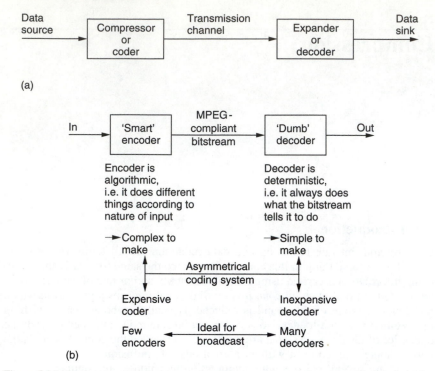

Figure 5.1 In (a) a compression system consists of compressor or coder, a transmission channel and a matching expander or decoder. The combination of coder and decoder is known as a codec. (b) MPEG is asymmetrical since the encoder is much more complex than the decoder.

do many others. The encoder needs to be algorithmic or adaptive whereas the decoder is 'dumb' and carries out fixed actions. This is advantageous in applications such as broadcasting where the number of expensive complex encoders is small but the number of simple inexpensive decoders is large. In point-to-point applications the advantage of asymmetrical coding is not so great. In MPEG audio coding the encoder is typically two or three times as complex as the decoder.

Figure 5.2 shows the use of a codec with a recorder. The playing time of the medium is extended in proportion to the compression factor. In the case of tapes, the access time is improved because the length of tape needed for a given recording is reduced and so it can be rewound more quickly. In some cases, compression may be used to improve the recorder quality. A lossless coder with a very light compression factor can be used to give a sixteen-bit DAT recorder eighteen- or twenty-bit performance.

In communications, the cost of data links is often roughly proportional to the data rate and so there is simple economic pressure to use a high compression factor. The use of heavy compression to allow audio to be sent over the Internet is an example of this.

In workstations designed for audio editing, the source material may be stored on hard disks for rapid access. Whilst top-grade systems may function without compression, many systems use compression to offset the high cost of disk storage.

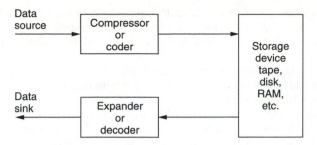

Figure 5.2 Compression can be used around a recording medium. The storage capacity may be increased or the access time reduced according to the application.

When a workstation is used for *off-line* editing, a high compression factor can be used and artifacts will be audible. This is of no consequence as these are only heard by the editor who uses the system to make an EDL (edit decision list) which is no more than a list of actions and the timecodes at which they occur. The original uncompressed material is then *conformed* to the EDL to obtain a high-quality edited work. When *on-line* editing is being performed, the output of the workstation is the finished product and clearly a lower compression factor will have to be used.

The cost of digital storage continues to fall and the pressure to use compression for recording purposes falls with it. Perhaps it is in broadcasting and the Internet where the use of compression will have its greatest impact. There is only one electromagnetic spectrum and pressure from other services such as cellular telephones makes efficient use of bandwidth mandatory. Analog broadcasting is an old technology and makes very inefficient use of bandwidth. Its replacement by a compressed digital transmission will be inevitable for the practical reason that the bandwidth is needed elsewhere.

Fortunately in broadcasting there is a mass market for decoders and these can be implemented as low-cost integrated circuits. Fewer encoders are needed and so it is less important if these are expensive. Whilst the cost of digital storage goes down year on year, the cost of electromagnetic spectrum goes up. Consequently in the future the pressure to use compression in recording will ease whereas the pressure to use it in radio communications will increase.

5.2 Lossless and perceptive coding

Although there are many different audio coding tools, all of them fall into one or other of these categories. In *lossless* coding, the data from the expander are identical bit-for-bit with the original source data. The so-called 'stacker' programs which increase the apparent capacity of disk drives in personal computers use lossless codecs. Clearly with computer programs the corruption of a single bit can be catastrophic. Lossless coding is generally restricted to compression factors of around 2:1.

It is important to appreciate that a lossless coder cannot guarantee a particular compression factor and the communications link or recorder used with it must be able to handle the variable output data rate. Audio material which results in poor compression factors on a given codec is described as *difficult*. It should be pointed out that the difficulty is often a function of the codec. In other words audio which one codec finds difficult may not be found difficult by another. Lossless codecs can be included in bit-error-rate testing schemes. It is also possible to cascade or *concatenate* lossless codecs without any special precautions.

In *lossy* coding, data from the decoder are not identical bit-for-bit with the source data and as a result comparing the input with the output is bound to reveal differences. Clearly lossy codecs are not suitable for computer data, but are used in many audio coders, MPEG included, as they allow greater compression factors than lossless codecs. The most successful lossy codecs are those in which the errors are arranged so that the listener finds them subjectively difficult to detect. Thus lossy codecs must be based on an understanding of psychoacoustic perception and are often called *perceptive* codes.

Perceptive coding relies on the principle of auditory masking, which was considered in Chapter 2. Masking causes the ear/brain combination to be less sensitive to sound at one frequency in the presence of another at a nearby frequency. If a first tone is present in the input, then it will mask signals of lower level at nearby frequencies. The quantizing of the first tone and of further tones at those frequencies can be made coarser. Fewer bits are needed and a coding gain results. The increased quantizing distortion is allowable if it is masked by the presence of the first tone.

In perceptive coding, the greater the compression factor required, the more accurately must the human senses be modelled. Perceptive coders can be forced to operate at a fixed compression factor. This is convenient for practical transmission applications where a fixed data rate is easier to handle than a variable rate. However, the result of a fixed compression factor is that the subjective quality can vary with the 'difficulty' of the input material. Perceptive codecs should not be concatenated indiscriminately especially if they use different algorithms. As the reconstructed signal from a perceptive codec is not bit-for-bit accurate, clearly such a codec cannot be included in any bit error rate testing system as the coding differences would be indistinguishable from real errors.

5.3 Compression principles

In a PCM audio system the bit rate is the product of the sampling rate and the number of bits in each sample and this is generally constant. Nevertheless the *information* rate of a real signal varies. In all real signals, part of the signal is obvious from what has gone before or what may come later and a suitable decoder can predict that part so that only the true information actually has to be sent. If the characteristics of a predicting decoder are known, the transmitter can omit parts of the message in the knowledge that the decoder has the ability to recreate it. Thus all encoders must contain a model of the decoder.

In a predictive codec there are two identical predictors, one in the coder and one in the decoder. Their job is to examine a run of previous data values and to extrapolate forward to estimate or predict what the next value will be. This is subtracted from the *actual* next value at the encoder to produce a prediction error or *residual* which is transmitted. The decoder then adds the prediction error to its own prediction to obtain the output code value again. Predictive coding can be applied to any type of information. In audio coders the information may be PCM samples, transform coefficients or even side-chain data such as scale factors.

Predictive coding has the advantage that provided the residual is transmitted intact, there is no loss of information.

One definition of information is that it is the unpredictable or surprising element of data. Newspapers are a good example of information because they only mention items which are surprising. Newspapers never carry items about individuals who have *not* been involved in an accident as this is the normal case. Consequently the phrase 'no news is good news' is remarkably true because if an information channel exists but nothing has been sent then it is most likely that nothing remarkable has happened.

The difference between the information rate and the overall bit rate is known as the redundancy. Compression systems are designed to eliminate as much of that redundancy as practicable or perhaps affordable. One way in which this can be done is to exploit statistical predictability in signals. The information content or *entropy* of a sample is a function of how different it is from the predicted value. Most signals have some degree of predictability. A sine wave is highly predictable, because all cycles look the same. According to Shannon's theory, any signal which is totally predictable carries no information. In the case of the sine wave this is clear because it represents a single frequency and so has no bandwidth.

At the opposite extreme a signal such as noise is completely unpredictable and as a result all codecs find noise *difficult*. There are two consequences of this characteristic. First, a codec which is designed using the statistics of real material should not be tested with random noise because it is not a representative test. Second, a codec which performs well with clean source material may perform badly with source material containing superimposed noise such as analog tape hiss. Practical compression units may require some form of pre-processing before the compression stage proper and appropriate noise reduction should be incorporated into the pre-processing if noisy signals are anticipated. It will also be necessary to restrict the degree of compression applied to noisy signals.

All real signals fall part-way between the extremes of total predictability and total unpredictability or noisiness. If the bandwidth (set by the sampling rate) and the dynamic range (set by the wordlength) of the transmission system are used to delineate an area, this sets a limit on the information capacity of the system. Figure 5.3(a) shows that most real signals only occupy part of that area. The signal may not contain all frequencies, or it may not have full dynamics at certain frequencies.

Entropy can be thought of as a measure of the actual area occupied by the signal. This is the area that *must* be transmitted if there are to be no subjective differences or *artifacts* in the received signal. The remaining area is called the *redundancy* because it adds nothing to the information conveyed. Thus an ideal coder could be imagined which miraculously sorts out the entropy from the redundancy and only sends the former. An ideal decoder would then recreate the original impression of the information quite perfectly.

As the ideal is approached, the coder complexity and the latency (delay) both rise. Figure 5.3(b) shows how complexity increases with compression factor. Figure 5.3(c) shows how increasing the codec latency can improve the compression factor. Obviously we would have to provide a channel which could accept whatever entropy the coder extracts in order to have transparent quality. As a result moderate coding gains which only remove redundancy need not in principle cause artifacts and can result in systems which are described as

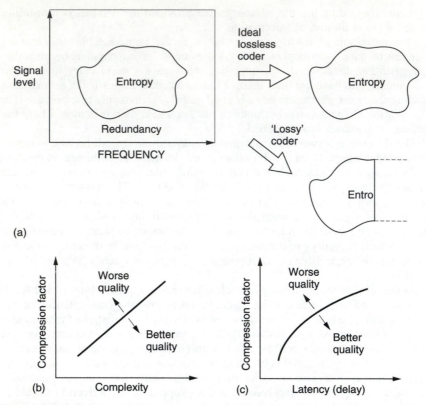

Figure 5.3 (a) A perfect coder removes only the redundancy from the input signal and results in subjectively lossless coding. If the remaining entropy is beyond the capacity of the channel some of it must be lost and the codec will then be lossy. An imperfect coder will also be lossy as it fails to keep all entropy. (b) As the compression factor rises, the complexity must also rise to maintain quality. (c) High compression factors also tend to increase latency or delay through the system.

subjectively lossless. This assumes that such systems are well engineered, which may not be the case in actual hardware.

If the channel capacity is not sufficient for that, then the coder will have to discard some of the entropy and with it useful information. Larger coding gains which remove some of the entropy must result in artifacts. It will also be seen from Figure 5.3 that an imperfect coder will fail to separate the redundancy and may discard entropy instead, resulting in artifacts at a suboptimal compression factor.

A single variable rate transmission channel is inconvenient and unpopular with channel providers because it is difficult to police. The requirement can be overcome by combining several compressed channels into one constant rate transmission in a way which flexibly allocates data rate between the channels. Provided the material is unrelated, the probability of all channels reaching peak entropy at once is very small and so those channels which are at one instant passing easy material will free up transmission capacity for those channels which are handling difficult material. This is the principle of statistical multiplexing.

Where the same type of source material is used consistently, e.g. English text, then it is possible to perform a statistical analysis on the frequency with which particular letters are used. Variable-length coding is used in which frequently used letters are allocated short codes and letters which occur infrequently are allocated long codes. This results in a lossless code. The well-known Morse code used for telegraphy is an example of this approach. The letter e is the most frequent in English and is sent with a single dot. An infrequent letter such as z is allocated a long complex pattern. It should be clear that codes of this kind which rely on a prior knowledge of the statistics of the signal are only effective with signals actually having those statistics. If Morse code is used with another language, the transmission becomes significantly less efficient because the statistics are quite different; the letter z, for example, is quite common in Czech.

The Huffman code[3] is one which is designed for use with a data source having known statistics and shares the same principles with the Morse code. The probability of the different code values to be transmitted is studied, and the most frequent codes are arranged to be transmitted with short wordlength symbols. As the probability of a code value falls, it will be allocated longer wordlength. The Huffman code is used in conjunction with a number of compression techniques and is shown in Figure 5.4.

The input or *source* codes are assembled in order of descending probability. The two lowest probabilities are distinguished by a single code bit and their probabilities are combined. The process of combining probabilities is continued until unity is reached and at each stage a bit is used to distinguish the path. The bit will be a zero for the most probable path and one for the least. The compressed output is obtained by reading the bits which describe which path to take going from right to left.

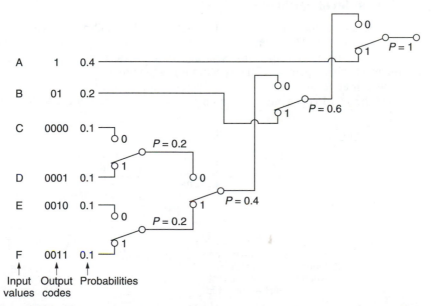

Figure 5.4 The Huffman code achieves compression by allocating short codes to frequent values. To aid deserializing the short codes are not prefixes of longer codes.

In the case of computer data, there is no control over the data statistics. Data to be recorded could be instructions, images, tables, text files and so on; each having their own code value distributions. In this case a coder relying on fixed source statistics will be completely inadequate. Instead a system is used which can learn the statistics as it goes along. The Lempel–Ziv–Welch (LZW) lossless codes are in this category. These codes build up a conversion table between frequent long source data strings and short transmitted data codes at both coder and decoder and initially their compression factor is below unity as the contents of the conversion tables are transmitted along with the data. However, once the tables are established, the coding gain more than compensates for the initial loss. In some applications, a continuous analysis of the frequency of code selection is made and if a data string in the table is no longer being used with sufficient frequency it can be deselected and a more common string substituted.

Lossless codes are less common in audio coding where perceptive codes are more popular. The perceptive codes often obtain a coding gain by shortening the wordlength of the data representing the signal waveform. This must increase the level of quantizing distortion and for good perceived quality the encoder must ensure that the resultant distortion is placed at frequencies where human senses are least able to perceive it. As a result although the received signal is measurably different from the source data, it can *appear* the same to the human listener under certain conditions. As these codes rely on the characteristics of human hearing, they can only fully be tested subjectively.

The compression factor of such codes can be set at will by choosing the wordlength of the compressed data. Whilst mild compression may be undetectable, with greater compression factors, artifacts become noticeable. Figure 5.3 shows that this is inevitable from entropy considerations.

5.4 Codec level calibration

The functioning of the ear is noticeably level dependent and perceptive coders take this into account. However, all signal processing takes place in the electrical or digital domain with respect to electrical or numerical levels whereas the hearing mechanism operates with respect to true sound pressure level. Figure 5.5 shows that in an ideal system the overall gain of the microphones and ADCs is such that the PCM codes have a relationship with sound pressure which is the

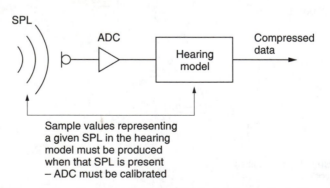

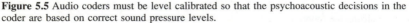

Figure 5.5 Audio coders must be level calibrated so that the psychoacoustic decisions in the coder are based on correct sound pressure levels.

same as that assumed by the model in the codec. Equally the overall gain of the DAC and loudspeaker system should be such that the sound pressure levels which the codec assumes are those actually heard. Clearly the gain control of the microphone and the volume control of the reproduction system must be calibrated if the hearing model is to function properly. If, for example, the microphone gain was too low and this was compensated by advancing the loudspeaker gain, the overall gain would be the same but the codec would be fooled into thinking that the sound pressure level was less than it really was and the masking model would not then be appropriate.

The above should come as no surprise as analog audio codecs such as the various Dolby systems have required and implemented line-up procedures and suitable tones. However obvious the need to calibrate coders may be, the degree to which this is recognized in the industry is almost negligible to date and this can only result in suboptimal performance.

5.5 Quality measurement

As has been seen, one way in which coding gain is obtained is to requantize sample values to reduce the wordlength. Since the resultant requantizing error is a distortion mechanism it results in energy moving from one frequency to another. The masking model is essential to estimate how audible the effect will be. The greater the degree of compression required, the more precise the model must be. If the masking model is inaccurate, then equipment based upon it may produce audible artifacts under some circumstances. Artifacts may also result if the model is not properly implemented. As a result, development of audio compression units requires careful listening tests with a wide range of source material[4,5] and precision loudspeakers. The presence of artifacts at a given compression factor indicates only that performance is below expectations; it does not distinguish between the implementation and the model. If the implementation is verified, then a more detailed model must be sought. Naturally comparative listening tests are only valid if all the codecs have been level calibrated and if the loudspeakers cause less loss of information than any of the codecs, a requirement which is frequently overlooked.

Properly conducted listening tests are expensive and time consuming, and alternative methods have been developed which can be used objectively to evaluate the performance of different techniques. The noise to masking ratio (NMR) is one such measurement.[6] Figure 5.6 shows how NMR is measured. Input audio signals are fed simultaneously to a data-reduction coder and decoder in tandem and to a compensating delay whose length must be adjusted to match the codec delay. At the output of the delay, the coding error is obtained by subtracting the codec output from the original. The original signal is spectrum-analysed into critical bands in order to derive the masking threshold of the input audio, and this is compared with the critical band spectrum of the error. The NMR in each critical band is the ratio between the masking threshold and the quantizing error due to the codec. An average NMR for all bands can be computed. A positive NMR in any band indicates that artifacts are potentially audible. Plotting the average NMR against time is a powerful technique, as with an ideal codec the NMR should be stable with different types of program material. If this is not the case the codec could perform quite differently as a function of the source material. NMR excursions can be correlated with the

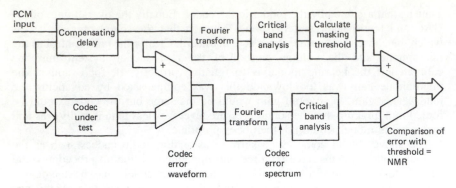

Figure 5.6 The noise-to-masking ratio is derived as shown here.

waveform of the audio input to analyse how the extra noise was caused and to redesign the codec to eliminate it.

Practical systems should have a finite NMR in order to give a degree of protection against difficult signals which have not been anticipated and against the use of post-codec equalization or several tandem codecs which could change the masking threshold. There is a strong argument that devices used for audio production should have a greater NMR than consumer or program delivery devices.

5.6 The limits

There are, of course, limits to all technologies. Eventually artifacts will be heard as the amount of compression is increased which no amount of detailed modelling will remove. The ear is only able to perceive a certain proportion of the information in a given sound. This could be called the perceptual entropy,[7] and all additional sound is redundant or irrelevant. Compression works by removing the redundancy, and clearly an ideal system would remove all of it, leaving only the entropy. Once this has been done, the masking capacity of the ear has been reached and the NMR has reached zero over the whole band.

Assuming an ideal masking model, further reduction of the data rate must cause the level of distortion products to rise above the masking level equally at all frequencies suddenly rendering them audible. The result is that the perceived quality of a codec suddenly falls at a critical bit rate. Figure 5.7 shows this effect which is variously known as a crash knee, graceless degradation or the cliff-edge effect. It is a simple consequence of human perception that a coder which keeps to the left of the crash knee 99 per cent of the time will still be marked down because the sudden failure for one per cent of the time causes irritation out of proportion to its duration.

To meet a particular bit rate constraint, the audio bandwidth may have to be reduced in order to keep the distortion level acceptable. For example, in MPEG-1, pre-filtering allows data from higher sub-bands to be neglected. MPEG-2 has also introduced some low sampling rate options for this purpose. Thus there is a limit to the degree of compression which can be achieved even with an ideal coder. Systems which go beyond that limit are not appropriate for high-quality music, but are relevant in news gathering and communications where speech intelligibility is the criterion.

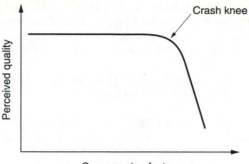

Crash knee

Figure 5.7 It is a characteristic of compression systems that failure is sudden.

Interestingly, the data rate out of a coder is virtually independent of the input sampling rate unless the sampling rate is very low. This is because the entropy of the sound is in the waveform, not in the number of samples carrying it.

It follows from the above that to obtain the highest audio quality for a given bit rate, every redundancy in the input signal must be explored. The more lossless coding tools which can be used, the less will be the extent to which the lossy tools operate. For example, MPEG Layers I and II audio coding don't employ prediction or buffering whereas Layer III uses buffering. MPEG-2 AAC[8] uses both prediction and buffering and can thus obtain better quality at a given bit rate or the same quality at a lower bit rate.

The compression factor of a coder is only part of the story. All codecs cause delay, and in general the greater the compression, the longer the delay. In some applications, such as telephony, a short delay is required.[9] In many applications, the compressed channel will have a constant bit rate, and so a constant compression factor is required. In real program material, the entropy varies and so the NMR will fluctuate. If greater delay can be accepted, as in a recording application, memory buffering can be used to allow the coder to operate at constant NMR and instantaneously variable data rate. The memory absorbs the instantaneous data rate differences of the coder and allows a constant rate in the channel. A higher effective compression factor will then be obtained. Near-constant quality can also be achieved using statistical multiplexing.

5.7 Some guidelines

Although compression techniques themselves are complex, there are some simple rules which can be used to avoid disappointment. Used wisely, audio compression has a number of advantages. Used in an inappropriate manner, disappointment is almost inevitable and the technology could get a bad name. The next few points are worth remembering.

- Compression technology may be exciting, but if it is not necessary it should not be used.
- If compression is to be used, the degree of compression should be as small as possible; i.e. use the highest practical bit rate.

- Cascaded compression systems cause loss of quality and the lower the bit rates, the worse this gets. Quality loss increases if any post-production steps are performed between compressions. Compression systems cause delay.
- Compression systems work best with clean source material. Noisy signals give poor results.
- Compressed data are generally more prone to transmission errors than non-compressed data. The choice of a compression scheme must consider the error characteristics of the channel.
- Audio codecs need to be level calibrated so that when sound-pressure-level-dependent decisions are made in the coder those levels actually exist at the microphone.
- Low bit rate coders should only be used for the final delivery of post-produced signals to the end user.
- Compression quality can only be assessed subjectively on precision loud-speakers. Codecs often sound fine on cheap speakers when in fact they are not.
- Compression works best in mono and less well in stereo and surround-sound systems where the imaging, ambience and reverb are frequently not well reproduced.
- Don't be browbeaten by the technology. You don't have to understand it to assess the results. Your ears are as good as anyone's so don't be afraid to criticize artifacts.

5.8 Audio compression tools

There are many different techniques available for audio compression, each having advantages and disadvantages. Real compressors will combine several techniques or tools in various ways to achieve different combinations of cost and complexity. Here it is intended to examine the tools separately before seeing how they are used in actual compression systems.

The simplest coding tool is companding which is a digital parallel of the noise reducers used in analog tape recording. Figure 5.8(a) shows that in companding the input signal level is monitored. Whenever the input level falls below maximum, it is amplified at the coder. The gain which was applied at the coder is added to the data stream so that the decoder can apply an equal attenuation. The advantage of companding is that the signal is kept as far away from the noise floor as possible. In analog noise reduction this is used to maximize the SNR of a tape recorder, whereas in digital compression it is used to keep the signal level as far as possible above the distortion introduced by various coding steps.

One common way of obtaining coding gain is to shorten the wordlength of samples so that fewer bits need to be transmitted. Figure 5.8(b) shows that when this is done, the distortion will rise by 6 dB for every bit removed. This is because removing a bit halves the number of quantizing intervals which then must be twice as large, doubling the error amplitude.

Clearly if this step follows the compander of (a), the audibility of the distortion will be minimized. As an alternative to shortening the wordlength, the uniform quantized PCM signal can be converted to a non-uniform format. In non-uniform coding, shown at (c), the size of the quantizing step rises with the magnitude of the sample so that the distortion level is greater when higher levels exist.

Companding is a relative of floating-point coding shown in Figure 5.9 where the sample value is expressed as a mantissa and a binary exponent which

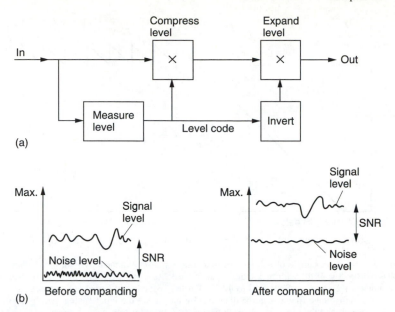

Figure 5.8 Digital companding. In (a) the encoder amplifies the input to maximum level and the decoder attenuates by the same amount. (b) In a companded system, the signal is kept as far as possible above the noise caused by shortening the sample wordlength.

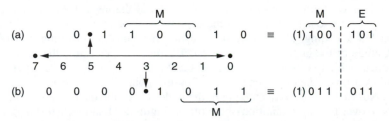

Figure 5.9 In this example of floating point notation, the radix point can have eight positions determined by the exponent E. The point is placed to the left of the first '1', and the next four bits to the right form the mantissa M. As the MSB of the mantissa is always 1, it need not always be stored.

determines how the mantissa needs to be shifted to have its correct absolute value on a PCM scale. The exponent is the equivalent of the gain setting or scale factor of a compander.

Clearly in floating point the signal-to-noise ratio is defined by the number of bits in the mantissa, and as shown in Figure 5.10, this will vary as a sawtooth function of signal level, as the best value, obtained when the mantissa is near overflow, is replaced by the worst value when the mantissa overflows and the exponent is incremented. Floating-point notation is used within DSP chips as it eases the computational problems involved in handling long wordlengths. For example, when multiplying floating point numbers, only the mantissae need to be multiplied. The exponents are simply added.

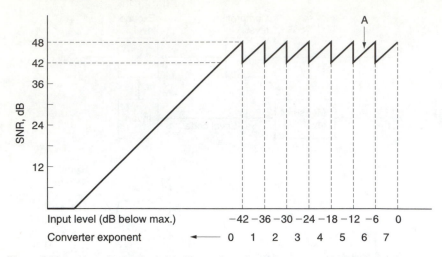

Figure 5.10 In this example of an eight-bit mantissa, three-bit exponent system, the maximum SNR is 6 dB × 8 = 48 dB with maximum input of 0 dB. As input level falls by 6 dB, the convertor noise remains the same, so SNR falls to 42 dB. Further reduction in signal level causes the convertor to shift range (point A in the diagram) by increasing the input analog gain by 6 dB. The SNR is restored, and the exponent changes from 7 to 6 in order to cause the same gain change at the receiver. The noise modulation would be audible in this simple system. A longer mantissa word is needed in practice.

A floating-point system requires one exponent to be carried with each mantissa and this is wasteful because in real audio material the level does not change so rapidly and there is redundancy in the exponents. A better alternative is floating-point block coding, also known as near-instantaneous companding, where the magnitude of the largest sample in a block is used to determine the value of an exponent which is valid for the whole block. Sending one exponent per block requires a lower data rate than in true floating point.[10]

In block coding the requantizing in the coder raises the quantizing error, but it does so over the entire duration of the block. Figure 5.11 shows that if a transient occurs towards the end of a block, the decoder will reproduce the waveform correctly, but the quantizing noise will start at the beginning of the block and may result in a burst of distortion products (also called pre-noise or pre-echo) which is audible before the transient. Temporal masking may be used to make this inaudible. With a 1 ms block, the artifacts are too brief to be heard.

Another solution is to use a variable time window according to the transient content of the audio waveform. When musical transients occur, short blocks are necessary and the coding gain will be low.[11] At other times the blocks become longer allowing a greater coding gain.

Whilst the above systems used alone do allow coding gain, the compression factor has to be limited because little benefit is obtained from masking. This is because the techniques above produce distortion which may be found anywhere over the entire audio band. If the audio input spectrum is narrow, this noise will not be masked.

Sub-band coding[12] splits the audio spectrum into many different frequency bands. Once this has been done, each band can be individually processed. In real audio signals many bands will contain lower-level signals than the loudest one.

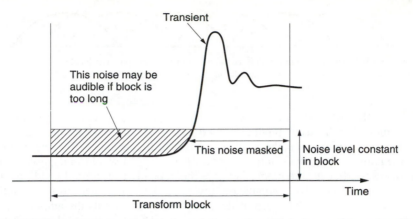

Figure 5.11 If a transient occurs towards the end of a transform block, the quantizing noise will still be present at the beginning of the block and may result in a pre-echo where the noise is audible before the transient.

Individual companding of each band will be more effective than broadband companding. Sub-band coding also allows the level of distortion products to be raised selectively so that distortion is created only at frequencies where spectral masking will be effective.

Following any perceptive coding steps, the resulting data may be further subjected to lossless binary compression tools such as prediction, Huffman coding or a combination of both.

Audio is usually considered to be a time-domain waveform as this is what emerges from a microphone. As has been seen in Chapter 3, spectral analysis allows any periodic waveform to be represented by a set of harmonically related components of suitable amplitude and phase. In theory it is perfectly possible to decompose a periodic input waveform into its constituent frequencies and phases, and to record or transmit the transform. The transform can then be inverted and the original waveform will be precisely recreated.

Although one can think of exceptions, the transform of a typical audio waveform changes relatively slowly much of the time. The slow speech of an organ pipe or a violin string or the slow decay of most musical sounds allow the rate at which the transform is sampled to be reduced, and a coding gain results. At some frequencies the level will be below maximum and a shorter wordlength can be used to describe the coefficient. Further coding gain will be achieved if the coefficients describing frequencies which will experience masking are quantized more coarsely.

In practice there are some difficulties, real sounds are not periodic, but contain transients which transformation cannot accurately locate in time. The solution to this difficulty is to cut the waveform into short segments and then to transform each individually. The delay is reduced, as is the computational task, but there is a possibility of artifacts arising because of the truncation of the waveform into rectangular time windows. A solution is to use window functions, and to overlap the segments as shown in Figure 5.12. Thus every input sample appears in just two transforms, but with variable weighting depending upon its position along the time axis.

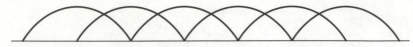

Figure 5.12 Transform coding can only be practically performed on short blocks. These are overlapped using window functions in order to handle continuous waveforms.

The DFT (discrete frequency transform) does not produce a continuous spectrum, but instead produces coefficients at discrete frequencies. The frequency resolution (i.e. the number of different frequency coefficients) is equal to the number of samples in the window. If overlapped windows are used, twice as many coefficients are produced as are theoretically necessary. In addition, the DFT requires intensive computation, owing to the requirement to use complex arithmetic to render the phase of the components as well as the amplitude. An alternative is to use discrete cosine transforms (DCT) or the modified discrete cosine transform (MDCT) which has the ability to eliminate the overhead of coefficients due to overlapping the windows and return to the critically sampled domain.[13] Critical sampling is a term which means that the number of coefficients does not exceed the number which would be obtained with non-overlapping windows.

5.9 Sub-band coding

Sub-band coding takes advantage of the fact that real sounds do not have uniform spectral energy. The wordlength of PCM audio is based on the dynamic range required and this is generally constant with frequency although any pre-emphasis will affect the situation. When a signal with an uneven spectrum is conveyed by PCM, the whole dynamic range is occupied only by the loudest spectral component, and all the other components are coded with excessive headroom. In its simplest form, sub-band coding works by splitting the audio signal into a number of frequency bands and companding each band according to its own level. Bands in which there is little energy result in small amplitudes which can be transmitted with short wordlength. Thus each band results in variable-length samples, but the sum of all the sample wordlengths is less than that of PCM and so a coding gain can be obtained. Sub-band coding is not restricted to the digital domain; the analog Dolby noise-reduction systems use it extensively.

The number of sub-bands to be used depends upon what other compression tools are to be combined with the sub-band coding. If it is intended to optimize compression based on auditory masking, the sub-bands should preferably be narrower than the critical bands of the ear, and therefore a large number will be required. This requirement is frequently not met: ISO/MPEG Layers I and II use only 32 sub-bands. Figure 5.13 shows the critical condition where the masking tone is at the top edge of the sub-band. It will be seen that the narrower the sub-band, the higher the requantizing 'noise' that can be masked. The use of an excessive number of sub-bands will, however, raise complexity and the coding delay, as well as risking pre-ringing on transients which may exceed the temporal masking.

On the other hand, if used in conjunction with predictive sample coding, relatively few bands are required. The apt-X100 system, for example, uses only four sub-bands as simulations showed that a greater number gave diminishing returns.[14]

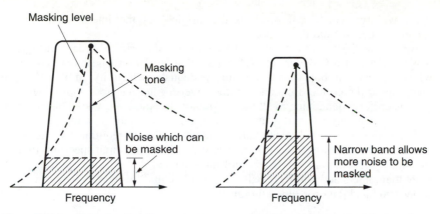

Figure 5.13 In sub-band coding the worst case occurs when the masking tone is at the top edge of the sub-band. The narrower the band, the higher the noise level which can be masked.

The bandsplitting process is complex and requires a lot of computation. One bandsplitting method which is useful is quadrature mirror filtering.[15] The QMF is a kind of twin FIR filter which converts a PCM sample stream into to two sample streams of half the input sampling rate, so that the output data rate equals the input data rate. The frequencies in the lower half of the audio spectrum are carried in one sample stream, and the frequencies in the upper half of the spectrum are carried in the other. Whilst the lower-frequency output is a PCM band-limited representation of the input waveform, the upper frequency output isn't. A moment's thought will reveal that it could not be so because the sampling rate is not high enough. In fact the upper half of the input spectrum has been heterodyned down to the same frequency band as the lower half by the clever use of aliasing. The waveform is unrecognizable, but when heterodyned back to its correct place in the spectrum in an inverse step, the correct waveform will result once more.

Sampling theory states that the sampling rate needed must be at least twice the bandwidth in the signal to be sampled. If the signal is band limited, the sampling rate need only be more than twice the signal *bandwidth* not the signal *frequency*. Downsampled signals of this kind can be reconstructed by a reconstruction or *synthesis* filter having a bandpass response rather than a low pass response. As only signals within the passband can be output, it is clear from Figure 5.14 that

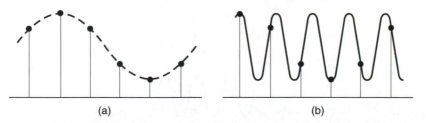

(a)   (b)

Figure 5.14 The sample stream shown would ordinarily represent the waveform shown in (a), but if it is known that the original signal could exist only between two frequencies then the waveform in (b) must be the correct one. A suitable bandpass reconstruction filter, or synthesis filter, will produce the waveform in (b).

the waveform which will result is the original as the intermediate aliased waveform lies outside the passband.

An inverse QMF will recombine the bands into the original broadband signal. It is a feature of a QMF/inverse QMF pair that any energy near the band edge which appears in both bands due to inadequate selectivity in the filtering reappears at the correct frequency in the inverse filtering process provided that there is uniform quantizing in all the sub-bands. In practical coders, this criterion is not met, but any residual artifacts are sufficiently small to be masked.

The audio band can be split into as many bands as required by cascading QMFs in a tree. However, each stage can only divide the input spectrum in half. In some coders certain sub-bands will have passed through one splitting stage more than others and will have half their bandwidth.[16] A delay is required in the wider sub-band data for time alignment.

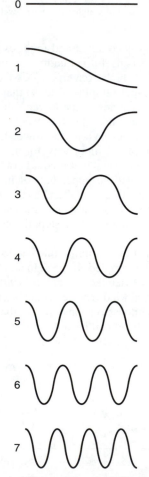

Figure 5.15 A table of basis functions for an eight-point DCT. If these waveforms are added together in various proportions, any original waveform can be reconstructed. In practice these waveforms are stored as samples, but after reconstruction to the analog domain they would appear as shown here.

5.10 Transform coding

Many transform coders use the discrete cosine transform described in section 3.14. The DCT works on blocks of samples which are windowed. For simplicity the following example uses a very small block of only eight samples whereas a real encoder might use several hundred.

Figure 5.15 shows the table of basis functions or *wave table* for an eight-point DCT. Adding these two-dimensional waveforms together in different proportions will give any combination of the original eight PCM audio samples. The coefficients of the DCT simply control the proportion of each wave which is added in the inverse transform. The top-left wave has no modulation at all because it conveys the DC component of the block. Increasing the DC coefficient adds a constant amount to every sample in the block.

Moving to the right the coefficients represent increasing frequencies. All these coefficients are bipolar, where the polarity indicates whether the original waveform at that frequency was inverted.

Figure 5.16 shows an example of an inverse transform. The DC coefficient produces a constant level throughout the sample block. The remaining waves in the table are AC coefficients. A zero coefficient would result in no modulation, leaving the DC level unchanged. The wave next to the DC component represents the lowest frequency in the transform which is half a cycle per block. A positive coefficient would increase the signal voltage at the left side of the block whilst reducing it on the right, whereas a negative coefficient would do the opposite. The magnitude of the coefficient determines the amplitude of the wave which is added. Figure 5.16 also shows that the next wave has a frequency of one cycle per block, i.e. the waveform is made more positive at both sides and more negative in the middle.

Consequently an inverse DCT is no more than a process of mixing various waveforms from the wave table where the relative amplitudes and polarity of these patterns are controlled by the coefficients. The original transform is simply

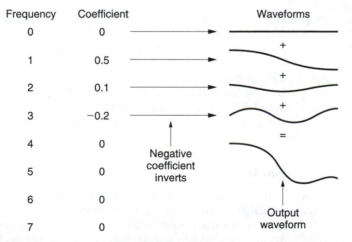

Figure 5.16 An example of an inverse DCT. The coefficients determine the amplitudes of the waves from the table in Figure 5.15 which are to be added together. Note that coefficient 3 is negative so that the wave is inverted.

a mechanism which finds the coefficient amplitudes from the original PCM sample block.

The DCT itself achieves no compression at all. The number of coefficients which are output always equals the number of audio samples in the block. However, in typical program material, not all coefficients will have significant values; there will often be a few dominant coefficients. The coefficients representing the higher frequencies will often be zero or of small value, due to the typical energy distribution of audio.

Coding gain (the technical term for reduction in the number of bits needed) is achieved by transmitting the low-valued coefficients with shorter wordlengths. The zero-valued coefficients need not be transmitted at all. Thus it is not the DCT which compresses the audio, it is the subsequent processing. The DCT simply expresses the audio samples in a form which makes the subsequent processing easier.

Higher compression factors require the coefficient wordlength to be further reduced using requantizing. Coefficients are divided by some factor which increases the size of the quantizing step. The smaller number of steps which results permits coding with fewer bits, but of course with an increased quantizing error. The coefficients will be multiplied by a reciprocal factor in the decoder to return to the correct magnitude.

Further redundancy in transform coefficients can also be identified. This can be done in various ways. Within a transform block, the coefficients may be transmitted using differential coding so that the first coefficient is sent in an absolute form whereas the remainder are transmitted as differences with respect to the previous one. Some coders attempt to predict the value of a given coefficient using the value of the same coefficient in typically the two previous blocks. The prediction is subtracted from the actual value to produce a prediction error or residual which is transmitted to the decoder. Another possibility is to use prediction within the transform block. The predictor scans the coefficients from, say, the low-frequency end upwards and tries to predict the value of the next coefficient in the scan from the values of the earlier coefficients. Again a residual is transmitted.

Inter-block prediction works well for stationary material, whereas intra-block prediction works well for transient material. An intelligent coder may select a prediction technique using the input entropy in the same way that it selects the window size.

Inverse transforming a requantized coefficient means that the frequency it represents is reproduced in the output with the wrong amplitude. The difference between the original and the reconstructed amplitude is considered to be a noise added to the wanted data. The audibility of such noise depends on the degree of masking prevailing.

5.11 Compression formats

There are numerous formats intended for audio compression and these can be divided into international standards and proprietary designs.

The ISO (International Standards Organization) and the IEC (International Electrotechnical Commission) recognized that compression would have an important part to play and in 1988 established the ISO/IEC/MPEG (Moving Picture Experts Group) to compare and assess various coding schemes in order to arrive at an international standard for compressing video. The terms of reference were extended the same year to include audio and the MPEG/Audio group was formed.

MPEG audio coding is used for DAB (digital audio broadcasting) and for the audio content of digital television broadcasts to the DVB standard.

In the USA, it has been proposed to use an alternative compression technique for the audio content of ATSC (advanced television systems committee) digital television broadcasts. This is the AC-3[17] system developed by Dolby Laboratories. The MPEG transport stream structure has also been standardized to allow it to carry AC-3 coded audio. The digital video disk (DVD) can also carry AC-3 or MPEG audio coding. Other popular proprietary codes include apt-X which is a mild compression factor/short delay codec and ATRAC which is the codec used in MiniDisc.

5.12 MPEG Layer I

Figure 5.17 shows a block diagram of a Layer I coder which is a simplified version of that used in the MUSICAM system. A polyphase filter divides the audio spectrum into 32 equal sub-bands. The output of the filter bank is critically

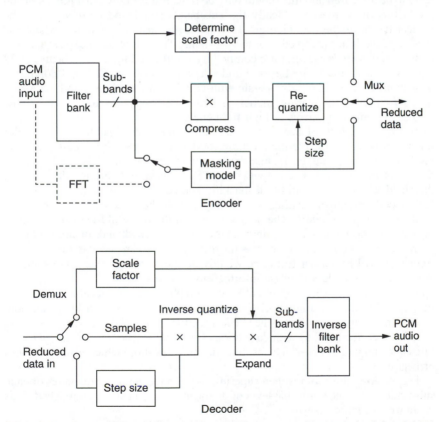

Figure 5.17 A simple sub-band coder. The bit allocation may come from analysis of the sub-band energy, or, for greater reduction, from a spectral analysis in a side chain.

sampled. In other words the output data rate is no higher than the input rate because each band has been heterodyned to a frequency range from zero upwards.

Sub-band compression takes advantage of the fact that real sounds do not have uniform spectral energy. The wordlength of PCM audio is based on the dynamic range required and this is generally constant with frequency although any pre-emphasis will affect the situation. When a signal with an uneven spectrum is conveyed by PCM, the whole dynamic range is occupied only by the loudest spectral component, and all the other components are coded with excessive headroom. In its simplest form, sub-band coding works by splitting the audio signal into a number of frequency bands and companding each band according to its own level. Bands in which there is little energy result in small amplitudes which can be transmitted with short wordlength. Thus each band results in variable-length samples, but the sum of all the sample wordlengths is less than that of the PCM input and so a degree of coding gain can be obtained.

A Layer I-compliant encoder, i.e. one whose output can be understood by a standard decoder, can be made which does no more than this. Provided the syntax of the bitstream is correct, the decoder is not concerned with how the coding decisions were made. However, higher compression factors require the distortion level to be increased and this should only be done if it is known that the distortion products will be masked. Ideally the sub-bands should be narrower than the critical bands of the ear. Figure 5.13 showed the critical condition where the masking tone is at the top edge of the sub-band. The use of an excessive number of sub-bands will, however, raise complexity and the coding delay. The use of 32 equal sub-bands in MPEG Layers I and II is a compromise. Efficient polyphase bandsplitting filters can only operate with equal-width sub-bands and the result, in an octave-based hearing model, is that sub-bands are too wide at low frequencies and too narrow at high frequencies.

To offset the lack of accuracy in the sub-band filter a parallel fast Fourier transform is used to drive the masking model. The standard suggests masking models, but compliant bitstreams can result from other models. In Layer I a 512-point FFT is used. The output of the FFT is used to determine the masking threshold which is the sum of all masking sources. Masking sources include at least the threshold of hearing which may locally be raised by the frequency content of the input audio. The degree to which the threshold is raised depends on whether the input audio is sinusoidal or atonal (broadband, or noise-like).

In the case of a sine wave, the magnitude and phase of the FFT at each frequency will be similar from one window to the next, whereas if the sound is atonal the magnitude and phase information will be chaotic.

The masking threshold is effectively a graph of just noticeable noise as a function of frequency. Figure 5.18(a) shows an example. The masking threshold is calculated by convolving the FFT spectrum with the cochlea spreading function (see section 2.5) with corrections for tonality. The level of the masking threshold cannot fall below the absolute masking threshold which is the threshold of hearing.

The masking threshold is then superimposed on the actual frequencies of each sub-band so that the allowable level of distortion in each can be established. This is shown in Figure 5.18(b).

Constant-size input blocks are used, containing 384 samples. At 48 kHz, 384 samples correspond to a period of 8 ms. After the sub-band filter each band

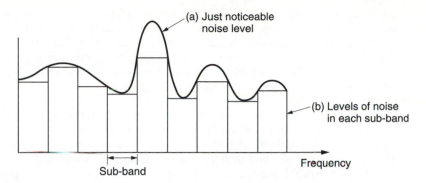

Figure 5.18 A continuous curve (a) of the just-noticeable noise level is calculated by the masking model. The levels of noise in each sub-band (b) must be set so as not to exceed the level of the curve.

contains 12 samples per block. The block size is too long to avoid the pre-masking phenomenon of Figure 5.11. Consequently the masking model must ensure that heavy requantizing is not used in a block which contains a large transient following a period of quiet. This can be done by comparing parameters of the current block with those of the previous block as a significant difference will indicate transient activity.

The samples in each sub-band block or *bin* are companded according to the peak value in the bin. A six-bit scale factor is used for each sub-band which applies to all 12 samples. The gain step is 2 dB and so with a six-bit code over 120 dB of dynamic range is available.

A fixed-output bit rate is employed, and as there is no buffering the size of the coded output block will be fixed. The wordlengths in each bin will have to be such that the sum of the bits from all the sub-bands equals the size of the coded block. Thus some sub-bands can have long wordlength coding if others have short wordlength coding. The process of determining the requantization step size, and hence the wordlength in each sub-band, is known as bit allocation. In Layer I all sub-bands are treated in the same way and fourteen different requantization classes are used. Each one has an odd number of quantizing intervals so that all codes are referenced to a precise zero level.

Where masking takes place, the signal is quantized more coarsely until the distortion level is raised to just below the masking level. The coarse quantization requires shorter wordlengths and allows a coding gain. The bit allocation may be iterative as adjustments are made to obtain an equal NMR across all sub-bands. If the allowable data rate is adequate, a positive NMR will result and the decoded quality will be optimal. However, at lower bit rates and in the absence of buffering a temporary increase in bit rate is not possible. The coding distortion cannot be masked and the best the encoder can do is to make the (negative) NMR equal across the spectrum so that artifacts are not emphasized unduly in any one sub-band. It is possible that in some sub-bands there will be no data at all, either because such frequencies were absent in the program material or because the encoder has discarded them to meet a low bit rate.

The samples of differing wordlength in each bin are then assembled into the output coded block. Unlike a PCM block, which contains samples of fixed wordlength, a coded block contains many different wordlengths which may vary from one sub-band to the next. In order to deserialize the block into samples of various wordlengths and demultiplex the samples into the appropriate frequency bins, the decoder has to be told what bit allocations were used when it was packed, and some synchronizing means is needed to allow the beginning of the block to be identified.

The compression factor is determined by the bit-allocation system. It is trivial to change the output block size parameter to obtain a different compression factor. If a larger block is specified, the bit allocator simply iterates until the new block size is filled. Similarly the decoder need only deserialize the larger block correctly into coded samples and then the expansion process is identical except for the fact that expanded words contain less noise. Thus codecs with varying degrees of compression are available which can perform different bandwidth/ performance tasks with the same hardware.

Figure 5.19(a) shows the format of the Layer I elementary stream. The frame begins with a sync pattern to reset the phase of deserialization, and a header which describes the sampling rate and any use of pre-emphasis. Following this is a block of 32 four-bit allocation codes. These specify the wordlength used in each sub-band and allow the decoder to deserialize the sub-band sample block. This is followed by a block of 32 six-bit scale factor indices, which specify the gain given to each band during companding. The last block contains 32 sets of 12 samples. These samples vary in wordlength from one block to the next, and can be from 0 to 15 bits long. The deserializer has to use the 32 allocation information codes to work out how to deserialize the sample block into individual samples of variable length.

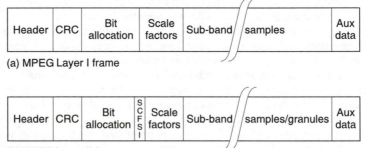

(a) MPEG Layer I frame

(b) MPEG Layer II frame

Figure 5.19 (a) The MPEG Layer I data frame has a simple structure. (b) In the Layer II frame, the compression of the scale factors requires the additional SCFSI code described in the text.

The Layer I MPEG decoder is shown in Figure 5.20. The elementary stream is deserialized using the sync pattern and the variable-length samples are assembled using the allocation codes. The variable-length samples are returned to fifteen-bit wordlength by adding zeros. The scale factor indices are then used to determine multiplication factors used to return the waveform in each sub-band to its original amplitude. The 32 sub-band signals are then merged into one spectrum by the synthesis filter. This is a set of bandpass filters which

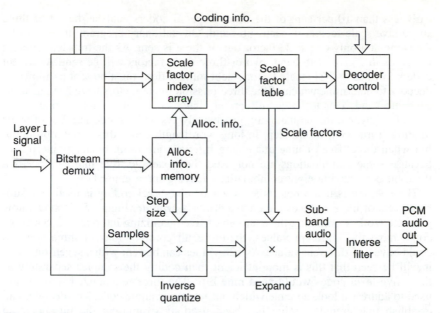

Figure 5.20 The Layer I decoder. See text for details.

heterodynes every sub-band to the correct place in the audio spectrum and then adds them to produce the audio output.

5.13 MPEG Layer II

MPEG Layer II audio coding is identical to MUSICAM. The same 32-band filterbank and the same block companding scheme as Layer I is used. In order to give the masking model better spectral resolution, the side-chain FFT has 1024 points. The FFT drives the masking model which may be the same as is suggested for Layer I. The block length is increased to 1152 samples. This is three times the block length of Layer I, corresponding to 24 ms at 48 kHz.

Figure 5.19(b) shows the Layer II elementary stream structure. Following the sync pattern the bit-allocation data are sent. The requantizing process of Layer II is more complex than in Layer I. The sub-bands are categorized into three frequency ranges, low, medium and high, and the requantizing in each range is different. Low-frequency samples can be quantized into 15 different wordlengths, mid-frequencies into seven different wordlengths and high frequencies into only three different wordlengths. Accordingly the bit-allocation data use words of four, three and two bits depending on the sub-band concerned. This reduces the amount of allocation data to be sent. In each case one extra combination exists in the allocation code. This is used to indicate that no data are being sent for that sub-band.

The 1152-sample block of Layer II is divided into three blocks of 384 samples so that the same companding structure as Layer I can be used. The 2 dB step size in the scale factors is retained. However, not all the scale factors are transmitted, because they contain a degree of redundancy. In real program material, the difference between scale factors in successive blocks in the same band exceeds

2 dB less than 10 per cent of the time. Layer II coders analyse the set of three successive scale factors in each sub-band. On stationary program, these will be the same and only one scale factor out of three is sent. As the transient content increases in a given sub-band, two or three scale factors will be sent. A two-bit code known as SCFSI (scale factor select information) must be sent to allow the decoder to determine which of the three possible scale factors have been sent for each sub-band. This technique effectively halves the scale factor bit rate.

As for Layer I, the requantizing process always uses an odd number of steps to allow a true centre zero step. In long wordlength codes this is not a problem, but when three, five or nine quantizing intervals are used, binary is inefficient because some combinations are not used. For example, five intervals needs a three-bit code having eight combinations leaving three unused.

The solution is that when three-, five- or nine-level coding is used in a sub-band, sets of three samples are encoded into a *granule*. Figure 5.21 shows how granules work. Continuing the example of five quantizing intervals, each sample could have five different values, therefore all combinations of three samples could have 125 different values. As 128 values can be sent with a seven-bit code, it will be seen that this is more efficient than coding the samples separately as three five-level codes would need nine bits. The three requantized samples are used to address a look-up table which outputs the granule code. The decoder can establish that granule coding has been used by examining the bit-allocation data.

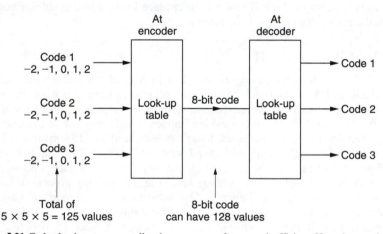

Figure 5.21 Codes having ranges smaller than a power of two are inefficient. Here three codes with a range of five values which would ordinarily need 3 × 3 bits can be carried in a single eight-bit word.

The requantized samples/granules in each sub-band, bit allocation data, scale factors and scale factor select codes are multiplexed into the output bitstream.

The Layer II decoder is shown in Figure 5.22. This is not much more complex than the Layer I decoder. The demultiplexing will separate the sample data from the side information. The bit-allocation data will specify the wordlength or granule size used so that the sample block can be deserialized and the granules decoded. The scale factor select information will be used to decode the

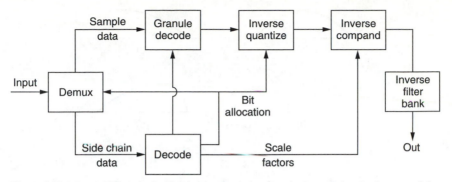

Figure 5.22 A Layer II decoder is slightly more complex than the Layer I decoder because of the need to decode granules and scale factors.

compressed scale factors to produce one scale factor per block of 384 samples. Inverse quantizing and inverse sub-band filtering takes place as for Layer I.

5.14 MPEG Layer III

Layer III is the most complex layer, and is only really necessary when the most severe data rate constraints must be met. It is also known as MP3 in its application of music delivery over the Internet. It is a transform code based on the ASPEC system with certain modifications to give a degree of commonality with Layer II. The original ASPEC coder used a direct MDCT on the input samples. In Layer III this was modified to use a hybrid transform incorporating the existing polyphase 32-band QMF of Layers I and II and retaining the block size of 1152 samples. In Layer III, the 32 sub-bands from the QMF are further processed by a critically sampled MDCT.

The windows overlap by two to one. Two window sizes are used to reduce pre-echo on transients. The long window works with 36 sub-band samples corresponding to 24 ms at 48 kHz and resolves 18 different frequencies, making 576 frequencies altogether. Coding products are spread over this period which is acceptable in stationary material but not in the vicinity of transients. In this case the window length is reduced to 8 ms. Twelve sub-band samples are resolved into six different frequencies making a total of 192 frequencies. This is the Heisenberg inequality: by increasing the time resolution by a factor of three, the frequency resolution has fallen by the same factor.

Figure 5.23 shows the available window types. In addition to the long and short symmetrical windows there is a pair of transition windows, know as start and stop windows which allow a smooth transition between the two window sizes. In order to use critical sampling, MDCTs must resolve into a set of frequencies which is a multiple of four. Switching between 576 and 192 frequencies allows this criterion to be met. Note that an 8 ms window is still too long to eliminate pre-echo. Pre-echo is eliminated using buffering. The use of a short window minimizes the size of the buffer needed.

Layer III provides a suggested (but not compulsory) pychoacoustic model which is more complex than that suggested for Layers I and II, primarily because of the need for window switching. Pre-echo is associated with the entropy in the

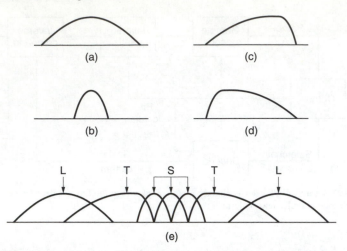

Figure 5.23 The window functions of Layer III coding. At (a) is the normal long window, whereas (b) shows the short window used to handle transients. Switching between window sizes requires transition windows (c) and (d). An example of switching using transition windows is shown in (e).

audio rising above the average value and this can be used to switch the window size. The perceptive model is used to take advantage of the high-frequency resolution available from the DCT which allows the noise floor to be shaped much more accurately than with the 32 sub-bands of Layers I and II. Although the MDCT has high-frequency resolution, it does not carry the phase of the waveform in an identifiable form and so is not useful for discriminating between tonal and atonal inputs. As a result a side FFT which gives conventional amplitude and phase data is still required to drive the masking model.

Non-uniform quantizing is used, in which the quantizing step size becomes larger as the magnitude of the coefficient increases. The quantized coefficients are then subject to Huffman coding. This is a technique where the most common code values are allocated the shortest wordlength. Layer III also has a certain amount of buffer memory so that pre-echo can be avoided during entropy peaks despite a constant output bit rate.

Figure 5.24 shows a Layer III encoder. The output from the sub-band filter is 32 continuous band-limited sample streams. These are subject to 32 parallel MDCTs. The window size can be switched individually in each sub-band as required by the characteristics of the input audio. The parallel FFT drives the masking model which decides on window sizes as well as producing the masking threshold for the coefficient quantizer. The distortion control loop iterates until the available output data capacity is reached with the most uniform NMR.

The available output capacity can vary owing to the presence of the buffer. Figure 5.25 shows that the buffer occupancy is fed back to the quantizer. During stationary program material, the buffer contents are deliberately run down by slight coarsening of the quantizing. The buffer empties because the output rate is fixed but the input rate has been reduced. When a transient arrives, the large coefficients which result can be handled by filling the buffer, avoiding raising the output bit rate whilst also avoiding the pre-echo which would result if the coefficients were heavily quantized.

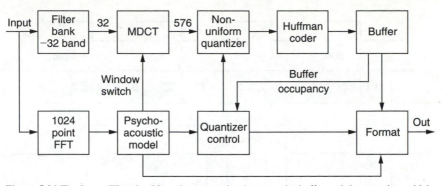

Figure 5.24 The Layer III coder. Note the connection between the buffer and the quantizer which allows different frames to contain different amounts of data.

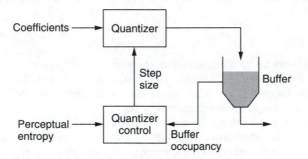

Figure 5.25 The variable rate coding of Layer III. An approaching transient via the perceptual entropy signal causes the coder to quantize more heavily in order to empty the buffer. When the transient arrives, the quantizing can be made more accurate and the increased data can be accepted by the buffer.

In order to maintain synchronism between encoder and decoder in the presence of buffering, headers and side information are sent synchronously at frame rate. However, the position of boundaries between the main data blocks which carry the coefficients can vary with respect to the position of the headers in order to allow a variable frame size. Figure 5.26 shows that the frame begins with an unique sync pattern which is followed by the side information. The side information contains a parameter called *main data begin* which specifies where the main data for the present frame began in the transmission. This parameter allows the decoder to find the coefficient block in the decoder buffer. As the frame headers are at fixed locations, the main data blocks may be interrupted by the headers.

5.15 MPEG-2 AAC

The MPEG standards system subsequently developed an enhanced system known as advanced audio coding (AAC).[8,18] This was intended to be a standard which delivered the highest possible performance using newly developed tools that could not be used in any standard which was backward compatible. AAC will also form the core of the audio coding of MPEG-4.

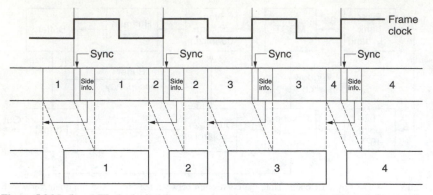

Figure 5.26 In Layer III, the logical frame rate is constant and is transmitted by equally spaced sync patterns. The data blocks do not need to coincide with sync. A pointer after each sync pattern specifies where the data block starts. In this example block 2 is smaller whereas 1 and 3 have enlarged.

AAC supports up to 48 audio channels with default support of monophonic, stereo and 5.1 channel (3/2) audio. The AAC concept is based on a number of coding tools known as *modules* which can be combined in different ways to produce bitstreams at three different profiles.

The main profile requires the most complex encoder which makes use of all the coding tools. The low-complexity (LC) profile omits certain tools and restricts the power of others to reduce processing and memory requirements. The remaining tools in LC profile coding are identical to those in main profile such that a main profile decoder can decode LC profile bitstreams.

The scaleable sampling rate (SSR) profile splits the input audio into four equal frequency bands each of which results in a self-contained bitstream. A simple decoder can decode only one, two or three of these bitstreams to produce a reduced bandwidth output. Not all the AAC tools are available to SSR profile.

The increased complexity of AAC allows the introduction of lossless coding tools. These allow a lower bit rate for the same quality or improved quality at a given bit rate where the reliance on lossy coding is reduced. There is greater attention given to the interplay between time-domain and frequency-domain precision in the human hearing system.

Figure 5.27 shows a block diagram of an AAC main profile encoder. The audio signal path is straight through the centre. The formatter assembles any side-chain data along with the coded audio data to produce a compliant bitstream. The input signal passes to the filter bank and the perceptual model in parallel.

The filter bank consists of a 50 per cent overlapped critically sampled MDCT which can be switched between block lengths of 2048 and 256 samples. At 48 kHz the filter allows resolutions of 23 Hz and 21 ms or 187 Hz and 2.6 ms. As AAC is a multichannel coding system, block length switching cannot be done indiscriminately as this would result in loss of block phase between channels. Consequently if short blocks are selected, the coder will remain in short block mode for integer multiples of eight blocks. This is illustrated in Figure 5.28 which also shows the use of transition windows between the block sizes as was done in Layer III.

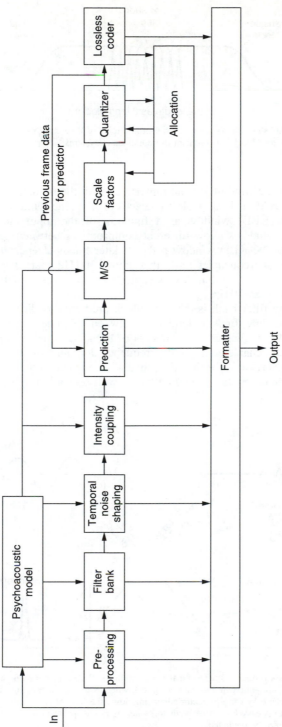

Figure 5.27 The AAC encoder. Signal flow is from left to right whereas side-chain data flow is vertical.

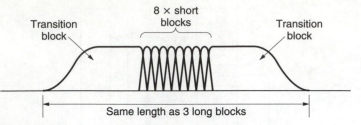

Figure 5.28 In AAC short blocks must be used in multiples of 8 so that the long block phase is undisturbed. This keeps block synchronism in multichannel systems.

The shape of the window function interferes with the frequency selectivity of the MDCT. In AAC it is possible to select either a sine window or a Kaiser–Bessel-derived (KBD) window as a function of the input audio spectrum. Different filter windows allow different compromises between bandwidth and rate of roll-off. The KBD window rolls off later but is steeper and thus gives better rejection of frequencies more than about 200 Hz apart whereas the sine window rolls off earlier but less steeply and so gives better rejection of frequencies less than 70 Hz.

Following the filter bank is the intra-block predictive coding module. When enabled this module finds redundancy between the coefficients within one transform block. In Chapter 3 the concept of transform duality was introduced, in which a certain characteristic in the frequency domain would be accompanied by a dual characteristic in the time domain and vice versa. Figure 5.29 shows that in the time domain, predictive coding works well on stationary signals but fails

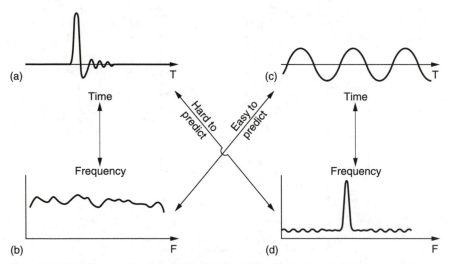

Figure 5.29 Transform duality suggests that predictability will also have a dual characteristic. A time predictor will not anticipate the transient in (a), whereas the broad spectrum of signal (a), shown in (b), will be easy for a predictor advancing down the frequency axis. In contrast, the stationary signal (c) is easy for a time predictor, whereas in the spectrum of (c) shown at (d) the spectral spike will not be predicted.

on transients. The dual of this characteristic is that in the frequency domain, predictive coding works well on transients but fails on stationary signals.

Equally, a predictive coder working in the time domain produces an error spectrum which is related to the input spectrum. The dual of this characteristic is that a predictive coder working in the frequency domain produces a prediction error which is related to the input time-domain signal.

This explains the use of the term *temporal noise shaping* (TNS) used in the AAC documents.[19] When used during transients, the TNS module produces a distortion which is time-aligned with the input such that pre-echo is avoided. The use of TNS also allows the coder to use longer blocks more of the time. This module is responsible for a significant amount of the increased performance of AAC.

Figure 5.30 shows that the coefficients in the transform block are serialized by a commutator. This can run from the lowest frequency to the highest or in reverse. The prediction method is a conventional forward predictor structure in which the result of filtering a number of earlier coefficients (20 in main profile) is used to predict the current one. The prediction is subtracted from the actual value to produce a prediction error or residual which is transmitted. At the decoder, an identical predictor produces the same prediction from earlier coefficient values and the error in this is cancelled by adding the residual.

Following the intra-block prediction, an optional module known as the intensity/coupling stage is found. This is used for very low bit rates where spatial information in stereo and surround formats is discarded to keep down the level of distortion. Effectively over at least part of the spectrum a mono signal is transmitted along with amplitude codes which allow the signal to be panned in the spatial domain at the decoder.

The next stage is the inter-block prediction module. Whereas the intra-block predictor is most useful on transients, the inter-block predictor module explores the redundancy between successive blocks on stationary signals.[20] This prediction only operates on coefficients below 16 kHz. For each DCT coefficient in a given block, the predictor uses the quantized coefficients from the same locations in two previous blocks to estimate the present value. As before, the prediction is subtracted to produce a residual which is transmitted. Note that the use of quantized coefficients to drive the predictor is necessary because this is what the decoder will have to do.

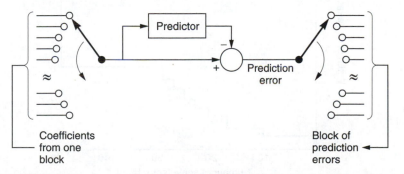

Figure 5.30 Predicting along the frequency axis is performed by running along the coefficients in a block and attempting to predict the value of the current coefficient from the values of some earlier ones. The prediction error is transmitted.

The predictor is adaptive and calculates its own coefficients from the signal history. The decoder uses the same algorithm so that the two predictors always track. The predictors run all the time whether prediction is enabled or not in order to keep the prediction coefficients adapted to the signal.

Audio coefficients are associated into sets known as *scale factor bands* for later companding. Within each scale factor band inter-block prediction can be turned on or off depending on whether a coding gain results.

Protracted use of prediction makes the decoder prone to bit errors and drift and removes decoding entry points from the bitstream. Consequently the prediction process is reset cyclically. The predictors are assembled into groups of 30 and after a certain number of frames a different group is reset until all have been reset. Predictor reset codes are transmitted in the side data. Reset will also occur if short frames are selected.

In stereo and 3/2 surround formats there is less redundancy because the signals also carry spatial information. The effecting of masking may be up to 20 dB less when distortion products are at a different location in the stereo image from the masking sounds. As a result stereo signals require much higher bit rate than two mono channels, particularly on transient material which is rich in spatial clues.

In some cases a better result can be obtained by converting the signal to a mid-side (M/S) or sum/difference format before quantizing. In surround-sound the M/S coding can be applied to the front L/R pair and the rear L/R pair of signals. The M/S format can be selected on a block-by-block basis for each scale factor band.

Next comes the lossy stage of the coder where distortion is selectively introduced as a function of frequency as determined by the masking threshold. This is done by a combination of amplification and requantizing. As mentioned, coefficients (or residuals) are grouped into scale factor bands. As Figure 5.31

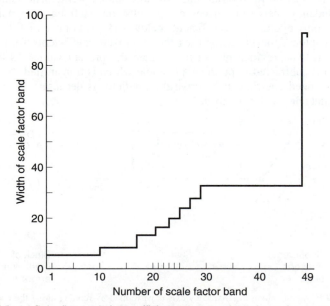

Figure 5.31 In AAC the fine-resolution coefficients are grouped together to form scale factor bands. The size of these varies to loosely mimic the width of critical bands.

shows, the number of coefficients varies in order to divide the coefficients into approximate critical bands. Within each scale factor band, all coefficients will be multiplied by the same scale factor prior to requantizing. Coefficients which have been multiplied by a large scale factor will suffer less distortion by the requantizer whereas those which have been multiplied by a small scale factor will have more distortion. Using scale factors, the psychoacoustic model can shape the distortion as a function of frequency so that it remains masked. The scale factors allow gain control in 1.5 dB steps over a dynamic range equivalent to 24-bit PCM and are transmitted as part of the side data so that the decoder can re-create the correct magnitudes.

The scale factors are differentially coded with respect to the first one in the block and the differences are then Huffman coded.

The requantizer uses non-uniform steps which give better coding gain and has a range of ±8191. The global step size (which applies to all scale factor bands) can be adjusted in 1.5 dB steps. Following requantizing the coefficients are Huffman coded.

There are many ways in which the coder can be controlled and any which results in a compliant bitstream is acceptable although the highest performance may not be reached. The requantizing and scale factor stages will need to be controlled in order to make best use of the available bit rate and the buffering. This is non-trivial because of the use of Huffman coding after the requantizer makes it impossible to predict the exact amount of data which will result from a given step size. This means that the process must iterate.

Whatever bit rate is selected, a good encoder will produce consistent quality by selecting window sizes, intra-or inter-frame prediction and using the buffer to handle entropy peaks. This suggests a connection between buffer occupancy and the control system. The psychoacoustic model will analyse the incoming audio entropy and during periods of average entropy it will empty the buffer by slightly raising the quantizer step size so that the bit rate entering the buffer falls. By running the buffer down, the coder can temporarily support a higher bit rate to handle transients or difficult material.

Simply stated, the scale factor process is controlled so that the distortion spectrum has the same shape as the masking threshold and the quantizing step size is controlled to make the level of the distortion spectrum as low as possible within the allowed bit rate. If the bit rate allowed is high enough, the distortion products will be masked.

References

1. ISO/IEC JTC1/SC29/WG11 MPEG, International standard ISO 11172–3, Coding of moving pictures and associated audio for digital storage media up to 1.5 Mbits/s, Part 3: Audio (1992)
2. MPEG Video Standard: ISO/IEC 13818–2: Information technology – generic coding of moving pictures and associated audio information: Video (1996) (aka ITU-T Rec. H-262) (1996)
3. Huffman, D.A. A method for the construction of minimum redundancy codes. *Proc. IRE*, **40**, 1098–1101 (1952)
4. Grewin, C. and Ryden, T., Subjective assessments on low bit-rate audio codecs. *Proc. 10th. Int. Audio Eng. Soc. Conf.*, 91–102, New York: Audio Engineeringse Society (1991)
5. Gilchrist, N.H.C., Digital sound: the selection of critical programme material and preparation of the recordings for CCIR tests on low bit rate codecs. *BBC Research Dept Report*, RD 1993/1
6. Colomes, C. and Faucon, G., A perceptual objective measurement system (POM) for the quality assessment of perceptual codecs. Presented at the 96th Audio Engineering Society Convention (Amsterdam, 1994), Preprint No. 3801 (P4.2)

7. Johnston, J., Estimation of perceptual entropy using noise masking criteria. *ICASSP*, 2524–2527 (1988)
8. ISO/iec 13818–7, Information Technology – Generic coding of moving pictures and associated audio, Part 7: Advanced audio coding (1997)
9. Gilchrist, N.H.C., Delay in broadcasting operations. Presented at the 90th Audio Engineering Society Convention (1991), Preprint 3033
10. Caine, C.R., English, A.R. and O'Clarey, J.W.H., NICAM-3: near-instantaneous companded digital transmission for high-quality sound programmes. *J. IERE*, **50**, 519–530 (1980)
11. Davidson, G.A. and Bosi, M., AC-2: High quality audio coding for broadcast and storage, in *Proc. 46th Ann. Broadcast Eng. Conf.*, Las Vegas, 98–105 (1992)
12. Crochiere, R.E., Sub-band coding. *Bell System Tech. J.*, **60**, 1633–1653 (1981)
13. Princen, J.P., Johnson, A. and Bradley, A.B., Sub-band/transform coding using filter bank designs based on time domain aliasing cancellation. *Proc. ICASSP*, 2161–2164 (1987)
14. Smyth, S.M.F. and McCanny, J.V., 4-bit Hi-Fi: High quality music coding for ISDN and broadcasting applications. *Proc. ASSP*, 2532–2535 (1988)
15. Jayant, N.S. and Noll, P., *Digital Coding of Waveforms: Principles and applications to speech and video*, Englewood Cliffs: Prentice Hall (1984)
16. Theile, G., Stoll, G. and Link, M., Low bit rate coding of high quality audio signals: an introduction to the MASCAM system. *EBU Tech. Review*, No. 230, 158–181 (1988)
17. Davis, M.F., The AC-3 multichannel coder. Presented at the 95th Audio Engineering Society Convention, Preprint 2774.
18. Bosi, M. *et al.*, ISO/IEC MPEG-2 Advanced Audio Coding. *JAES*, **45**, 789–814 (1997)
19. Herre, J. and Johnston, J.D., Enhancing the performance of perceptual audio coders by using temporal noise shaping (TNS). Presented at the 101st Audio Engineering Society Convention, Preprint 4384 (1996)
20. Fuchs, H., Improving MPEG audio coding by backward adaptive linear stereo prediction. Presented at the 99th Audio Engineering Society Convention (1995), Preprint 4086

6

Digital coding principles

Recording and communication are quite different tasks, but they have a great deal in common. Digital transmission consists of converting data into a waveform suitable for the path along which they are to be sent. Digital recording is basically the process of recording a digital transmission waveform on a suitable medium. In this chapter the fundamentals of digital recording and transmission are introduced along with descriptions of the coding and error-correction techniques used in practical applications.

6.1 Introduction

Data can be recorded on many different media and conveyed using many forms of transmission. Once audio is converted to data all of these freedoms become available to it.

The generic term for the path down which the information is sent is the *channel*. In a transmission application, the channel may be a point-to-point cable, a network stage or a radio link. In a recording application the channel will include the record head, the medium and the replay head.

In digital circuitry there is a great deal of noise immunity because the signal has only two states, which are widely separated compared with the amplitude of noise. In both digital recording and transmission this is not always the case. In magnetic recording, noise immunity is a function of track width and reduction of the working SNR of a digital track allows the same information to be carried in a smaller area of the medium, improving economy of operation. In broadcasting, the noise immunity is a function of the transmitter power and reduction of working SNR allows lower power to be used with consequent economy. These reductions also increase the random error rate, but, as was seen in Chapter 1, an error-correction system may already be necessary in a practical system and it is simply made to work harder.

In real channels, the signal may *originate* with discrete states which change at discrete times, but the channel will treat it as an analog waveform and so it will not be *received* in the same form. Various frequency-dependent loss mechanisms will reduce the amplitude of the signal. Noise will be picked up as a result of stray electric fields or magnetic induction and in radio receivers the circuitry will contribute some noise. As a result, the received voltage will have an infinitely varying state along with a degree of uncertainty due to the noise. Different frequencies can propagate at different speeds in the channel; this is the phenomenon of group delay. An alternative way of considering group delay is that there will be frequency-dependent phase shifts in the signal and these will result in uncertainty in the timing of pulses.

In digital circuitry, the signals are generally accompanied by a separate clock signal which reclocks the data to remove jitter as was shown in Chapter 1. In contrast, it is generally not feasible to provide a separate clock in recording and transmission applications. In the transmission case, a separate clock line would not only raise cost, but is impractical because at high frequency it is virtually impossible to ensure that the clock cable propagates signals at the same speed as the data cable except over short distances. In the recording case, provision of a separate clock track is impractical at high density because mechanical tolerances cause phase errors between the tracks. The result is the same; timing differences between parallel channels which are known as skew.

The solution is to use a self-clocking waveform and the generation of this is a further essential function of the coding process. Clearly, if data bits are simply clocked serially from a shift register in so-called direct recording or transmission this characteristic will not be obtained. If all the data bits are the same, for example all zeros, there is no clock when they are serialized.

It is not the channel which is digital; instead the term describes the way in which the received signals are *interpreted*. When the receiver makes discrete decisions from the input waveform it attempts to reject the uncertainties in voltage and time. The technique of channel coding is one where transmitted waveforms are restricted to those which still allow the receiver to make discrete decisions despite the degradations caused by the analog nature of the channel.

6.2 Types of transmission channel

Transmission can be by electrical conductors, radio or optical fibre. Although these appear to be completely different, they are in fact just different examples of electromagnetic energy travelling from one place to another. If the energy is made time-variant, information can be carried.

Electromagnetic energy propagates in a manner which is a function of frequency, and our partial understanding requires it to be considered as electrons, waves or photons so that we can predict its behaviour in given circumstances.

At DC and at the low frequencies used for power distribution, electromagnetic energy is called electricity and needs to be transported completely inside conductors. It has to have a complete circuit to flow in, and the resistance to current flow is determined by the cross-sectional area of the conductor. The insulation around the conductor and the spacing between the conductors has no effect on the ability of the conductor to pass current. At DC an inductor appears to be a short circuit, and a capacitor appears to be an open circuit.

As frequency rises, resistance is exchanged for impedance. Inductors display increasing impedance with frequency, capacitors show falling impedance. Electromagnetic energy increasingly tends to leave the conductor. The first symptom is the skin effect: the current flows only in the outside layer of the conductor effectively causing the resistance to rise.

As the energy is starting to leave the conductors, the characteristics of the space between them become important. This determines the impedance. A change of impedance causes reflections in the energy flow and some of it heads back towards the source. Constant impedance cables with fixed conductor spacing are necessary, and these must be suitably terminated to prevent reflections. The most important characteristic of the insulation is its thickness as this determines the spacing between the conductors.

As frequency rises still further, the energy travels less in the conductors and more in the insulation between them, and their composition becomes important and they begin to be called dielectrics. A poor dielectric like PVC absorbs high-frequency energy and attenuates the signal. So-called low-loss dielectrics such as PTFE are used, and one way of achieving low loss is to incorporate as much air into the dielectric as possible by making it in the form of a foam or extruding it with voids.

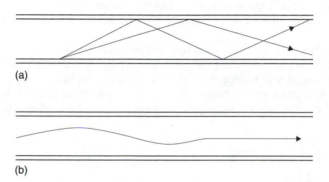

(a)

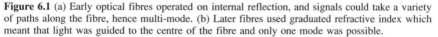

(b)

Figure 6.1 (a) Early optical fibres operated on internal reflection, and signals could take a variety of paths along the fibre, hence multi-mode. (b) Later fibres used graduated refractive index which meant that light was guided to the centre of the fibre and only one mode was possible.

High-frequency signals can also be propagated without a medium, and are called radio. As frequency rises further the electromagnetic energy is termed 'light' which can also travel without a medium, but can be also be guided through a suitable medium. Figure 6.1(a) shows an early type of optical fibre in which total internal reflection is used to guide the light. It will be seen that the length of the optical path is a function of the angle at which the light is launched. Thus at the end of a long fibre sharp transitions would be smeared by this effect. Later optical fibres are made with a radially varying refractive index such that light diverging from the axis is automatically refracted back into the fibre. Figure 6.1(b) shows that in single-mode fibre light can only travel down one path and so the smearing of transitions is avoided.

6.3 Transmission lines

Frequency-dependent behaviour is the most important factor in deciding how best to harness electromagnetic energy flow for information transmission. It is obvious that the higher the frequency, the greater the possible information rate, but in general, losses increase with frequency, and flat frequency response is elusive. The best that can be managed is that over a narrow band of frequencies, the response can be made reasonably constant with the help of equalization. Unfortunately raw data when serialized have an unconstrained spectrum. Runs of identical bits can produce frequencies much lower than the bit rate would suggest. One of the essential steps in a transmission system is to modify the spectrum of the data into something more suitable.

At moderate bit rates, say a few megabits per second, and with moderate cable lengths, say a few metres, the dominant effect will be the capacitance of the cable due to the geometry of the space between the conductors and the dielectric between. The capacitance behaves under these conditions as if it were a single capacitor connected across the signal. The effect of the series source resistance and the parallel capacitance is that signal edges or transitions are turned into exponential curves as the capacitance is effectively being charged and discharged through the source impedance.

As cable length increases, the capacitance can no longer be lumped as if it were a single unit; it has to be regarded as being distributed along the cable. With rising frequency, the cable inductance also becomes significant, and it too is distributed.

The cable is now a transmission line and pulses travel down it as current loops which roll along as shown in Figure 6.2. If the pulse is positive, as it is launched along the line, it will charge the dielectric locally as at (a). As the pulse moves along, it will continue to charge the local dielectric as at (b). When the driver finishes the pulse, the trailing edge of the pulse follows the leading edge along the

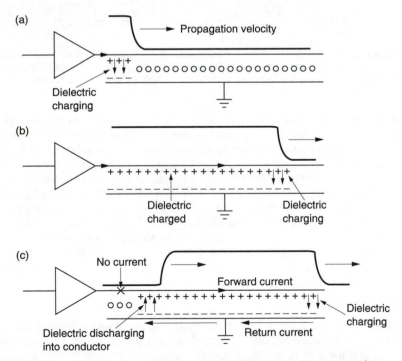

Figure 6.2 A transmission line conveys energy packets which appear with respect to the dielectric. In (a) the driver launches a pulse which charges the dielectric at the beginning of the line. As it propagates the dielectric is charged further along as in (b). When the drive ends the pulse, the charged dielectric discharges into the line. A current loop is formed where the current in the return loop flows in the opposite direction to the current in the 'hot' wire.

line. The voltage of the dielectric charged by the leading edge of the pulse is now higher than the voltage on the line, and so the dielectric discharges into the line as at (c). The current flows forward as it is in fact the same current which is flowing into the dielectric at the leading edge. There is thus a loop of current rolling down the line flowing forward in the 'hot' wire and backwards in the return.

The constant to-ing and fro-ing of charge in the dielectric results in dielectric loss of signal energy. Dielectric loss increases with frequency and so a long transmission line acts as a filter. Thus the term 'low-loss' cable refers primarily to the kind of dielectric used.

Transmission lines which transport energy in this way have a characteristic impedance caused by the interplay of the inductance along the conductors with the parallel capacitance. One consequence of that transmission mode is that correct termination or matching is required between the line and both the driver and the receiver. When a line is correctly matched, the rolling energy rolls straight out of the line into the load and the maximum energy is available. If the impedance presented by the load is incorrect, there will be reflections from the mismatch. An open circuit will reflect all the energy back in the same polarity as the original, whereas a short circuit will reflect all the energy back in the opposite polarity. Thus impedances above or below the correct value will have a tendency towards reflections whose magnitude depends upon the degree of mismatch and whose polarity depends upon whether the load is too high or too low. In practice it is the need to avoid reflections which is the most important reason to terminate correctly.

A perfectly square pulse contains an indefinite series of harmonics, but the higher ones suffer progressively more loss. A square pulse at the driver becomes less and less square with distance as Figure 6.3 shows. The harmonics are progressively lost until in the extreme case all that is left is the fundamental. A transmitted square wave is received as a sine wave. Fortunately data can still be recovered from the fundamental signal component.

Once all the harmonics have been lost, further losses cause the amplitude of the fundamental to fall. The effect worsens with distance and it is necessary to ensure that data recovery is still possible from a signal of unpredictable level.

6.4 Types of recording medium

Digital media do not need to have linear transfer functions, nor do they need to be noise-free or continuous. All they need to do is to allow the player to be able to distinguish the presence or absence of replay events, such as the generation of pulses, with reasonable (rather than perfect) reliability. In a magnetic medium, the event will be a flux change from one direction of magnetization to another. In an optical medium, the event must cause the pickup to perceive a change in the intensity of the light falling on the sensor. This may be obtained through selective absorption of light by dyes, or by phase contrast (see Chapter 11). In magneto-optical disks the recording itself is magnetic, but it is made and read using light.

6.5 Magnetic recording

Magnetic recording relies on the hysteresis of certain magnetic materials. After an applied magnetic field is removed, the material remains magnetized in the same direction. By definition, the process is non-linear.

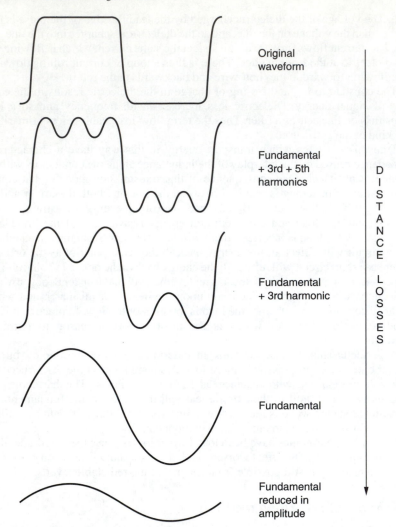

Figure 6.3 A signal may be square at the transmitter, but losses increase with frequency, and as the signal propagates, more of the harmonies are lost until only the fundamental remains. The amplitude of the fundamental then falls with further distance.

Figure 6.4 shows the construction of a typical digital record head. A magnetic circuit carries a coil through which the record current passes and generates flux. A non-magnetic gap forces the flux to leave the magnetic circuit of the head and to penetrate the medium. The current through the head must be set to suit the coercivity of the tape, and is arranged to almost saturate the track. The amplitude of the current is constant, and recording is performed by reversing the direction of the current with respect to time. As the track passes the head, this is converted to the reversal of the magnetic field left on the tape with respect to distance. The magnetic recording is therefore bipolar. Figure 6.5 shows that the recording is actually made just after the trailing pole of the record head where the flux

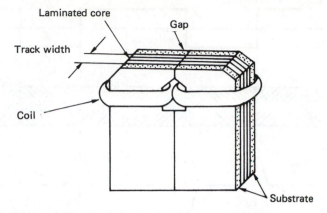

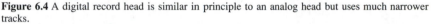

Figure 6.4 A digital record head is similar in principle to an analog head but uses much narrower tracks.

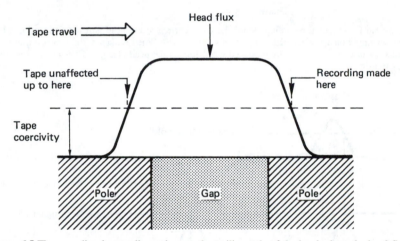

Figure 6.5 The recording is actually made near the trailing pole of the head where the head flux falls below the coercivity of the tape.

strength from the gap is falling. As in analog recorders, the width of the gap is generally made quite large to ensure that the full thickness of the magnetic coating is recorded, although this cannot be done if the same head is intended to replay.

Figure 6.6 shows what happens when a conventional inductive head, i.e. one having a normal winding, is used to replay the bipolar track made by reversing the record current. The head output is proportional to the rate of change of flux and so only occurs at flux reversals. In other words, the replay head differentiates the flux on the track. The polarity of the resultant pulses alternates as the flux changes and changes back. A circuit is necessary which locates the peaks of the pulses and outputs a signal corresponding to the original record current waveform. There are two ways in which this can be done.

The amplitude of the replay signal is of no consequence and often an AGC system is used to keep the replay signal constant in amplitude. What matters is

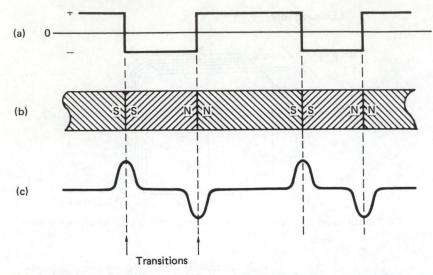

Figure 6.6 Basic digital recording, At (a) the write current in the head is reversed from time to time, leaving a binary magnetization pattern shown at (b). When replayed, the waveform at (c) results because an output is only produced when flux in the head changes. Changes are referred to as transitions.

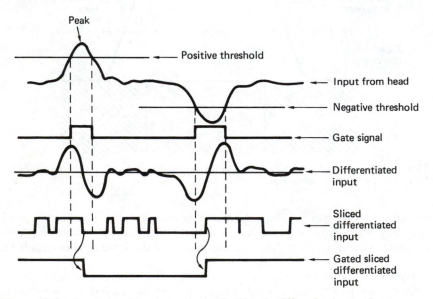

Figure 6.7 Gated peak detection rejects noise by disabling the differentiated output between transitions.

the time at which the write current, and hence the flux stored on the medium, reverses. This can be determined by locating the peaks of the replay impulses, which can conveniently be done by differentiating the signal and looking for zero crossings. Figure 6.7 shows that this results in noise between the peaks. This problem is overcome by the gated peak detector, where only zero crossings from

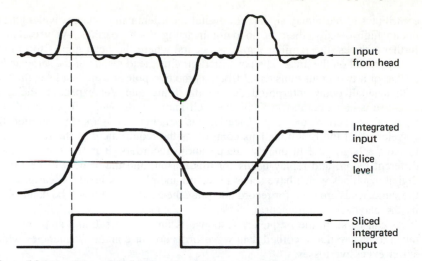

Figure 6.8 Integration method for recreating write-current waveform.

a pulse which exceeds the threshold will be counted. The AGC system allows the thresholds to be fixed. As an alternative, the record waveform can also be restored by integration, which opposes the differentiation of the head as in Figure 6.8.[1]

The head shown in Figure 6.4 has a frequency response shown in Figure 6.9. At DC there is no change of flux and no output. As a result, inductive heads are at a disadvantage at very low speeds. The output rises with frequency until the rise is halted by the onset of thickness loss. As the frequency rises, the recorded wavelength falls and flux from the shorter magnetic patterns cannot be picked up so far away. At some point, the wavelength becomes so short that flux from the back of the tape coating cannot reach the head and a decreasing thickness of tape

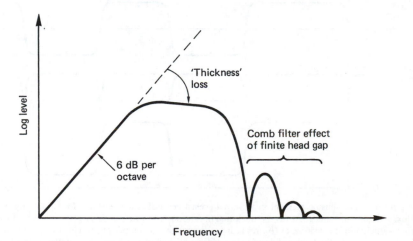

Figure 6.9 The major mechanisms defining magnetic channel bandwidth.

contributes to the replay signal.[2] In digital recorders using short wavelengths to obtain high density, there is no point in using thick coatings. As wavelength further reduces, the familiar gap loss occurs, where the head gap is too big to resolve detail on the track. The construction of the head results in the same action as that of a two-point transversal filter, as the two poles of the head see the tape with a small delay interposed due to the finite gap. As expected, the head response is like a comb filter with the well-known nulls where flux cancellation takes place across the gap. Clearly, the smaller the gap, the shorter the wavelength of the first null. This contradicts the requirement of the record head to have a large gap. In quality analog audio recorders, it is the norm to have different record and replay heads for this reason, and the same will be true in digital machines which have separate record and playback heads. Clearly, where the same heads are used for record and play, the head gap size will be determined by the playback requirement.

As can be seen, the frequency response is far from ideal, and steps must be taken to ensure that recorded data waveforms do not contain frequencies which suffer excessive losses.

A more recent development is the magneto-resistive (M-R) head. This is a head which measures the flux on the tape rather than using it to generate a signal directly. Flux-measurement works down to DC and so offers advantages at low tape speeds. Unfortunately flux-measuring heads are not polarity-conscious but sense the modulus of the flux and if used directly they respond to positive and negative flux equally, as shown in Figure 6.10. This is overcome by using a small extra winding in the head carrying a constant current. This creates a steady bias field which adds to the flux from the tape. The flux seen by the head is now unipolar and changes between two levels and a more useful output waveform

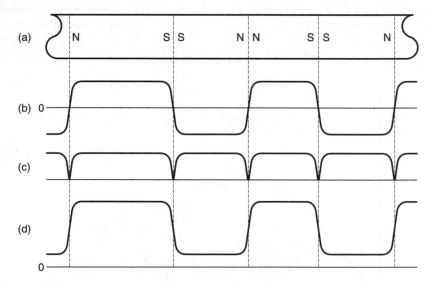

Figure 6.10 The sensing element in a magneto-resistive head is not sensitive to the polarity of the flux, only the magnitude. At (a) the track magnetization is shown and this causes a bidirectional flux variation in the head as at (b), resulting in the magnitude output at (c). However, if the flux in the head due to the track is biased by an additional field, it can be made unipolar as at (d) and the correct waveform is obtained.

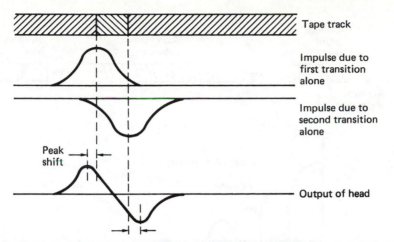

Figure 6.11 Readout pulses from two closely recorded transitions are summed in the head and the effect is that the peaks of the waveform are moved outwards. This is known as peak-shift distortion and equalization is necessary to reduce the effect.

results. Recorders which have low head-to-medium speed use M–R heads, whereas recorders with high head-to-medium speed, such as ADAT, use inductive heads.

Heads designed for use with tape work in actual contact with the magnetic coating. The tape is tensioned to pull it against the head. There will be a wear mechanism and need for periodic cleaning. In the hard disk, the rotational speed is high in order to reduce access time, and the drive must be capable of staying on-line for extended periods. In this case the heads do not contact the disk surface, but are supported on a boundary layer of air. The presence of the air film causes spacing loss, which restricts the wavelengths at which the head can replay. This is the penalty of rapid access.

Digital media must operate at high density for economic reasons. This implies that the shortest possible wavelengths will be used. Figure 6.11 shows that when two flux changes, or transitions, are recorded close together, they affect each other on replay. The amplitude of the composite signal is reduced, and the position of the peaks is pushed outwards. This is known as inter-symbol interference, or peak-shift distortion and it occurs in all magnetic media.

The effect is primarily due to high-frequency loss and it can be reduced by equalization on replay, as is done in most tapes, or by precompensation on record as is done in hard disks.

6.6 Azimuth recording and rotary heads

Figure 6.12(a) shows that in azimuth recording, the transitions are laid down at an angle to the track by using a head which is tilted. Machines using azimuth recording must always have an even number of heads, so that adjacent tracks can be recorded with opposite azimuth angle. The two track types are usually referred to as A and B. Figure 6.12(b) shows the effect of playing a track with the wrong type of head. The playback process suffers from an enormous azimuth error. The effect of azimuth error can be understood by imagining the tape track to be made

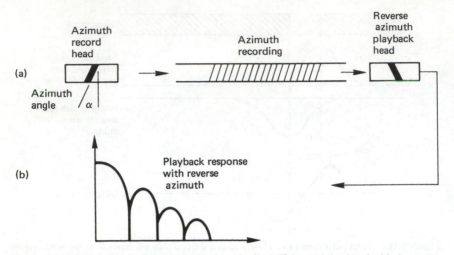

Figure 6.12 In azimuth recording (a), the head gap is tilted. If the track is played with the same head, playback is normal, but the response of the reverse azimuth head is attenuated (b).

from many identical parallel strips. In the presence of azimuth error, the strips at one edge of the track are played back with a phase shift relative to strips at the other side. At some wavelengths, the phase shift will be 180°, and there will be no output; at other wavelengths, especially long wavelengths, some output will reappear. The effect is rather like that of a comb filter, and serves to attenuate crosstalk due to adjacent tracks so that no guard bands are required. Since no tape is wasted between the tracks, more efficient use is made of the tape. The term 'guard-bandless recording' is often used instead of, or in addition to, 'azimuth recording'. The failure of the azimuth effect at long wavelengths is a characteristic of azimuth recording, and it is necessary to ensure that the spectrum of the signal to be recorded has a small low-frequency content. The signal will need to pass through a rotary transformer to reach the heads, and cannot therefore contain a DC component.

In some rotary head recorders there is no separate erase process, and erasure is achieved by overwriting with a new waveform. Overwriting is only successful when there are no long wavelengths in the earlier recording, since these penetrate deeper into the tape, and the short wavelengths in a new recording will not be able to erase them. In this case the ratio between the shortest and longest wavelengths recorded on tape should be limited.

Restricting the spectrum of the code to allow erasure by overwrite also eases the design of the rotary transformer.

6.7 Optical and magneto-optical disks

Optical recorders have the advantage that light can be focused at a distance whereas magnetism cannot. This means that there need be no physical contact between the pickup and the medium and no wear mechanism.

In the same way that the recorded wavelength of a magnetic recording is limited by the gap in the replay head, the density of optical recording is limited by the size of light spot which can be focused on the medium. This is controlled

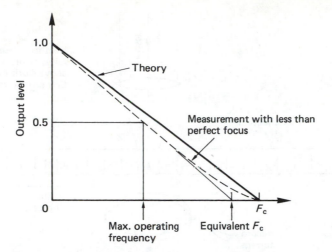

Figure 6.13 Frequency response of laser pickup. Maximum operating frequency is about half of cut-off frequency F_c.

by the wavelength of the light used and by the aperture of the lens. When the light spot is as small as these limits allow, it is said to be diffraction limited. The frequency response of an optical disk is shown in Figure 6.13. The response is best at DC and falls steadily to the optical cut-off frequency. Although the optics work down to DC, this cannot be used for the data recording. DC and low frequencies in the data would interfere with the focus and tracking servos and, as will be seen, difficulties arise when attempting to demodulate a unipolar signal. In practice the signal from the pickup is split by a filter. Low frequencies go to the servos, and higher frequencies go to the data circuitry. As a result, the optical disk channel has the same inability to handle DC as does a magnetic recorder, and the same techniques are needed to overcome it.

When a magnetic material is heated above its Curie temperature, it becomes demagnetized, and on cooling will assume the magnetization of an applied field which would be too weak to influence it normally. This is the principle of magneto-optical recording. The heat is supplied by a finely focused laser, the field is supplied by a coil which is much larger.

Figure 6.14 shows that the medium is initially magnetized in one direction only. In order to record, the coil is energized with a current in the opposite direction. This is too weak to influence the medium in its normal state, but when it is heated by the recording laser beam the heated area will take on the magnetism from the coil when it cools. Thus a magnetic recording with very small dimensions can be made even though the magnetic circuit involved is quite large in comparison.

Readout is obtained using the Kerr effect or the Faraday effect, which are phenomena whereby the plane of polarization of light can be rotated by a magnetic field. The angle of rotation is very small and needs a sensitive pickup. The pickup contains a polarizing filter before the sensor. Changes in polarization change the ability of the light to get through the polarizing filter and results in an intensity change which once more produces a unipolar output.

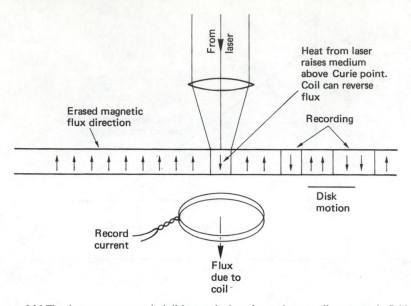

Figure 6.14 The thermomagneto-optical disk uses the heat from a laser to allow magnetic field to record on the disk.

The magneto-optic recording can be erased by reversing the current in the coil and operating the laser continuously as it passes along the track. A new recording can then be made on the erased track.

A disadvantage of magneto-optical recording is that all materials having a Curie point low enough to be useful are highly corrodible by air and need to be kept under an effectively sealed protective layer.

6.8 Equalization and data separation

The characteristics of most channels are that signal loss occurs which increases with frequency. This has the effect of slowing down rise times and thereby sloping off edges. If a signal with sloping edges is sliced, the time at which the waveform crosses the slicing level will be changed, and this causes jitter. Figure 6.15 shows that slicing a sloping waveform in the presence of baseline wander causes more jitter.

On a long cable, high-frequency roll-off can cause sufficient jitter to move a transition into an adjacent bit period. This is called inter-symbol interference and

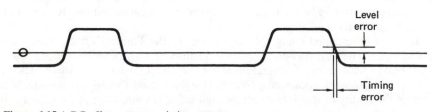

Figure 6.15 A DC offset can cause timing errors.

the effect becomes worse in signals which have greater asymmetry, i.e. short pulses alternating with long ones. The effect can be reduced by the application of equalization, which is typically a high-frequency boost, and by choosing a channel code which has restricted asymmetry.

Compensation for peak shift distortion in recording requires equalization of the channel,[3] and this can be done by a network after the replay head, termed an equalizer or pulse sharpener,[4] as in Figure 6.16(a). This technique uses transversal filtering to oppose the inherent transversal effect of the head. As an alternative, precompensation in the record stage can be used as shown in Figure 6.16(b). Transitions are written in such a way that the anticipated peak shift will move the readout peaks to the desired timing.

The important step of information recovery at the receiver or replay circuit is known as data separation. The data separator is rather like an analog-to-digital convertor because the two processes of sampling and quantizing are both present. In the time domain, the sampling clock is derived from the clock content of the channel waveform. In the voltage domain, the process of *slicing* converts the analog waveform from the channel back into a binary representation. The slicer

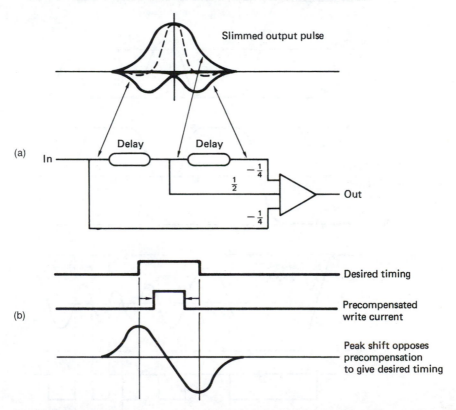

Figure 6.16 Peak-shift distortion is due to the finite width of replay pulses. The effect can be reduced by the pulse slimmer shown in (a) which is basically a transversal filter. The use of a linear operational amplifier emphasizes the analog nature of channels. Instead of replay pulse slimming, transitions can be written with a displacement equal and opposite to the anticipated peak shift as shown in (b).

is thus a form of quantizer which has only one-bit resolution. The slicing process makes a discrete decision about the voltage of the incoming signal in order to reject noise. The sampler makes discrete decisions along the time axis in order to reject jitter. These two processes will be described in detail.

6.9 Slicing and jitter rejection

The slicer is implemented with a comparator which has analog inputs but a binary output. In a cable receiver, the input waveform can be sliced directly. In an inductive magnetic replay system, the replay waveform is differentiated and must first pass through a peak detector (Figure 6.7) or an integrator (Figure 6.8). The signal voltage is compared with the midway voltage, known as the threshold, baseline or slicing level by the comparator. If the signal voltage is above the threshold, the comparator outputs a high level, if below, a low level results.

Figure 6.17 shows some waveforms associated with a slicer. At (a) the transmitted waveform has an uneven duty cycle. The DC component, or average

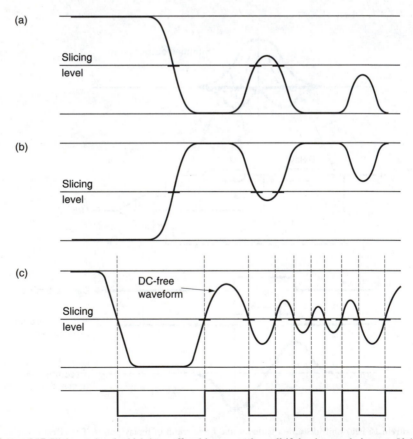

Figure 6.17 Slicing a signal which has suffered losses works well if the duty cycle is even. If the duty cycle is uneven, as at (a), timing errors will become worse until slicing fails. With the opposite duty cycle, the slicing fails in the opposite direction as at (b). If, however, the signal is DC free, correct slicing can continue even in the presence of serious losses, as (c) shows.

level, of the signal is received with high amplitude, but the pulse amplitude falls as the pulse gets shorter. Eventually the waveform cannot be sliced.

At (b) the opposite duty cycle is shown. The signal level drifts to the opposite polarity and once more slicing is impossible. The phenomenon is called baseline wander and will be observed with any signal whose average voltage is not the same as the slicing level.

At (c) it will be seen that if the transmitted waveform has a relatively constant average voltage, slicing remains possible up to high frequencies even in the presence of serious amplitude loss, because the received waveform remains symmetrical about the baseline.

It is clearly not possible simply to serialize data in a shift register for so-called direct transmission, because successful slicing can only be obtained if the number of ones is equal to the number of zeros; there is little chance of this happening consistently with real data. Instead, a modulation code or channel

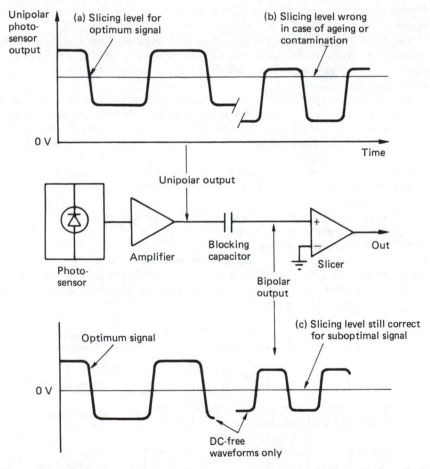

Figure 6.18 (a) Slicing a unipolar signal requires a non-zero threshold. (b) If the signal amplitude changes, the threshold will then be incorrect. (c) If a DC-free code is used, a unipolar waveform can be converted to a bipolar waveform using a series capacitor. A zero threshold can be used and slicing continues with amplitude variations.

code is necessary. This converts the data into a waveform which is DC-free or nearly so for the purpose of transmission.

The slicing threshold level is naturally zero in a bipolar system such as magnetic inductive replay or a cable. When the amplitude falls it does so symmetrically and slicing continues. The same is not true of M–R heads and optical pickups, which both respond to intensity and therefore produce a unipolar output. If the replay signal is sliced directly, the threshold cannot be zero, but must be some level approximately half the amplitude of the signal as shown in Figure 6.18(a). Unfortunately when the signal level falls it falls towards zero and not towards the slicing level. The threshold will no longer be appropriate for the signal as can be seen at (b). This can be overcome by using a DC-free coded waveform. If a series capacitor is connected to the unipolar signal from an optical pickup, the waveform is rendered bipolar because the capacitor blocks any DC component in the signal. The DC-free channel waveform passes through unaltered. If an amplitude loss is suffered, Figure 6.18(c) shows that the resultant bipolar signal now reduces in amplitude about the slicing level and slicing can continue.

The binary waveform at the output of the slicer will be a replica of the transmitted waveform, except for the addition of jitter or time uncertainty in the position of the edges due to noise, baseline wander, intersymbol interference and imperfect equalization.

Binary circuits reject noise by using discrete voltage levels which are spaced further apart than the uncertainty due to noise. In a similar manner, digital coding combats time uncertainty by making the time axis discrete using events, known as transitions, spaced apart at integer multiples of some basic time period, called a detent, which is larger than the typical time uncertainty. Figure 6.19 shows how this jitter-rejection mechanism works. All that matters is to identify the detent in which the transition occurred. Exactly where it occurred within the detent is of no consequence.

As ideal transitions occur at multiples of a basic period, an oscilloscope, which is repeatedly triggered on a channel-coded signal carrying random data, will show an eye pattern if connected to the output of the equalizer. Study of the eye pattern reveals how well the coding used suits the channel. In the case of transmission, with a short cable, the losses will be small, and the eye opening will be virtually square except for some edge-sloping due to cable capacitance. As

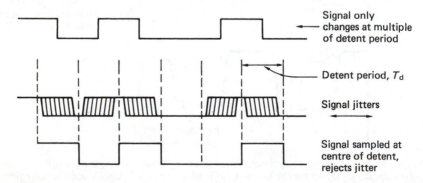

Figure 6.19 A certain amount of jitter can be rejected by changing the signal at multiples of the basic detent period T_d.

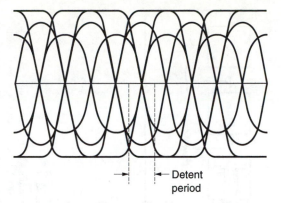

Detent
period

Figure 6.20 A transmitted waveform will appear like this on an oscilloscope as successive parts of the waveform are superimposed on the tube. When the waveform is rounded off by losses, diamond-shaped eyes are left in the centre, spaced apart by the detent period.

cable length increases, the harmonics are lost and the remaining fundamental gives the eyes a diamond shape. The same eye pattern will be obtained with a recording channel where it is uneconomic to provide bandwidth much beyond the fundamental.

Noise closes the eyes in a vertical direction, and jitter closes the eyes in a horizontal direction, as in Figure 6.20. If the eyes remain sensibly open, data separation will be possible. Clearly, more jitter can be tolerated if there is less noise, and vice versa. If the equalizer is adjustable, the optimum setting will be where the greatest eye opening is obtained.

In the centre of the eyes, the receiver must make binary decisions at the channel bit rate about the state of the signal, high or low, using the slicer output. As stated, the receiver is sampling the output of the slicer, and it needs to have a sampling clock in order to do that. In order to give the best rejection of noise and jitter, the clock edges which operate the sampler must be in the centre of the eyes.

As has been stated, a separate clock is not practicable in recording or transmission. A fixed-frequency clock at the receiver is of no use as even if it was sufficiently stable, it would not know what phase to run at.

The only way in which the sampling clock can be obtained is to use a phase-locked loop to regenerate it from the clock content of the self-clocking channel-coded waveform. In phase-locked loops, the voltage-controlled oscillator is driven by a phase error measured between the output and some reference, such that the output eventually has the same frequency as the reference. If a divider is placed between the VCO and the phase comparator, as in section 3.17, the VCO frequency can be made to be a multiple of the reference. This also has the effect of making the loop more heavily damped. If a channel-coded waveform is used as a reference to a PLL, the loop will be able to make a phase comparison whenever a transition arrives and will run at the channel bit rate. When there are several detents between transitions, the loop will *flywheel* at the last known frequency and phase until it can rephase at a subsequent transition. Thus a continuous clock is recreated from the clock content of the channel waveform. In a recorder, if the speed of the medium should change, the PLL will change

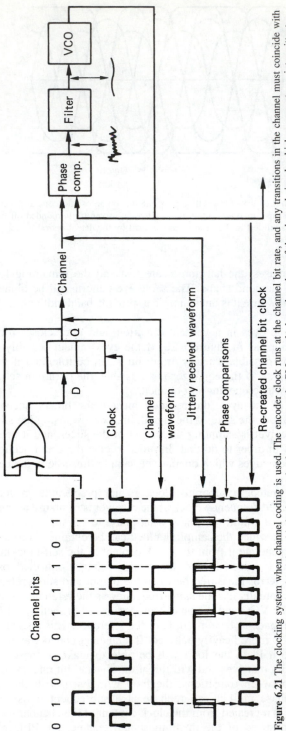

Figure 6.21 The clocking system when channel coding is used. The encoder clock runs at the channel bit rate, and any transitions in the channel must coincide with encoder clock edges. The reason for doing this is that, at the data separator, the PLL can lock to the edges of the channel signal, which represents an intermittent clock, and turn it into a continuous clock. The jitter in the edges of the channel signal causes noise in the phase error of the PLL, but the damping acts as a filter and the PLL runs at the average phase of the channel bits, rejecting the jitter.

frequency to follow. Once the loop is locked, clock edges will be phased with the average phase of the jittering edges of the input waveform. If, for example, rising edges of the clock are phased to input transitions, then falling edges will be in the centre of the eyes. If these edges are used to clock the sampling process, the maximum jitter and noise can be rejected. The output of the slicer when sampled by the PLL edge at the centre of an eye is the value of a channel bit. Figure 6.21 shows the complete clocking system of a channel code from encoder to data separator.

Clearly, data cannot be separated if the PLL is not locked, but it cannot be locked until it has seen transitions for a reasonable period. In recorders, which have discontinuous recorded blocks to allow editing, the solution is to precede each data block with a pattern of transitions whose sole purpose is to provide a timing reference for synchronizing the phase-locked loop. This pattern is known as a preamble. In interfaces, the transmission can be continuous and there is no difficulty remaining in lock indefinitely. There will simply be a short delay on first applying the signal before the receiver locks to it.

One potential problem area which is frequently overlooked is to ensure that the VCO in the receiving PLL is correctly centred. If it is not, it will be running with a static phase error and will not sample the received waveform at the centre of the eyes. The sampled bits will be more prone to noise and jitter errors. VCO centring can simply be checked by displaying the control voltage. This should not change significantly when the input is momentarily interrupted.

6.10 Channel coding

It is not practicable simply to serialize raw data in a shift register for the purpose of recording or for transmission except over relatively short distances. Practical systems require the use of a modulation scheme, known as a channel code, which expresses the data as waveforms which are self-clocking in order to reject jitter, separate the received bits and to avoid skew on separate clock lines. The coded waveforms should further be DC-free or nearly so to enable slicing in the presence of losses and have a narrower spectrum than the raw data both for economy and to make equalization easier.

Jitter causes uncertainty about the time at which a particular event occurred. The frequency response of the channel then places an overall limit on the spacing of events in the channel. Particular emphasis must be placed on the interplay of bandwidth, jitter and noise, which will be shown here to be the key to the design of a successful channel code.

Figure 6.22 shows that a channel coder is necessary prior to the record stage, and that a decoder, known as a data separator, is necessary after the replay stage. The output of the channel coder is generally a logic-level signal which contains a 'high' state when a transition is to be generated. The waveform generator produces the transitions in a signal whose level and impedance is suitable for driving the medium or channel. The signal may be bipolar or unipolar as appropriate.

Some codes eliminate DC entirely, which is advantageous for cable transmission, optical media and rotary head recording. Some codes can reduce the channel bandwidth needed by lowering the upper spectral limit. This permits higher linear density, usually at the expense of jitter rejection. Other codes narrow the spectrum, by raising the lower limit. A code with a narrow spectrum

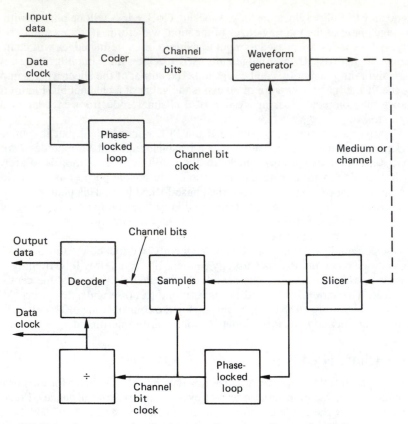

Figure 6.22 The major components of a channel coding system. See text for details.

has a number of advantages. The reduction in asymmetry will reduce peak shift and data separators can lock more readily because the range of frequencies in the code is smaller. In theory the narrower the spectrum, the less noise will be suffered, but this is only achieved if filtering is employed. Filters can easily cause phase errors which will nullify any gain.

A convenient definition of a channel code (for there are certainly others) is 'A method of modulating real data such that they can be reliably received despite the shortcomings of a real channel, while making maximum economic use of the channel capacity'. The basic time periods of a channel-coded waveform are called positions or detents, in which the transmitted voltage will be reversed or stay the same. The symbol used for the units of channel time is T_d.

One of the fundamental parameters of a channel code is the density ratio (DR). One definition of density ratio is that it is the worst-case ratio of the number of data bits recorded to the number of transitions in the channel. It can also be thought of as the ratio between the Nyquist rate of the data (one-half the bit rate) and the frequency response required in the channel. The storage density of data recorders has steadily increased due to improvements in medium and transducer technology, but modern storage densities are also a function of improvements in channel coding.

As jitter is such an important issue in digital recording and transmission, a parameter has been introduced to quantify the ability of a channel code to reject time instability. This parameter, the jitter margin, also known as the window margin or phase margin (T_w), is defined as the permitted range of time over which a transition can still be received correctly, divided by the data bit-cell period (T). Equalization is often difficult in practice, a code which has a large jitter margin will sometimes be used because it resists the effects of inter-symbol interference well. Such a code may achieve a better performance in practice than a code with a higher density ratio but poor jitter performance.

A more realistic comparison of code performance will be obtained by taking into account both density ratio and jitter margin. This is the purpose of the figure of merit (FoM), which is defined as DR $\times$ T_w.

6.11 Simple codes

In the Non-Return to Zero (NRZ) code shown in Figure 6.23(a), the record current does not cease between bits, but flows at all times in one direction or the other dependent on the state of the bit to be recorded. This results in a replay pulse only when the data bits change from state to another. As a result, if one pulse was missed, the subsequent bits would be inverted. This was avoided by adapting the coding such that the record current would change state or invert whenever a data one occurred, leading to the term Non-Return to Zero Invert or NRZI shown in Figure 6.23(b). In NRZI a replay pulse occurs whenever there is a data one. Clearly, neither NRZ or NRZI are self-clocking, but require a separate clock track. Skew between tracks can only be avoided by working at low density and so the system cannot be used directly for digital audio. However, virtually all the codes used for magnetic recording are based on the principle of reversing the record current to produce a transition.

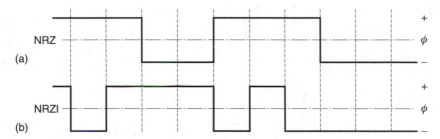

Figure 6.23 In the NRZ code (a) a missing replay pulse inverts every following bit. This was overcome in the NRZI code (b) which reverses write current on a data one.

The FM code, also known as Manchester code or bi-phase mark code, shown in Figure 6.24(a) was the first practical self-clocking binary code and it is suitable for both transmission and recording. It is DC-free and very easy to encode and decode. It is the code specified for the AES/EBU digital audio interconnect standard. In the field of recording it remains in use today only where density is not of prime importance, for example in SMPTE/EBU timecode for professional audio and video recorders.

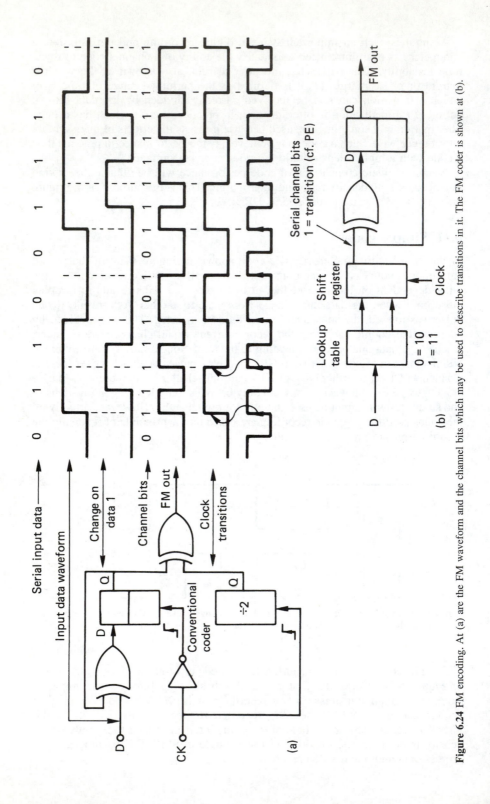

Figure 6.24 FM encoding. At (a) are the FM waveform and the channel bits which may be used to describe transitions in it. The FM coder is shown at (b).

In FM there is always a transition at the bit-cell boundary which acts as a clock. For a data one, there is an additional transition at the bit-cell centre. Figure 6.24(a) shows that each data bit can be represented by two channel bits. For a data zero, they will be 10, and for a data one they will be 11. Since the first bit is always one, it conveys no information, and is responsible for the density ratio of only one-half. Since there can be two transitions for each data bit, the jitter margin can only be half a bit, and the resulting FoM is only 0.25. The high clock content of FM does, however, mean that data recovery is possible over a wide range of speeds; hence the use of timecode. The lowest frequency in FM is due to a stream of zeros and is equal to half the bit rate. The highest frequency is due to a stream of ones, and is equal to the bit rate. Thus the fundamentals of FM are within a band of one octave. Effective equalization is generally possible over such a band. FM is not polarity-conscious and can be inverted without changing the data.

Figure 6.24(b) shows how an FM coder works. Data words are loaded into the input shift register which is clocked at the data bit rate. Each data bit is converted to two channel bits in the codebook or look-up table. These channel bits are loaded into the output register. The output register is clocked twice as fast as the input register because there are twice as many channel bits as data bits. The ratio of the two clocks is called the code rate, in this case it is a rate one-half code. Ones in the serial channel bit output represent transitions whereas zeros represent no change. The channel bits are fed to the waveform generator which is a one-bit delay, clocked at the channel bit rate, and an exclusive-OR gate. This changes state when a channel bit one is input. The result is a coded FM waveform where there is always a transition at the beginning of the data bit period, and a second optional transition whose presence indicates a one.

In modified frequency modulation (MFM) also known as Miller code,[5] the highly redundant clock content of FM was reduced by the use of a phase-locked loop in the receiver which could flywheel over missing clock transitions. This technique is implicit in all the more advanced codes. Figure 6.25(a) shows that

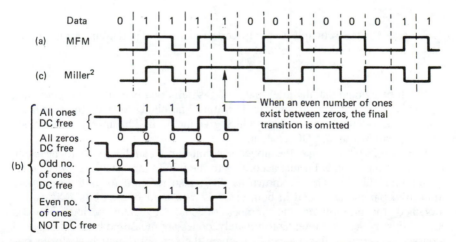

Figure 6.25 MFM or Miller code is generated as shown here. The minimum transition spacing is twice that of FM or PE. MFM is not always DC-free as shown at (b). This can be overcome by the modification of (c) which results in the Miller^2 code.

the bit-cell centre transition on a data one was retained, but the bit-cell boundary transition is now required only between successive zeros. There are still two channel bits for every data bit, but adjacent channel bits will never be one, doubling the minimum time between transitions, and giving a DR of 1. Clearly, the coding of the current bit is now influenced by the preceding bit. The maximum number of prior bits which affect the current bit is known as the constraint length L_c, measured in data-bit periods. For MFM $L_c = T$. Another way of considering the constraint length is that it assesses the number of data bits which may be corrupted if the receiver misplaces one transition. If L_c is long, all errors will be burst errors.

MFM doubled the density ratio compared to FM and PE without changing the jitter performance; thus the FoM also doubles, becoming 0.5. It was adopted for many rigid disks at the time of its development, and remains in use on double-density floppy disks. It is not, however, DC-free. Figure 6.25(b) shows how MFM can have DC content under certain conditions.

The Miller[2] code is derived from MFM, and Figure 6.25(c) shows that the DC content is eliminated by a slight increase in complexity.[6,7] Wherever an even number of ones occurs between zeros, the transition at the last one is omitted. This creates two additional, longer run lengths and increases the T_{max} of the code. The decoder can detect these longer run lengths in order to re-insert the suppressed ones. The FoM of Miller[2] is 0.5 as for MFM.

6.12 Group codes

Further improvements in coding rely on converting patterns of real data to patterns of channel bits with more desirable characteristics using a conversion table known as a codebook. If a data symbol of m bits is considered, it can have 2^m different combinations. As it is intended to discard undesirable patterns to improve the code, it follows that the number of channel bits n must be greater than m. The number of patterns which can be discarded is:

$$2^n - 2^m$$

One name for the principle is group code recording (GCR), and an important parameter is the code rate, defined as:

$$R = m/n$$

It will be evident that the jitter margin T_w is numerically equal to the code rate, and so a code rate near to unity is desirable. The choice of patterns which are used in the codebook will be those which give the desired balance between clock content, bandwidth and DC content.

Figure 6.26 shows that the upper spectral limit can be made to be some fraction of the channel bit rate according to the minimum distance between ones in the channel bits. This is known as T_{min}, also referred to as the minimum transition parameter M and in both cases is measured in data bits T. It can be obtained by multiplying the number of channel detent periods between transitions by the code rate. Unfortunately, codes are measured by the number of consecutive zeros in the channel bits, given the symbol d, which is always one less than the number of detent periods. In fact T_{min} is numerically equal to the density ratio.

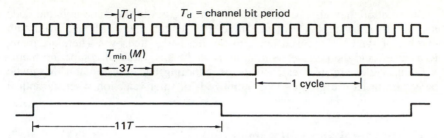

Figure 6.26 A channel code can control its spectrum by placing limits on T_{min} (M) and T_{max} which define upper and lower frequencies. The ratio of T_{max}/T_{min} determines the asymmetry of waveform and predicts DC content and peak shift. Example shown is EFM.

$$T_{min} = M = DR = \frac{(d + 1) \times m}{n}$$

It will be evident that choosing a low code rate could increase the density ratio, but it will impair the jitter margin. The figure of merit is:

$$FoM = DR \times T_w = \frac{(d + 1) \times m^2}{n^2}$$

since $T_w = m/n$.

Figure 6.26 also shows that the lower spectral limit is influenced by the maximum distance between transitions T_{max}. This is also obtained by multiplying the maximum number of detent periods between transitions by the code rate. Again, codes are measured by the maximum number of zeros between channel ones, k, and so:

$$T_{max} = \frac{(k + 1) \times m}{n}$$

and the maximum/minimum ratio P is:

$$p = \frac{(k + 1)}{(d + 1)}$$

The length of time between channel transitions is known as the *run length*. Another name for this class is the run-length-limited (RLL) codes.[8] Since m data bits are considered as one symbol, the constraint length L_c will be increased in RLL codes to at least m. It is, however, possible for a code to have run-length limits without it being a group code.

In practice, the junction of two adjacent channel symbols may violate run-length limits, and it may be necessary to create a further codebook of symbol size $2n$ which converts violating code pairs to acceptable patterns. This is known as merging and follows the golden rule that the substitute $2n$ symbol must finish with a pattern which eliminates the possibility of a subsequent violation. These patterns must also differ from all other symbols.

Substitution may also be used to different degrees in the same nominal code in order to allow a choice of maximum run length, e.g. 3PM. The maximum number of symbols involved in a substitution is denoted by r. There are many RLL codes and the parameters d, k, m, n, and r are a way of comparing them.

Group codes are used extensively in recording and transmission. Magnetic tapes and disks use group codes optimized for jitter rejection whereas optical disks use group codes optimized for density ratio.

6.13 Randomizing and encryption

Randomizing is not a channel code, but a technique which can be used in conjunction with almost any channel code. It is widely used in digital audio and video broadcasting and in a number of recording and transmission formats. The randomizing system is arranged outside the channel coder. Figure 6.27 shows that, at the encoder, a pseudo-random sequence is added modulo-2 to the serial data. This process makes the signal spectrum in the channel more uniform, drastically reduces T_{max} and reduces DC content. At the receiver the transitions are converted back to a serial bitstream to which the same pseudo-random sequence is again added modulo-2. As a result, the random signal cancels itself out to leave only the serial data, provided that the two pseudo-random sequences are synchronized to bit accuracy.

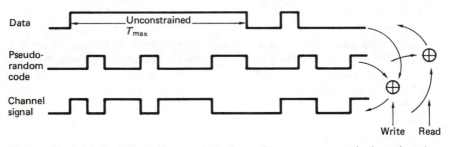

Figure 6.27 Modulo-2 addition with a pseudo-random code removes unconstrained runs in real data. Identical process must be provided on replay.

Many channel codes, especially group codes, display pattern sensitivity because some waveforms are more sensitive to peak shift distortion than others. Pattern sensitivity is only a problem if a sustained series of sensitive symbols needs to be recorded. Randomizing ensures that this cannot happen because it breaks up any regularity or repetition in the data. The data randomizing is performed by using the exclusive-OR function of the data and a pseudo-random sequence as the input to the channel coder. On replay the same sequence is generated, synchronized to bit accuracy, and the exclusive-OR of the replay bitstream and the sequence is the original data.

The generation of randomizing polynomials was described in section 3.15. Clearly, the sync pattern cannot be randomized, since this causes a Catch-22 situation where it is not possible to synchronize the sequence for replay until the sync pattern is read, but it is not possible to read the sync pattern until the sequence is synchronized!

In recorders, the randomizing is block based, since this matches the block structure on the medium. Where there is no obvious block structure, convolutional or endless randomizing can be used. In convolutional randomizing, the signal sent down the channel is the serial data waveform which has been convolved with the impulse response of a digital filter. On reception the signal is deconvolved to restore the original data.

In a randomized transmission, if the receiver is not able to recreate the pseudo-random sequence, the data cannot be decoded. This can be used as the basis for encryption in which only authorized users can decode transmitted data. In an encryption system, the goal is security whereas in a channel-coding system the goal is simplicity. Channel coders use pseudo-random sequences because these are economical to create using feedback shift registers. However, there are a limited number of pseudo-random sequences and it would be too easy to try them all until the correct one was found. Encryption systems use the same processes, but the key sequence which is added to the data at the encoder is truly random. This makes it much harder for unauthorized parties to access the data. Only a receiver in possession of the correct sequence can decode the channel signal. If the sequence is made long enough, the probability of stumbling across the sequence by trial and error can be made sufficiently small. Security systems of this kind can be compromised if the delivery of the key to the authorized user is intercepted.

6.14 Synchronizing

Once the PLL in the data separator has locked to the clock content of the transmission, a serial channel bitstream and a channel bit clock will emerge from the sampler. In a group code, it is essential to know where a group of channel bits begins in order to assemble groups for decoding to data bit groups. In a randomizing system it is equally vital to know at what point in the serial data stream the words or samples commence. In serial transmission and in recording, channel bit groups or randomized data words are sent one after the other, one bit at a time, with no spaces in between, so that although the designer knows that a data block contains, say, 128 bytes, the receiver simply finds 1024 bits in a row. If the exact position of the first bit is not known, then it is not possible to put all the bits in the right places in the right bytes; a process known as deserializing. The effect of sync slippage is devastating, because a one-bit disparity between the bit count and the bitstream will corrupt every symbol in the block.

The synchronization of the data separator and the synchronization to the block format are two distinct problems, which are often solved by the same sync pattern. Deserializing requires a shift register which is fed with serial data and read out once per word. The sync detector is simply a set of logic gates which are arranged to recognize a specific pattern in the register. The sync pattern is either identical for every block or has a restricted number of versions and it will be recognized by the replay circuitry and used to reset the bit count through the block. Then by counting channel bits and dividing by the group size, groups can be deserialized and decoded to data groups. In a randomized system, the pseudo-random sequence generator is also reset. Then counting derandomized bits from the sync pattern and dividing by the wordlength enables the replay circuitry to deserialize the data words.

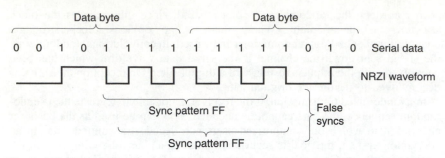

Figure 6.28 Concatenation of two words can result in the accidental generation of a word which is reserved for synchronizing.

Even if a specific code were excluded from the recorded data so that it could be used for synchronizing, this cannot ensure that the same pattern cannot be falsely created at the junction between two allowable data words. Figure 6.28 shows how false synchronizing can occur due to concatenation. It is thus not practical to use a bit pattern which is a data code value in a simple synchronizing recognizer. The problem is overcome in some synchronous systems by using the fact that sync patterns occur exactly once per block and therefore contain redundancy. If the pattern is seen by the recognizer at block rate, a genuine sync condition exists. Sync patterns seen at other times must be false. Such systems take a few milliseconds before sync is achieved, but once achieved it should not be lost unless the transmission is interrupted.

In run-length-limited codes false syncs are not a problem. The sync pattern is no longer a data bit pattern but is a specific waveform. If the sync waveform contains run lengths which violate the normal coding limits, there is no way that these run lengths can occur in encoded data, nor any possibility that they will be interpreted as data. They can, however, be readily detected by the replay circuitry.

In a group code there are many more combinations of channel bits than there are combinations of data bits. Thus after all data bit patterns have been allocated group patterns, there are still many unused group patterns which cannot occur in the data. With care, group patterns can be found which cannot occur due to the concatenation of any pair of groups representing data. These are then unique and can be used for synchronizing.

6.15 Basic error correction

There are many different types of recording and transmission channel and consequently there will be many different mechanisms which may result in errors. Bit errors in audio cause impulsive noise or transients whose effect depends upon the significance of the affected bit. Errors in compressed audio data are more serious as they may cause the decoder to lose sync.

In magnetic recording, data can be corrupted by mechanical problems such as media dropout and poor tracking or head contact, or Gaussian thermal noise in replay circuits and heads. In optical recording, contamination of the medium interrupts the light beam. When group codes are used, a single defect in a group changes the group symbol and may cause errors up to the size of the group.

Single-bit errors are therefore less common in group-coded channels. Inside equipment, data are conveyed on short wires and the noise environment is under the designer's control. With suitable design techniques, errors can be made effectively negligible whereas in communication systems, there is considerably less control of the electromagnetic environment.

Irrespective of the cause, all these mechanisms cause one of two effects. There are large isolated corruptions, called error bursts, where numerous bits are corrupted all together in an area which is otherwise error-free, and there are random errors affecting single bits or symbols. Whatever the mechanism, the result will be that the received data will not be exactly the same as those sent. In binary the discrete bits will be each either right or wrong. If a binary digit is known to be wrong, it is only necessary to invert its state and then it must be right. Thus error correction itself is trivial; the hard part is working out *which* bits need correcting.

There are a number of terms which have idiomatic meanings in error correction. The raw BER (bit error rate) is the error rate of the medium, whereas the residual or uncorrected BER is the rate at which the error-correction system fails to detect or miscorrects errors. In practical digital systems, the residual BER is negligibly small. If the error correction is turned off, the two figures become the same.

Error correction works by adding some bits to the data which are calculated from the data. This creates an entity called a codeword which spans a greater length of time than one bit alone. The statistics of noise means that whilst one bit may be lost in a codeword, the loss of the rest of the codeword because of noise is highly improbable. As will be described later in this chapter, codewords are designed to be able to correct totally a finite number of corrupted bits. The greater the timespan over which the coding is performed, or, on a recording medium, the greater area over which the coding is performed, the greater will be the reliability achieved, although this does mean that an encoding delay will be experienced on recording, and a similar or greater decoding delay on reproduction.

Shannon[9] disclosed that a message can be sent to any desired degree of accuracy provided that it is spread over a sufficient timespan. Engineers have to compromise, because an infinite coding delay in the recovery of an error-free signal is not acceptable. Digital interfaces such as AES/EBU do not employ error correction because the build-up of coding delays in large production systems is unacceptable.

If error correction is necessary as a practical matter, it is then only a small step to put it to maximum use. All error correction depends on adding bits to the original message, and this, of course, increases the number of bits to be recorded, although it does not increase the information recorded. It might be imagined that error correction is going to reduce storage capacity, because space has to be found for all the extra bits. Nothing could be further from the truth. Once an error-correction system is used, the signal-to-noise ratio of the channel can be reduced, because the raised BER of the channel will be overcome by the error-correction system. Reduction of the SNR by 3 dB in a magnetic track can be achieved by halving the track width, provided that the system is not dominated by head or preamplifier noise. This doubles the recording density, making the storage of the additional bits needed for error correction a trivial matter. By a similar argument, the power of a digital transmitter can be reduced if error correction is used. In short, error

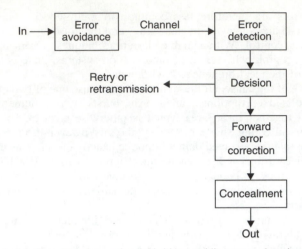

Figure 6.29 Error-handling strategies can be divided into avoiding errors, detecting errors and deciding what to do about them. Some possibilities are shown here. Of all these the detection is the most critical, as nothing can be done if the error is not detected.

correction is not a nuisance to be tolerated; it is a vital tool needed to maximize the efficiency of storage devices and transmission. Convergent systems would not be economically viable without it.

Figure 6.29 shows the broad subdivisions of error handling. The first stage might be called error avoidance and includes such measures as creating bad block files on hard disks or using verified media. Properly terminating network cabling is also in this category. Placing the audio blocks near to the centre of the tape in DVTRs is a further example. The data pass through the channel, which causes whatever corruptions it feels like. On receipt of the data the occurrence of errors is first detected, and this process must be extremely reliable, as it does not matter how effective the correction or how good the concealment algorithm, if it is not known that they are necessary! The detection of an error then results in a course of action being decided.

In the case of a file transfer, real-time operation is not required. If a disk drive detects a read error a retry is easy as the disk is turning at several thousand rpm and will quickly re-present the data. An error due to a dust particle may not occur on the next revolution. A packet in error in a network will result in a retransmission. Many magnetic tape systems have *read after write*. During recording, offtape data are immediately checked for errors. If an error is detected, the tape may abort the recording, reverse to the beginning of the current block and erase it. The data from that block may then be recorded further down the tape. This is the recording equivalent of a retransmission in a communications system.

In many cases of digital video or audio replay a retry or retransmission is not possible because the data are required in real time. In this case the solution is to encode the message using a system which is sufficiently powerful to correct the errors in real time. These are called forward error-correcting schemes (FEC). The term 'forward' implies that the transmitter does not need to take any action in the case of an error; the receiver will perform the correction.

6.16 Concealment by interpolation

There are some practical differences between data recording for video and the computer data recording application. Although video or audio recorders seldom have time for retries, they have the advantage that there is a certain amount of redundancy in the information conveyed. Thus if an error cannot be corrected, then it can be concealed. If a sample is lost, it is possible to obtain an approximation to it by interpolating between samples in the vicinity of the missing one. Clearly, concealment of any kind cannot be used with computer instructions or compressed data, although concealment can be applied after compressed signals have been decoded.

If there is too much corruption for concealment, the only course in video is repeat the previous field or frame in a freeze as it is unlikely that the corrupt picture is watchable. In audio the equivalent is muting.

In general, if use is to be made of concealment on replay, the data must generally be reordered or shuffled prior to recording. To take a simple example, odd-numbered samples are recorded in a different area of the medium from even-numbered samples. On playback, if a gross error occurs on the medium, depending on its position, the result will be either corrupted odd samples or corrupted even samples, but it is most unlikely that both will be lost. Interpolation is then possible if the power of the correction system is exceeded. The concealment technique described here is only suitable for PCM recording. If compression has been employed, different concealment techniques will be needed.

It should be stressed that corrected data are indistinguishable from the original and thus there can be no visible or audible artifacts. In contrast, concealment is only an approximation to the original information and could be detectable. In practical equipment, concealment occurs infrequently unless there is a defect requiring attention, and its presence is difficult to see.

6.17 Parity

The error-detection and error-correction processes are closely related and will be dealt with together here. The actual correction of an error is simplified tremendously by the adoption of binary. As there are only two symbols, 0 and 1, it is enough to know that a symbol is wrong, and the correct value is obvious. Figure 6.30 shows a minimal circuit required for correction once the bit in error has been identified. The XOR (exclusive-OR) gate shows up extensively in error

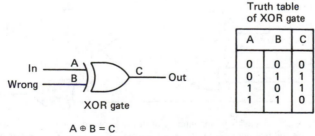

Truth table of XOR gate

A	B	C
0	0	0
0	1	1
1	0	1
1	1	0

$A \oplus B = C$

Figure 6.30 Once the position of the error is identified, the correction process in binary is easy.

correction and the figure also shows the truth table. One way of remembering the characteristics of this useful device is that there will be an output when the inputs are different. Inspection of the truth table will show that there is an even number of ones in each row (zero is an even number) and so the device could also be called an even parity gate. The XOR gate is also an adder in modulo-2.

Parity is a fundamental concept in error detection. In Figure 6.31, the example is given of a four-bit data word which is to be protected. If an extra bit is added to the word which is calculated in such a way that the total number of ones in the five-bit word is even, this property can be tested on receipt. The generation of the parity bit can be performed by a number of the ubiquitous XOR gates configured into what is known as a parity tree. In the figure, if a bit is corrupted, the received message will be seen no longer to have an even number of ones. If two bits are

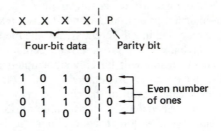

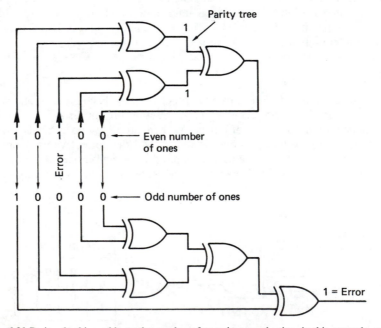

Figure 6.31 Parity checking adds up the number of ones in a word using, in this example, parity trees. One error bit and odd numbers of errors are detected. Even numbers of errors cannot be detected.

corrupted, the failure will be undetected. This example can be used to introduce much of the terminology of error correction. The extra bit added to the message carries no information of its own, since it is calculated from the other bits. It is therefore called a *redundant* bit.

The addition of the redundant bit gives the message a special property, i.e. the number of ones is even. A message having some special property *irrespective of the actual data content* is called a *codeword*. All error correction relies on adding redundancy to real data to form codewords for transmission. If any corruption occurs, the intention is that the received message will not have the special property; in other words, if the received message is not a codeword there has definitely been an error. The receiver can check for the special property without any prior knowledge of the data content. Thus the same check can be made on all received data. If the received message is a codeword, there probably has not been an error. The word 'probably' must be used because the figure shows that two bits in error will cause the received message to be a codeword, which cannot be discerned from an error-free message.

If it is known that generally the only failure mechanism in the channel in question is loss of a single bit, it is *assumed* that receipt of a codeword means that there has been no error. If there is a probability of two error bits, that becomes very nearly the probability of failing to detect an error, since all odd numbers of errors will be detected, and a four-bit error is much less likely. It is paramount in all error-correction systems that the protection used should be appropriate for the probability of errors to be encountered. An inadequate error-correction system is actually worse than not having any correction. Error correction works by trading probabilities. Error-free performance with a certain error rate is achieved at the expense of performance at higher error rates. Figure 6.32 shows the effect of an error-correction system on the residual BER for a given raw BER. It will be seen that there is a characteristic knee in the graph. If the expected raw BER has been misjudged, the consequences can be disastrous. Another result demonstrated by the example is that we can only guarantee to detect the same number of bits in error as there are redundant bits.

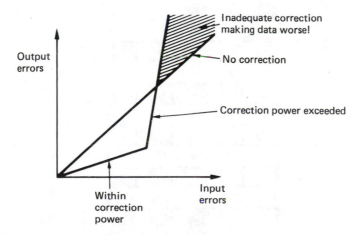

Figure 6.32 An error-correction system can only reduce errors at normal error rates at the expense of increasing errors at higher rates. It is most important to keep a system working to the left of the knee in the graph.

6.18 Block and convolutional codes

Figure 6.33(a) shows a strategy known as a crossword code, or product code. The data are formed into a two-dimensional array, in which each location can be a single bit or a multi-bit symbol. Parity is then generated on both rows and columns. If a single bit or symbol fails, one row parity check and one column parity check will fail, and the failure can be located at the intersection of the two failing checks. Although two symbols in error confuse this simple scheme, using more complex coding in a two-dimensional structure is very powerful, and further examples will be given throughout this chapter.

The example of Figure 6.33(a) assembles the data to be coded into a block of finite size and then each codeword is calculated by taking a different set of symbols. This should be contrasted with the operation of the circuit of Figure 6.33(b). Here the data are not in a block, but form an endless stream. A shift register allows four symbols simultaneously to be available to the encoder. The action of the encoder depends upon the delays. When symbol 3 emerges from the first delay, it will be added (modulo-2) to symbol 6. When this sum emerges from the second delay, it will be added to symbol 9 and so on. The codeword produced is shown in Figure 6.39(c) where it will be seen to be bent such that it has a vertical section and a diagonal section. Four symbols later the next codeword will be created one column further over in the data.

This is a convolutional code because the coder always takes parity on the same pattern of symbols which is convolved with the data stream on an endless basis. Figure 6.33(c) also shows that if an error occurs, it can be located because it will cause parity errors in two codewords. The error will be on the diagonal part of one codeword and on the vertical part of the other so that it can be located uniquely at the intersection and corrected by parity.

Comparison with the block code of Figure 6.33(a) will show that the convolutional code needs less redundancy for the same single-symbol location and correction performance as only a single redundant symbol is required for

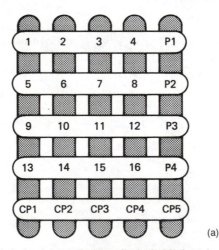

(a)

Figure 6.33 A block code is shown in (a). Each location in the block can be a bit or a word. Horizontal parity checks are made by adding P1, P2, etc., and cross-parity or vertical checks are made by adding CP1, CP2, etc. Any symbol in error will be at the intersection of the two failing codewords.

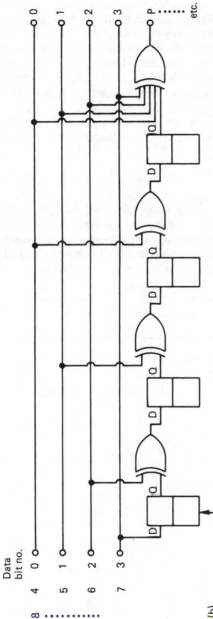

Data
bit no.

0 — 0
1 — 1
2 — 2
3 — 3
P — .... etc.

4 — 0
5 — 1
6 — 2
7 — 3
8 — ....

(b)

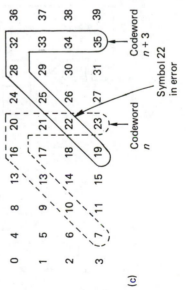

0	4	8	13	16	20	24	28	32	36
1	5	9	13	17	21	25	29	33	37
2	6	10	14	18	22	26	30	34	38
3	7	11	15	19	23	27	31	35	39

Codeword
n

Symbol 22
in error

Codeword
n + 3

(c)

Figure 6.33 (*Continued*) In (b) a convolutional coder is shown. Symbols entering are subject to different delays which result in the codewords in (c) being calculated. These have a vertical part and a diagonal part. A symbol in error will be at the intersection of the diagonal part of one code and the vertical part of another.

every four data symbols. Convolutional codes are computed on an endless basis which makes them inconvenient in recording applications where editing is anticipated. Here the block code is more appropriate as it allows edit gaps to be created between codes. In the case of uncorrectable errors, the convolutional principle causes the syndromes to be affected for some time afterwards and results in miscorrections of symbols which were not actually in error. This is called error propagation and is a characteristic of convolutional codes. Recording media tend to produce somewhat variant error statistics because media defects and mechanical problems cause errors which do not fit the classical additive noise channel. Convolutional codes can easily be taken beyond their correcting power if used with real recording media.

In transmission and broadcasting, the error statistics are more stable and the editing requirement is absent. As a result, convolutional codes tend to be used in digital broadcasting.

6.19 Cyclic codes

In digital recording applications, the data are stored serially on a track, and it is desirable to use relatively large data blocks to reduce the amount of the medium devoted to preambles, addressing and synchronizing. The principle of codewords having a special characteristic will still be employed, but they will be generated and checked algorithmically by equations. The syndrome will then be converted to the bit(s) in error by solving equations.

Where data can be accessed serially, simple circuitry can be used because the same gate will be used for many XOR operations. The circuit of Figure 6.34 is

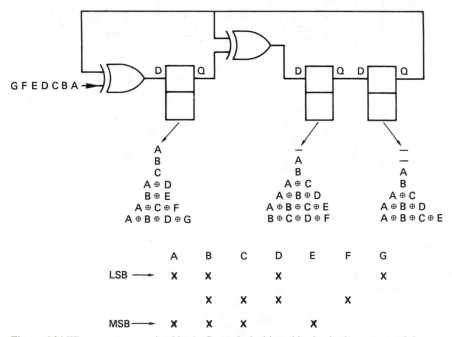

Figure 6.34 When seven successive bits A–G are clocked into this circuit, the contents of the three latches are shown for each clock. The final result is a parity-check matrix.

a kind of shift register, but with a particular feedback arrangement which leads it to be known as a twisted-ring counter. If seven message bits A–G are applied serially to this circuit, and each one of them is clocked, the outcome can be followed in the diagram. As bit A is presented and the system is clocked, bit A will enter the left-hand latch. When bits B and C are presented, A moves across to the right. Both XOR gates will have A on the upper input from the right-hand latch, the left one has D on the lower input and the right one has B on the lower input. When clocked, the left latch will thus be loaded with the XOR of A and D, and the right one with the XOR of A and B. The remainder of the sequence can be followed, bearing in mind that when the same term appears on both inputs of an XOR gate, it goes out, as the exclusive-OR of something with itself is nothing. At the end of the process, the latches contain three different expressions. Essentially, the circuit makes three parity checks through the message, leaving the result of each in the three stages of the register. In the figure, these expressions have been used to draw up a check matrix. The significance of these steps can now be explained.

The bits A B C and D are four data bits, and the bits E F and G are redundancy. When the redundancy is calculated, bit E is chosen so that there are an even number of ones in bits A B C and E; bit F is chosen such that the same applies to bits B C D and F, and similarly for bit G. Thus the four data bits and the three check bits form a seven-bit codeword. If there is no error in the codeword, when it is fed into the circuit shown, the result of each of the three parity checks will be zero and every stage of the shift register will be cleared. As the register has eight possible states, and one of them is the error-free condition, then there are seven remaining states, hence the seven-bit codeword. If a bit in the codeword is corrupted, there will be a non-zero result. For example, if bit D fails, the check on bits A B D and G will fail, and a one will appear in the left-hand latch. The check on bits B C D F will also fail, and the centre latch will set. The check on bits A B C E will not fail, because D is not involved in it, making the right-hand bit zero. There will be a syndrome of 110 in the register, and this will be seen from the check matrix to correspond to an error in bit D. Whichever bit fails, there will be a different three-bit syndrome which uniquely identifies the failed bit. As there are only three latches, there can be eight different syndromes. One of these is zero, which is the error-free condition, and so there are seven remaining error syndromes. The length of the codeword cannot exceed seven bits, or there would not be enough syndromes to correct all the bits. This can also be made to tie in with the generation of the check matrix. If fourteen bits, A to N, were fed into the circuit shown, the result would be that the check matrix repeated twice, and if a syndrome of 101 were to result, it could not be determined whether bit D or bit K failed. Because the check repeats every seven bits, the code is said to be a cyclic redundancy check (CRC) code.

It has been seen that the circuit shown makes a matrix check on a received word to determine if there has been an error, but the same circuit can also be used to generate the check bits. To visualize how this is done, examine what happens if only the data bits A B C and D are known, and the check bits E F and G are set to zero. If this message, ABCD000, is fed into the circuit, the left-hand latch will afterwards contain the XOR of A B C and zero, which is, of course, what E should be. The centre latch will contain the XOR of B C D and zero, which is what F should be and so on. This process is not quite ideal,

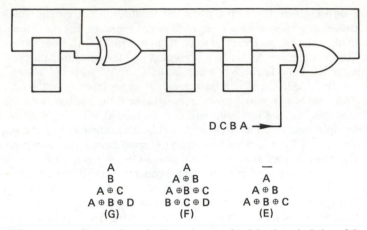

$$
\begin{array}{ccc}
\text{A} & \text{A} & \overline{\text{A}} \\
\text{B} & \text{A} \oplus \text{B} & \text{A} \\
\text{A} \oplus \text{C} & \text{A} \oplus \text{B} \oplus \text{C} & \text{A} \oplus \text{B} \\
\text{A} \oplus \text{B} \oplus \text{D} & \text{B} \oplus \text{C} \oplus \text{D} & \text{A} \oplus \text{B} \oplus \text{C} \\
\text{(G)} & \text{(F)} & \text{(E)}
\end{array}
$$

Figure 6.35 By moving the insertion point three places to the right, the calculation of the check bits is completed in only four clock periods and they can follow the data immediately. This is equivalent to premultiplying the data by x^3.

however, because it is necessary to wait for three clock periods after entering the data before the check bits are available. Where the data are simultaneously being recorded and fed into the encoder, the delay would prevent the check bits being easily added to the end of the data stream. This problem can be overcome by slightly modifying the encoder circuit as shown in Figure 6.35. By moving the position of the input to the right, the operation of the circuit is advanced so that the check bits are ready after only four clocks. The process can be followed in the diagram for the four data bits A B C and D. On the first clock, bit A enters the left two latches, whereas on the second clock, bit B will appear on the upper input of the left XOR gate, with bit A on the lower input, causing the centre latch to load the XOR of A and B and so on.

The way in which the cyclic codes work has been described in engineering terms, but it can be described mathematically if analysis is contemplated.

Just as the position of a decimal digit in a number determines the power of ten (whether that digit means one, ten or a hundred), the position of a binary digit determines the power of two (whether it means one, two or four). It is possible to rewrite a binary number so that it is expressed as a list of powers of two. For example, the binary number 1101 means 8 + 4 + 1, and can be written:

$$2^3 + 2^2 + 2^0$$

In fact, much of the theory of error correction applies to symbols in number bases other than 2, so that the number can also be written more generally as

$$x^3 + x^2 + 1 \ (2^0 = 1)$$

which also looks much more impressive. This expression, containing as it does various powers, is, of course, a polynomial, and the circuit of Figure 6.34 which has been seen to construct a parity-check matrix on a codeword can also be described as calculating the remainder due to dividing the input by a polynomial using modulo-2 arithmetic. In modulo-2 there are no borrows or carries, and addition and subtraction are replaced by the XOR function, which makes

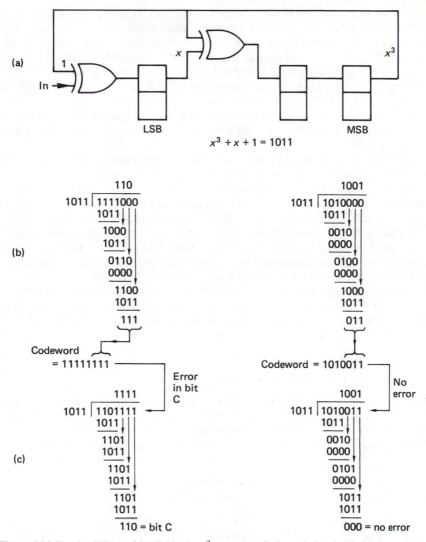

Figure 6.36 Circuit of Figure 6.34 divides by $x^3 + x + 1$ to find remainder. At (b) this is used to calculate check bits. At (c) right, zero syndrome, no error.

hardware implementation very easy. In Figure 6.36 it will be seen that the circuit of Figure 6.34 actually divides the codeword by a polynomial which is

$$x^3 + x + 1 \text{ or } 1011$$

This can be deduced from the fact that the right-hand bit is fed into two lower-order stages of the register at once. Once all the bits of the message have been clocked in, the circuit contains the remainder. In mathematical terms, the special property of a codeword is that it is a polynomial which yields a remainder of zero when divided by the generating polynomial. The receiver will make this division, and the result should be zero in the error-free case. Thus the codeword itself disappears from the division. If an error has occurred it is considered that this is due to an error polynomial which has been added to the codeword polynomial. If a codeword divided by the check polynomial is zero, a non-zero syndrome must represent the error polynomial divided by the check polynomial. Thus if the syndrome is multiplied by the check polynomial, the latter will be cancelled out and the result will be the error polynomial. If this is added modulo-2 to the received word, it will cancel out the error and leave the corrected data.

Some examples of modulo-2 division are given in Figure 6.36 which can be compared with the parallel computation of parity checks according to the matrix of Figure 6.34.

The process of generating the codeword from the original data can also be described mathematically. If a codeword has to give zero remainder when divided, it follows that the data can be converted to a codeword by adding the remainder when the data are divided. Generally speaking, the remainder would have to be subtracted, but in modulo-2 there is no distinction. This process is also illustrated in Figure 6.36. The four data bits have three zeros placed on the right-hand end, to make the wordlength equal to that of a codeword, and this word is then divided by the polynomial to calculate the remainder. The remainder is added to the zero-extended data to form a codeword. The modified circuit of Figure 6.35 can be described as premultiplying the data by x^3 before dividing.

CRC codes are of primary importance for detecting errors, and several have been standardized for use in digital communications. The most common of these are:

$$x^{16} + x^{15} + x^2 + 1 \text{ (CRC-16)}$$

$$x^{16} + x^{12} + x^5 + 1 \text{ (CRC-CCITT)}$$

The sixteen-bit cyclic codes have codewords of length $2^{16} - 1$ or 65 535 bits long. This may be too long for the application. Another problem with very long codes is that with a given raw BER, the longer the code, the more errors will occur in it. There may be enough errors to exceed the power of the code. The solution in

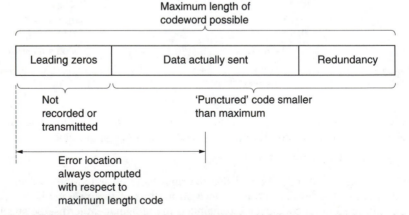

Figure 6.37 Codewords are often shortened, or punctured, which means that only the end of the codeword is actually transmitted. The only precaution to be taken when puncturing codes is that the computed position of an error will be from the beginning of the codeword, not from the beginning of the message.

both cases is to shorten or *puncture* the code. Figure 6.37 shows that in a punctured code, only the end of the codeword is used, and the data and redundancy are preceded by a string of zeros. It is not necessary to record these zeros, and, of course, errors cannot occur in them. Implementing a punctured code is easy. If a CRC generator starts with the register cleared and is fed with serial zeros, it will not change its state. Thus it is not necessary to provide the zeros, encoding can begin with the first data bit. In the same way, the leading zeros need not be provided during playback. The only precaution needed is that if a syndrome calculates the location of an error, this will be from the beginning of the codeword not from the beginning of the data. Where codes are used for detection only, this is of no consequence.

6.20 Introduction to the Reed–Solomon codes

The Reed–Solomon codes (Irving Reed and Gustave Solomon) are inherently burst correcting[10] because they work on multi-bit symbols rather than individual bits. The R–S codes are also extremely flexible in use. One code may be used both to detect and correct errors and the number of bursts which are correctable can be chosen at the design stage by the amount of redundancy. A further advantage of the R–S codes is that they can be used in conjunction with a separate error-detection mechanism in which case they perform the correction only by erasure. R–S codes operate at the theoretical limit of correcting efficiency. In other words, no more efficient code can be found.

In the simple CRC system described in section 6.19, the effect of the error is detected by ensuring that the codeword can be divided by a polynomial. The CRC codeword was created by adding a redundant symbol to the data. In the Reed–Solomon codes, several errors can be isolated by ensuring that the codeword will divide by a number of polynomials. Clearly, if the codeword must divide by, say, two polynomials, it must have two redundant symbols. This is the minimum case of an R–S code. On receiving an R–S-coded message there will be two syndromes following the division. In the error-free case, these will both be zero. If both are not zero, there is an error.

It has been stated that the effect of an error is to add an error polynomial to the message polynomial. The number of terms in the error polynomial is the same as the number of errors in the codeword. The codeword divides to zero and the syndromes are a function of the error only. There are two syndromes and two equations. By solving these simultaneous equations it is possible to obtain two unknowns. One of these is the position of the error, known as the *locator* and the other is the error bit pattern, known as the *corrector*. As the locator is the same size as the code symbol, the length of the codeword is determined by the size of the symbol. A symbol size of eight bits is commonly used because it fits in conveniently with both sixteen-bit audio samples and byte-oriented computers. An eight-bit syndrome results in a locator of the same wordlength. Eight bits have 2^8 combinations, but one of these is the error-free condition, and so the locator can specify one of only 255 symbols. As each symbol contains eight bits, the codeword will be $255 \times 8 = 2040$ bits long.

As further examples, five-bit symbols could be used to form a codeword 31 symbols long, and three-bit symbols would form a codeword seven symbols long. This latter size is small enough to permit some worked examples, and will be used further here. Figure 6.38 shows that in the seven-symbol codeword, five

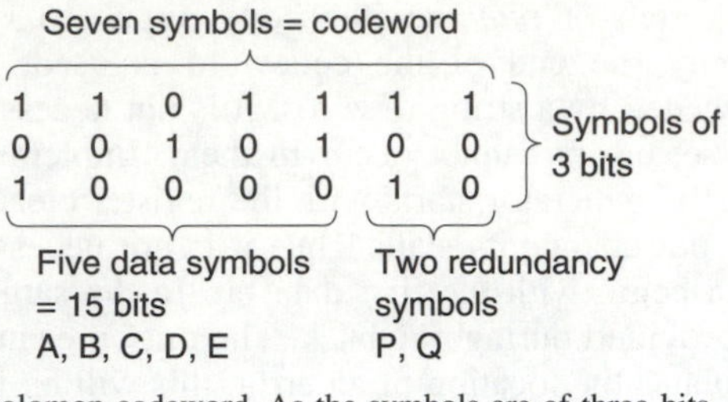

Seven symbols = codeword

$$\left.\begin{array}{ccccccc} 1 & 1 & 0 & 1 & 1 & 1 & 1 \\ 0 & 0 & 1 & 0 & 1 & 0 & 0 \\ 1 & 0 & 0 & 0 & 0 & 1 & 0 \end{array}\right\} \begin{array}{l} \text{Symbols of} \\ \text{3 bits} \end{array}$$

Five data symbols   Two redundancy
= 15 bits   symbols
A, B, C, D, E   P, Q

Figure 6.38 A Reed–Solomon codeword. As the symbols are of three bits, there can only be eight possible syndrome values. One of these is all zeros, the error-free case, and so it is only possible to point to seven errors; hence the codeword length of seven symbols. Two of these are redundant, leaving five data symbols.

symbols of three bits each, A–E, are the data, and P and Q are the two redundant symbols. This simple example will locate and correct a single symbol in error. It does not matter, however, how many bits in the symbol are in error.

The two check symbols are solutions to the following equations:

$$A \oplus B \oplus C \oplus D \oplus E \oplus P \oplus Q = 0 \; (\oplus = \text{XOR symbol})$$

$$a^7A \oplus a^6B \oplus a^5C \oplus a^4D \oplus a^3E \oplus a^2P \oplus aQ = 0$$

where a is a constant. The original data A–E followed by the redundancy P and Q pass through the channel.

The receiver makes two checks on the message to see if it is a codeword. This is done by calculating syndromes using the following expressions, where the ($'$) implies the received symbol which is not necessarily correct:

$$S_0 = A' \oplus B' \oplus C' \oplus D' \oplus E' \; P' \oplus Q'$$

(This is in fact a simple parity check.)

$$S_1 = a^7A' \oplus a^6B' \oplus a^5C' \oplus a^4D' \oplus a^3E' \oplus a^2P' \oplus aQ'$$

If two syndromes of all zeros are not obtained, there has been an error. The information carried in the syndromes will be used to correct the error. For the purpose of illustration, let it be considered that D$'$ has been corrupted before moving to the general case. D$'$ can be considered to be the result of adding an error of value E to the original value D such that $D' = D \oplus E$.

As $A \oplus B \oplus C \oplus D \oplus E \oplus P \oplus Q = 0$

then $A \oplus B \oplus C \oplus (D \oplus E) \oplus E \oplus P \oplus Q = E = S_0$

As $D' = D \oplus E$

then $D = D' \oplus E = D' \oplus S_0$

Thus the value of the corrector is known immediately because it is the same as the parity syndrome S_0. The corrected data symbol is obtained simply by adding S_0 to the incorrect symbol.

At this stage, however, the corrupted symbol has not yet been identified, but this is equally straightforward:

As $a^7A \oplus a^6B \oplus a^5C \oplus a^4D \oplus a^3E \oplus a^2P \oplus aQ = 0$

then:

$$a^7A \oplus a^6B \oplus a^5C \oplus a^4 (D \oplus E) \oplus a^3E \oplus a^2P \oplus aQ = a^4 E = S_1$$

Thus the syndrome S_1 is the error bit pattern E, but it has been raised to a power of a which is a function of the position of the error symbol in the block. If the position of the error is in symbol k, then k is the locator value and:

$$S_0 \times a^k = S_1$$

Hence:

$$a^k = \frac{S_1}{S_0}$$

The value of k can be found by multiplying S_0 by various powers of a until the product is the same as S_1. Then the power of a necessary is equal to k. The use of the descending powers of a in the codeword calculation is now clear because the error is then multiplied by a different power of a dependent upon its position, known as the locator, because it gives the position of the error. The process of finding the error position by experiment is known as a Chien search.

Whilst the expressions above show that the values of P and Q are such that the two syndrome expressions sum to zero, it is not yet clear how P and Q are calculated from the data. Expressions for P and Q can be found by solving the two R–S equations simultaneously. This has been done in Appendix 6.1. The following expressions must be used to calculate P and Q from the data in order to satisfy the codeword equations. These are:

$$P = a^6A \oplus aB \oplus a^2C \oplus a^5D \oplus a^3E$$

$$Q = a^2A \oplus a^3B \oplus a^6C \oplus a^4D \oplus aE$$

In both the calculation of the redundancy shown here and the calculation of the corrector and the locator it is necessary to perform numerous multiplications and raising to powers. This appears to present a formidable calculation problem at both the encoder and the decoder. This would be the case if the calculations involved were conventionally executed. However, the calculations can be simplified by using logarithms. Instead of multiplying two numbers, their logarithms are added. In order to find the cube of a number, its logarithm is added three times. Division is performed by subtracting the logarithms. Thus all the manipulations necessary can be achieved with addition or subtraction, which is straightforward in logic circuits.

The success of this approach depends upon simple implementation of log tables. As was seen in section 3.16, raising a constant, a, known as the *primitive element* to successively higher powers in modulo-2 gives rise to a Galois field. Each element of the field represents a different power n of a. It is a fundamental of the R–S codes that all the symbols used for data, redundancy and syndromes are considered to be elements of a Galois field. The number of bits in the symbol determines the size of the Galois field, and hence the number of symbols in the codeword.

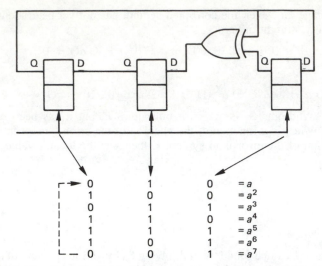

Figure 6.39 The bit patterns of a Galois field expressed as powers of the primitive element a. This diagram can be used as a form of log table in order to multiply binary numbers. Instead of an actual multiplication, the appropriate powers of a are simply added.

In Figure 6.39, the binary values of the elements are shown alongside the power of a they represent. In the R–S codes, symbols are no longer considered simply as binary numbers, but also as equivalent powers of a. In Reed–Solomon coding and decoding, each symbol will be multiplied by some power of a. Thus if the symbol is also known as a power of a it is only necessary to add the two powers. For example, if it is necessary to multiply the data symbol 100 by a^3, the calculation proceeds as follows, referring to Figure 6.39.

$$100 = a^2 \text{ so } 100 \times a^3 = a^{(2 + 3)} = a^5 = 111$$

Note that the results of a Galois multiplication are quite different from binary multiplication. Because all products must be elements of the field, sums of powers which exceed seven wrap around by having seven subtracted. For example:

$$a^5 \times a^6 = a^{11} = a^4 = 110$$

Figure 6.40 shows some examples of circuits which will perform this kind of multiplication. Note that they require a minimum amount of logic.

Figure 6.41 gives an example of the Reed–Solomon encoding process. The Galois field shown in Figure 6.39 has been used, having the primitive element a = 010. At the beginning of the calculation of P, the symbol A is multiplied by a^6. This is done by converting A to a power of a. According to Figure 6.39, 101 = a^6 and so the product will be $a^{(6+6)} = a^{12} = a^5 = 111$. In the same way, B is multiplied by a, and so on, and the products are added modulo-2. A similar process is used to calculate Q.

Figure 6.42 shows a circuit which can calculate P or Q. The symbols A–E are presented in succession, and the circuit is clocked for each one. On the first clock, a^6A is stored in the left-hand latch. If B is now provided at the input, the second GF multiplier produces aB and this is added to the output of the first latch

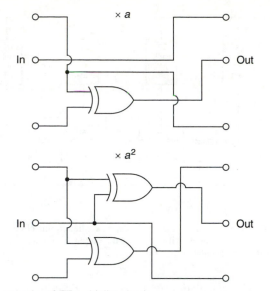

Figure 6.40 Some examples of GF multiplier circuits.

Input data	A	101	$a^6A = 111$	$a^2A = 010$
	B	100	$a\ B = 011$	$a^3B = 111$
	C	010	$a^2C = 011$	$a^6C = 001$
	D	100	$a^5D = 001$	$a^4D = 101$
	E	111	$a^3E = 010$	$a\ E = 101$
Check symbols	P	100 ←	————100	
	Q	100 ←		⌐100

Codeword	A	101	$a^7A = 101$
	B	100	$a^6B = 010$
	C	010	$a^5C = 101$
	D	100	$a^4D = 101$
	E	111	$a^3E = 010$
	P	100	$a^2P = 110$
	Q	100	$a\ Q = 011$
	$S_0 = \overline{000}$		$S_1 = \overline{000}$ ◄——— Both syndromes zero

Figure 6.41 Five data symbols A–E are used as terms in the generator polynomials derived in Appendix 6.1 to calculate two redundant symbols P and Q. An example is shown at the top. Below is the result of using the codeword symbols A–Q as terms in the checking polynomials. As there is no error, both syndromes are zero.

and when clocked will be stored in the second latch which now contains $a^6A + aB$. The process continues in this fashion until the complete expression for P is available in the right-hand latch. The intermediate contents of the right-hand latch are ignored.

The entire codeword now exists, and can be recorded or transmitted. Figure 6.41 also demonstrates that the codeword satisfies the checking equations. The modulo-2 sum of the seven symbols, S_0, is 000 because each column has an even

A, B, C, D, E

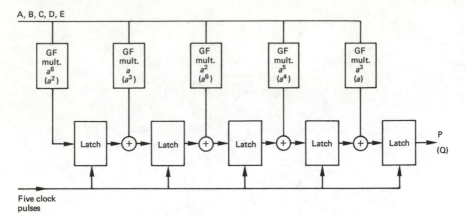

Five clock
pulses

Figure 6.42 If the five data symbols of Figure 6.41 are supplied to this circuit in sequence, after five clocks, one of the check symbols will appear at the output. Terms without brackets will calculate P, bracketed terms calculate Q.

number of ones. The calculation of S_1 requires multiplication by descending powers of a. The modulo-2 sum of the products is again zero. These calculations confirm that the redundancy calculation was properly carried out.

Figure 6.43 gives three examples of error correction based on this codeword. The erroneous symbol is marked with a dash. As there has been an error, the syndromes S_0 and S_1 will not be zero.

7	A	101	$a^7 A = 101$
6	B	100	$a^6 B = 010$
5	C	010	$a^5 C = 101$
4	D'	101	$a^4 D' = 011$
3	E	111	$a^3 E = 010$
2	P	100	$a^2 P = 110$
1	Q	100	$a\ Q = 011$

$S_0 = 001$ $S_1 = 110$

$$\frac{S_1}{S_0} = \frac{a^4}{1} = a^4$$

$k = 4$

$D' + S_0 = 101 + 001$
$D = 100$

7	A	101	$a^7 A = 101$
6	B	100	$a^6 B = 010$
5	C'	110	$a^5 C = 100$
4	D	100	$a^4 D = 101$
3	E	111	$a^3 E = 010$
2	P	100	$a^2 P = 110$
1	Q	100	$a\ Q = 011$

$S_0 = 100$ $S_1 = 001$

$$\frac{S_1}{S_0} = \frac{1}{a^2} = \frac{1}{a^2} \times \frac{a^5}{a^5} = a^5$$

$k = 5$

$C' + S_0 = 110 + 100$
$C = 010$

7	A'	111	$a^7 A = 111$
6	B	100	$a^6 B = 010$
5	C	010	$a^5 C = 101$
4	D	100	$a^4 D = 101$
3	E	111	$a^3 E = 010$
2	P	100	$a^2 P = 110$
1	Q	100	$a\ Q = 011$

$S_0 = 010$ $S_1 = 010$

$$\frac{S_1}{S_0} = \frac{a}{a} = 001 = a^7$$

$k = 7$

$A' + S_0 = 111 + 010$
$A = 101$

Figure 6.43 Three examples of error location and correction. The number of bits in error in a symbol is irrelevant; if all three were wrong, S_o would be 111, but correction is still possible.

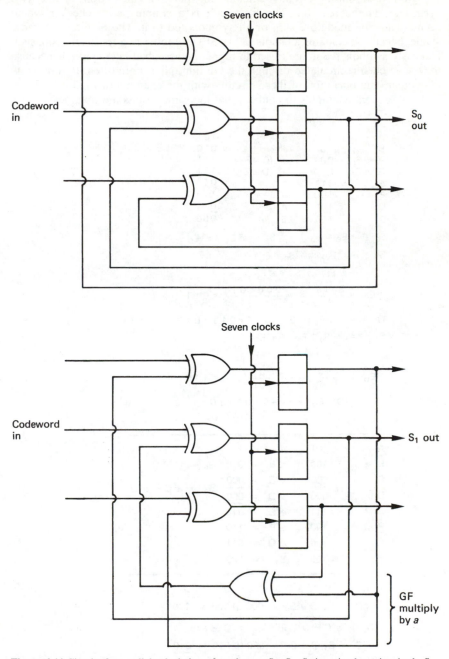

Figure 6.44 Circuits for parallel calculation of syndromes S_0, S_1, S_0 is a simple parity check. S_1 has a GF multiplication by a in the feedback, so that A is multiplied by a^7, B is multiplied by a^6, etc., and all are summed to give S_1.

Figure 6.44 shows circuits suitable for parallel calculation of the two syndromes at the receiver. The S_0 circuit is a simple parity checker which accumulates the modulo-2 sum of all symbols fed to it. The S_1 circuit is more subtle, because it contains a Galois field (GF) multiplier in a feedback loop, such that early symbols fed in are raised to higher powers than later symbols because they have been recirculated through the GF multiplier more often. It is possible to compare the operation of these circuits with the example of Figure 6.45 and with subsequent examples to confirm that the same results are obtained.

$$
\begin{array}{llll}
A & 101 & a^7 A = & 101 \\
B & 100 & a^6 B = & 010 \\
(C \oplus E_C) & 001 & a^5 (C \oplus E_C) & 111 \\
(D \oplus E_D) & 010 & a^4 (D \oplus E_D) & 111 \\
E & 111 & a^3 E = & 010 \\
P & 100 & a^2 P = & 110 \\
Q & 100 & a\, Q = & \underline{011} \\
S_1 \quad = & \overline{101} & S_1 \quad = & \overline{000}
\end{array}
$$

$$S_0 = E_C \oplus E_D \qquad S_1 = a^5 E_C \oplus a^4 E_D$$

$$S_1 = a^5 E_C \oplus a^4 (S_0 \oplus E_C)$$

$$\quad = a^5 E_C \oplus a^4 S_0 \oplus a^4 E_C$$

$$\therefore E_C = \frac{S_1 \oplus a^4 S_0}{a^5 \oplus a^4} = \frac{000 \oplus 011}{001} = 011$$

$$C = (C \oplus E_C) \oplus E_C = 001 \oplus 011 = \underline{010}$$

$$S_1 = a^5 (S_0 \ominus E_D) \oplus a^4 E_D$$

$$\quad = a^5 S_0 \oplus a^5 E_D \oplus a^4 E_D$$

$$\therefore E_D = \frac{S_1 \oplus a^5 S_0}{a^5 \oplus a^4} = \frac{000 \oplus 110}{001} = 110$$

$$D = (D \oplus E_D) + E_D = 010 \oplus 110 = \underline{100} \qquad \textbf{(a)}$$

$$
\begin{array}{llll}
A & 101 & a^7 A = 101 & \\
B & 100 & a^6 B = 010 & S_0 = C \oplus D \\
C & \underline{000} & a^5 C = \underline{000} & \\
D & \underline{000} & a^4 D = \underline{000} & S_1 = a^5 C \oplus a^4 D \\
E & 111 & a^3 E = 010 & \\
P & 100 & a^2 P = 110 & \\
Q & \underline{100} & a\, Q = \underline{011} & \\
S_0 & =100 & S_1 = 000 &
\end{array}
$$

$$S_1 = a^5 S_0 \oplus a^5 D \oplus a^4 D = a^5 S_0 \oplus D$$

$$\therefore D = S_1 \oplus a^5 S_0 = 000 \oplus 100 = \underline{100}$$

$$S_1 = a^5 C \oplus a^4 C \oplus a^4 S_0 = C \oplus a^4 S_0$$

$$\therefore C = S_1 \oplus a^4 S_0 = 000 \oplus 010 = \underline{010}$$

$$\textbf{(b)}$$

Figure 6.45 If the location of errors is known, then the syndromes are a known function of the two errors as shown in (a). It is, however, much simpler to set the incorrect symbols to zero, i.e. to *erase* them as in (b). Then the syndromes are a function of the wanted symbols and correction is easier.

6.21 Correction by erasure

In the examples of Figure 6.43, two redundant symbols P and Q have been used to locate and correct one error symbol. If the positions of errors are known by some separate mechanism (see product codes, section 6.23) the locator need not be calculated. The simultaneous equations may instead be solved for two correctors. In this case the number of symbols which can be corrected is equal to the number of redundant symbols. In Figure 6.45(a) two errors have taken place, and it is known that they are in symbols C and D. Since S_0 is a simple parity check, it will reflect the modulo-2 sum of the two errors. Hence $S_1 = EC \oplus ED$.

The two errors will have been multiplied by different powers in S_1, such that:

$$S_1 = a^5 EC \oplus a^4 ED$$

These two equations can be solved, as shown in the figure, to find EC and ED, and the correct value of the symbols will be obtained by adding these correctors to the erroneous values. It is, however, easier to set the values of the symbols in error to zero. In this way the nature of the error is rendered irrelevant and it does not enter the calculation. This setting of symbols to zero gives rise to the term erasure. In this case,

$$S_0 = C \oplus D$$
$$S_1 = a^5C + a^4D$$

Erasing the symbols in error makes the errors equal to the correct symbol values and these are found more simply as shown in Figure 6.45(b)

Practical systems will be designed to correct more symbols in error than in the simple examples given here. If it is proposed to correct by erasure an arbitrary number of symbols in error given by t, the codeword must be divisible by t different polynomials. Alternatively, if the errors must be located and corrected, $2t$ polynomials will be needed. These will be of the form $(x + a^n)$ where n takes all values up to t or $2t$. a is the primitive element.

Where four symbols are to be corrected by erasure, or two symbols are to be located and corrected, four redundant symbols are necessary, and the codeword polynomial must then be divisible by

$$(x + a^0) (x + a^1) (x + a^2) (x + a^3)$$

Upon receipt of the message, four syndromes must be calculated, and the four correctors or the two error patterns and their positions are determined by solving four simultaneous equations. This generally requires an iterative procedure, and a number of algorithms have been developed for the purpose.[11-13] Modern DVTR formats use eight-bit R–S codes and erasure extensively. The primitive polynomial commonly used with GF(256) is:

$$x^8 + x^4 + x^3 + x^2 + 1$$

The codeword will be 255 bytes long but will often be shortened by puncturing. The larger Galois fields require less redundancy, but the computational problem increases. LSI chips have been developed specifically for R–S decoding in many high-volume formats.

6.22 Interleaving

The concept of bit interleaving was introduced in connection with a single-bit correcting code to allow it to correct small bursts. With burst-correcting codes such as Reed–Solomon, bit interleave is unnecessary. In most channels, particularly high-density recording channels used for digital video or audio, the burst size may be many bytes rather than bits, and to rely on a code alone to correct such errors would require a lot of redundancy. The solution in this case is to employ symbol interleaving, as shown in Figure 6.46. Several codewords are encoded from input data, but these are not recorded in the order they were input, but are physically reordered in the channel, so that a real burst error is split into smaller bursts in several codewords. The size of the burst seen by each

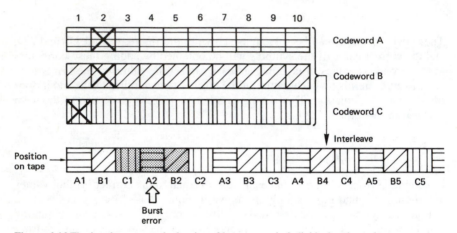

Figure 6.46 The interleave controls the size of burst errors in individual codewords.

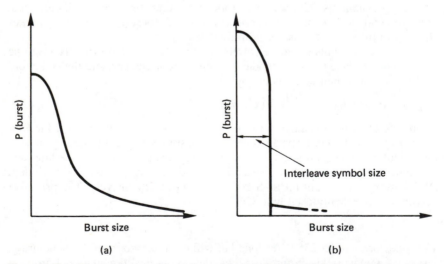

Figure 6.47 (a) The distribution of burst sizes might look like this. (b) Following interleave, the burst size within a codeword is controlled to that of the interleave symbol size, except for gross errors which have low probability.

codeword is now determined primarily by the parameters of the interleave, and Figure 6.47 shows that the probability of occurrence of bursts with respect to the burst length in a given codeword is modified. The number of bits in the interleave word can be made equal to the burst-correcting ability of the code in the knowledge that it will be exceeded only very infrequently.

There are a number of different ways in which interleaving can be performed. Figure 6.48 shows that in block interleaving, words are reordered within blocks which are themselves in the correct order. This approach is attractive for rotary-head recorders, because the scanning process naturally divides the tape up into blocks. The block interleave is achieved by writing samples into a memory in sequential address locations from a counter, and reading the memory with non-sequential addresses from a sequencer. The effect is to convert a one-dimensional sequence of samples into a two-dimensional structure having rows and columns.

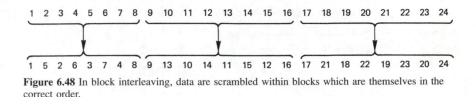

Figure 6.48 In block interleaving, data are scrambled within blocks which are themselves in the correct order.

The alternative to block interleaving is convolutional interleaving where the interleave process is endless. In Figure 6.49 symbols are assembled into short blocks and then delayed by an amount proportional to the position in the block. It will be seen from the figure that the delays have the effect of shearing the symbols so that columns on the left side of the diagram become diagonals on the right. When the columns on the right are read, the convolutional interleave will be obtained. Convolutional interleave works well in transmission applications such as DVB where there is no natural track break. Convolutional interleave has the advantage of requiring less memory to implement than a block code. This is because a block code requires the entire block to be written into the memory before it can be read, whereas a convolutional code requires only enough memory to cause the required delays.

6.23 Product codes

In the presence of burst errors alone, the system of interleaving works very well, but it is known that in most practical channels there are also uncorrelated errors of a few bits due to noise. Figure 6.50 shows an interleaving system where a dropout-induced burst error has occurred which is at the maximum correctable size. All three codewords involved are working at their limit of one symbol. A random error due to noise in the vicinity of a burst error will cause the correction power of the code to be exceeded. Thus a random error of a single bit causes a further entire symbol to fail. This is a weakness of an interleave solely designed to handle dropout-induced bursts. Practical high-density equipment must address the problem of noise-induced or random errors and burst errors occurring at the same time. This is done by forming codewords both before and after the

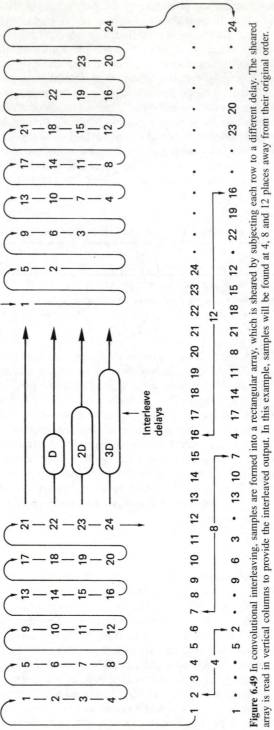

Figure 6.49 In convolutional interleaving, samples are formed into a rectangular array, which is sheared by subjecting each row to a different delay. The sheared array is read in vertical columns to provide the interleaved output. In this example, samples will be found at 4, 8 and 12 places away from their original order.

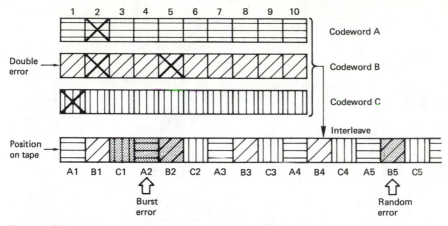

Figure 6.50 The interleave system falls down when a random error occurs adjacent to a burst.

interleave process. In block interleaving, this results in a *product code*, whereas in the case of convolutional interleave the result is called *cross-interleaving*.

Figure 6.51 shows that in a product code the redundancy calculated first and checked last is called the outer code, and the redundancy calculated second and checked first is called the inner code. The inner code is formed along tracks on the medium. Random errors due to noise are corrected by the inner code and do not impair the burst-correcting power of the outer code. Burst errors are declared uncorrectable by the inner code which flags the bad samples on the way into the de-interleave memory. The outer code reads the error flags in order to correct the flagged symbols by erasure. The error flags are also known as erasure flags. As it does not have to compute the error locations, the outer code needs half as much redundancy for the same correction power. Thus the inner code redundancy does not raise the code overhead. The combination of codewords with interleaving in several dimensions yields an error-protection strategy which is truly synergistic, in that the end result is more powerful than the sum of the parts. Needless to say, the technique is used extensively in modern storage formats.

Appendix 6.1 Calculation of Reed–Solomon generator polynomials

For a Reed–Solomon codeword over $GF(2^3)$, there will be seven three-bit symbols. For location and correction of one symbol, there must be two redundant symbols P and Q, leaving A–E for data.

The following expressions must be true, where a is the primitive element of $x^3 \oplus x \oplus 1$ and $\oplus$ is XOR throughout:

$$A \oplus B \oplus C \oplus D \oplus E \oplus P \oplus Q = 0 \tag{1}$$

$$a^7A \oplus a^6B \oplus a^5C \oplus a^4D \oplus a^3E \oplus a^2P \oplus aQ = 0 \tag{2}$$

Dividing equation (2) by a:

$$a^6A \oplus a^5B \oplus a^4C \oplus a^3D \oplus a^2E \oplus aP \oplus Q = 0$$

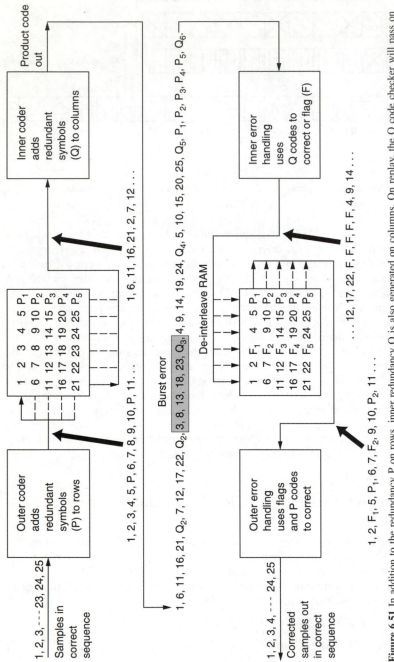

Figure 6.51 In addition to the redundancy P on rows, inner redundancy Q is also generated on columns. On replay, the Q code checker will pass on flags F if it finds an error too large to handle itself. The flags pass through the de-interleave process and are used by the outer error correction to identify which symbol in the row needs correcting with P redundancy. The concept of crossing two codes in this way is called a product code.

$$= A \oplus B \oplus C \oplus D \oplus E \oplus P \oplus Q$$

Cancelling Q, and collecting terms:

$$(a^6 \oplus 1)A \oplus (a^5 \oplus 1)B \oplus (a^4 \oplus 1)C \oplus (a^3 \oplus 1)D \oplus (a^2 \oplus 1)E$$
$$= (a \oplus 1)P$$

Using section 3.16 to calculate $(a^n + 1)$, e.g. $a^6 + 1 = 101 + 001 = 100 = a^2$:

$$a^2A \oplus a^4B \oplus a^5C \oplus aD \oplus a^6E = a^3P$$
$$a^6A \oplus aB \oplus a^2C \oplus a^5D \oplus a^3E = P$$

Multiplying equation (1) by a^2 and equating to equation (2):

$$a^2A \oplus a^2B \oplus a^2C \oplus a^2D \oplus a^2E \oplus a^2P \oplus a^2Q = 0$$
$$= a^7A \oplus a^6B \oplus a^5C \oplus a^4D \oplus a^3E \oplus a^2P \oplus aQ$$

Cancelling terms a^2P and collecting terms (remember $a^2 \oplus a^2 = 0$):

$$(a^7 \oplus a^2)A \oplus (a^6 \oplus a^2)B \oplus (a^5 \oplus a^2)C \oplus (a^4 \oplus a^2)D \oplus$$
$$(a^3 \oplus a^2)E = (a^2 \oplus a)Q$$

Adding powers according to section 3.16, e.g.

$$a^7 \oplus a^2 = 001 \oplus 100 = 101 = a^6:$$
$$a^6A \oplus B \oplus a^3C \oplus aD \oplus a^5E = a^4Q$$
$$a^2A \oplus a^3B \oplus a^6C \oplus a^4D \oplus aE = Q$$

References

1. Deeley, E.M., Integrating and differentiating channels in digital tape recording. *Radio Electron. Eng.*, **56**, 169–173 (1986)
2. Mee, C.D., *The Physics of Magnetic Recording*, Amsterdam and New York: Elsevier-North-Holland Publishing (1978)
3. Jacoby, G.V., Signal equalization in digital magnetic recording. *IEEE Trans. Magn.*, **MAG-11**, 302–305 (1975)
4. Schneider, R.C., An improved pulse-slimming method for magnetic recording. *IEEE Trans. Magn.*, **MAG-11**, 1240–1241 (1975)
5. Miller, A., US Patent. No.3 108 261
6. Mallinson, J.C. and Miller, J.W., Optimum codes for digital magnetic recording. *Radio and Electron. Eng.*, **47**, 172–176 (1977)
7. Miller, J.W., DC-free encoding for data transmission system. US Patent 4 027 335 (1977)
8. Tang, D.T., Run-length-limited codes. IEEE International Symposium on Information Theory (1969)
9. Shannon, C.E., A mathematical theory of communication. *Bell System Tech. J.*, **27**, 379 (1948)
10. Reed, I.S. and Solomon, G., Polynomial codes over certain finite fields. *J. Soc. Indust. Appl. Math.*, **8**, 300–304 (1960)
11. Berlekamp, E.R., *Algebraic Coding Theory*. New York: McGraw-Hill (1967). Reprint edition: Laguna Hills, CA: Aegean Park Press (1983)
12. Sugiyama, Y. *et al.*, An erasures and errors decoding algorithm for Goppa codes. *IEEE Trans. Inf. Theory*, **IT-22** (1976)
13. Peterson, W.W. and Weldon, E.J., *Error Correcting Codes*, 2nd edn, Cambridge MA: MIT Press (1972)

7
Transmission

7.1 Introduction

The distances involved in transmission vary from that of a short cable between adjacent units to communication anywhere on earth via data networks or radio communication. This chapter must consider a correspondingly wide range of possibilities. The importance of direct digital interconnection between audio devices was realized early, and this was standardized by the AES/EBU digital audio interface for professional equipment and the SPDIF interface for consumer equipment. These standards were extended to produce the MADI standard for multi-channel interconnects. All of these work on uncompressed PCM audio.

As digital audio and computers continue to converge, computer networks are increasingly used for audio purposes. Audio may be transmitted on networks such as Ethernet, ISDN, ATM and Internet. Here compression may or may not be used, and non-real-time transmission may also be found according to economic pressures.

Digital audio is now being broadcast in its own right as DAB, alongside traditional analog television as NICAM digital audio and as the sound channels of digital television broadcasts. All of these applications use compression.

Whatever the transmission medium, one universal requirement is a reliable synchronization system. In PCM systems, synchronization of the sampling rate between sources is necessary for mixing. In packet-based networks, synchronization allows the original sampling rate to be established at the receiver despite intermittent packet delivery. In digital television systems, synchronization between vision and sound is a further requirement.

7.2 The AES/EBU interface

The AES/EBU digital audio interface, originally published in 1985,[1] was proposed to embrace all the functions of existing formats in one standard. Alongside the professional format, Sony and Philips developed a similar format now known as SPDIF (Sony Philips Digital Interface) intended for consumer use. This offers different facilities to suit the application, yet retains sufficient compatibility with the professional interface so that, for many purposes, consumer and professional machines can be connected together.[2,3]

During the standardization process it was considered desirable to be able to use existing analog audio cabling for digital transmission. Existing professional analog signals use nominally 600 Ω impedance balanced line screened signalling, with one cable per audio channel, or in some cases one twisted pair per channel with a common screen. The 600 Ω standard came from telephony where long distances are involved in comparison with electrical audio wavelengths. The distances likely to be found within a studio complex are short compared to audio electrical wavelengths and as a result at audio frequency the impedance of cable is high and the 600 Ω figure is that of the source and termination. Such a cable has a different impedance at the frequencies used for digital audio, around 110 Ω.

If a single serial channel is to be used, the interconnect has to be self-clocking and self-synchronizing, i.e. the single signal must carry enough information to allow the boundaries between individual bits, words and blocks to be detected reliably. To fulfil these requirements, the AES/EBU and SPDIF interfaces use FM channel code (see Chapter 6) which is DC-free, strongly self-clocking and capable of working with a changing sampling rate. Synchronization of deserialization is achieved by violating the usual encoding rules.

The use of FM means that the channel frequency is the same as the bit rate when sending data ones. Tests showed that in typical analog audio cabling installations, sufficient bandwidth was available to convey two digital audio channels in one twisted pair. The standard driver and receiver chips for RS-422A[4] data communication (or the equivalent CCITT-V.11) are employed for professional use, but work by the BBC[5] suggested that equalization and transformer coupling were desirable for longer cable runs, particularly if several twisted pairs occupy a common shield. Successful transmission up to 350 m has been achieved with these techniques.[6] Figure 7.1 shows the standard configuration. The output impedance of the drivers will be about 110 Ω, and the impedance of the cable used should be similar at the frequencies of interest. The driver was specified in AES-3–1985 to produce between 3 and 10 V peak-to-peak into such an impedance but this was changed to between 2 and 7 volts in AES-3–1992 to better reflect the characteristics of actual RS-422 driver chips.

The original receiver impedance was set at a high 250 Ω with the intention that up to four receivers could be driven from one source. This was found to be inadvisable because of reflections caused by impedance mismatches and AES-3–1992 is now a point-to-point interface with source, cable and load impedance all set at 110 Ω. Whilst analog audio cabling was adequate for digital signalling,

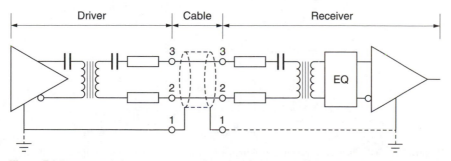

Figure 7.1 Recommended electrical circuit for use with the standard two-channel interface.

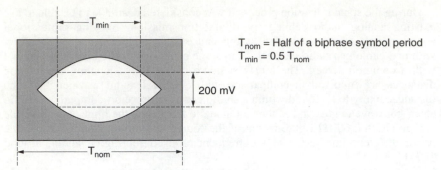

Figure 7.2 The minimum eye pattern acceptable for correct decoding of standard two-channel data.

cable manufacturers have subsequently developed cables which are more appropriate for new digital installations, having lower loss factors allowing greater transmission distances.

In Figure 7.2, the specification of the receiver is shown in terms of the minimum eye pattern (see Chapter 6) which can be detected without error. It will be noted that the voltage of 200 mV specifies the height of the eye opening at a width of half a channel bit period. The actual signal amplitude will need to be larger than this, and even larger if the signal contains noise. Figure 7.3 shows the recommended equalization characteristic which can be applied to signals received over long lines.

As an adequate connector in the shape of the XLR is already in wide service, the connector made to IEC 268 Part 12 has been adopted for digital audio use. Effectively, existing analog audio cables having XLR connectors can be used without alteration for digital connections. The AES/EBU standard does, however, require that suitable labelling should be used so that it is clear that the connections on a particular unit are digital.

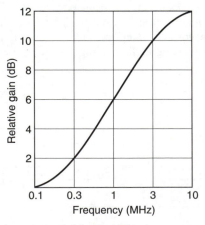

Figure 7.3 EQ characteristic recommended by the AES to improve reception in the case of long lines.

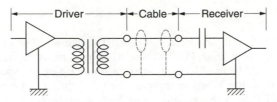

Figure 7.4 The consumer electrical interface.

The need to drive long cables does not generally arise in the domestic environment, and so a low-impedance balanced signal was not considered necessary. The electrical interface of the consumer format uses a 0.5 V peak single-ended signal, which can be conveyed down conventional audio-grade coaxial cable connected with RCA 'phono' plugs. Figure 7.4 shows the resulting consumer interface as specified by IEC 958.

There is the alternative[7] of a professional interface using coaxial cable and BNC connectors for distances of around 1000 m. This is simply the AES/EBU protocol but with a 75 Ω coaxial cable carrying a one-volt signal so that it can be handled by analog video distribution amplifiers. Impedance converting transformers are commercially available allowing balanced 110 Ω to unbalanced 75 Ω matching.

In Figure 7.5 the basic structure of the professional and consumer formats can be seen. One subframe consists of 32 bit-cells, of which four will be used by a synchronizing pattern. Subframes from the two audio channels, A and B, alternate on a time-division basis. Up to 24-bit sample wordlength can be used, which should cater for all conceivable future developments, but normally 20-bit maximum length samples will be available with four auxiliary data bits, which can be used for a voice-grade channel in a professional application. In a consumer DAT machine, subcode can be transmitted in bits 4–11, and the sixteen-bit audio in bits 12–27.

The format specifies that audio data must be in two's complement coding. Whilst pure binary could accept various alignments of different wordlengths with only a level change, this is not true of two's complement. If different wordlengths are used, the MSBs must always be in the same bit position otherwise the polarity will be misinterpreted. Thus the MSB has to be in bit 27 irrespective of

0	3	4	7	8		27	28		31
Preamble		L S B	L S B	Audio sample word		M S B	V	U C	P

```
Validity flag ────────┘
User data ──────────┘
Channel status ───────┘
Parity bit ──────────┘
```

Figure 7.5 The basic subframe structure of the AES/EBU format. Sample can be twenty bits with four auxiliary bits, or twenty-four bits. LSB is transmitted first.

wordlength. Shorter words are leading zero filled up to the twenty-bit capacity.

Four status bits accompany each subframe. The validity flag will be reset if the associated sample is reliable. Whilst there have been many aspirations regarding what the V bit could be used for, in practice a single bit cannot specify much, and if combined with other V bits to make a word, the time resolution is lost. AES-3–1992 described the V bit as indicating that the information in the associated subframe is 'suitable for conversion to an analog signal'. Thus it might be reset if the interface was being used for non-audio data as is done, for example, in CD-I players.

The parity bit produces even parity over the subframe, such that the total number of ones in the subframe is even. This allows for simple detection of an odd number of bits in error, but its main purpose is that it makes successive sync patterns have the same polarity, which can be used to improve the probability of detection of sync. The user and channel-status bits are discussed later.

Two of the subframes described above make one frame, which repeats at the sampling rate in use. The first subframe will contain the sample from channel A, or from the left channel in stereo working. The second subframe will contain the sample from channel B, or the right channel in stereo. At 48 kHz, the bit rate will be 3.072 MHz, but as the sampling rate can vary, the clock rate will vary in proportion.

In order to separate the audio channels on receipt the synchronizing patterns for the two subframes are different as Figure 7.6 shows. These sync patterns begin with a run length of 1.5 bits which violates the FM channel coding rules and so cannot occur due to any data combination. The type of sync pattern is denoted by the position of the second transition which can be 0.5, 1.0 or 1.5 bits away from the first. The third transition is designed to make the sync patterns DC-free.

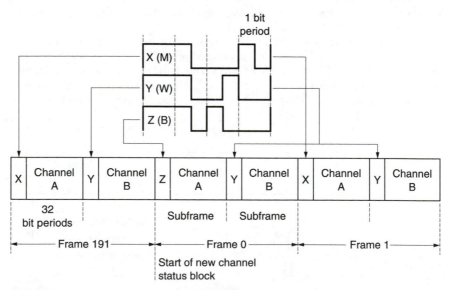

Figure 7.6 Three different preambles (X, Y and Z) are used to synchronize a receiver at the starts of subframes.

The channel status and user bits in each subframe form serial data streams with one bit of each per audio channel per frame. The channel status bits are given a block structure and synchronized every 192 frames, which at 48 kHz gives a block rate of 250 Hz, corresponding to a period of four milliseconds. In order to synchronize the channel-status blocks, the channel A sync pattern is replaced for one frame only by a third sync pattern which is also shown in Figure 7.6. The AES standard refers to these as X,Y and Z whereas IEC 958 calls them M,W and B. As stated, there is a parity bit in each subframe, which means that the binary level at the end of a subframe will always be the same as at the beginning. Since the sync patterns have the same characteristic, the effect is that sync patterns always have the same polarity and the receiver can use that information to reject noise. The polarity of transmission is not specified, and indeed an accidental inversion in a twisted pair is of no consequence, since it is only the transition that is of importance, not the direction.

When 24-bit resolution is not required, the four auxiliary bits can be used to provide talkback. This was proposed by broadcasters[8] to allow voice coordination between studios as well as program exchange on the same cables.

Twelve-bit samples of the talkback signal are taken at one third the main sampling rate. Each twelve-bit sample is then split into three four-bit nibbles which can be sent in the auxiliary data slot of three successive samples in the same audio channel. As there are 192 nibbles per channel status block period, there will be exactly 64 talkback samples in that period. The reassembly of the nibbles can be synchronized by the channel status sync pattern. Channel status byte 2 reflects the use of auxiliary data in this way.

7.3 Channel status

In both the professional and consumer formats, the sequence of channel-status bits over 192 subframes builds up a 24-byte channel-status block. However, the contents of the channel status data are completely different between the two applications. The professional channel status structure is shown in Figure 7.7. Byte 0 determines the use of emphasis and the sampling rate, with details in Figure 7.8. Byte 1 determines the channel usage mode, i.e. whether the data transmitted are a stereo pair, two unrelated mono signals or a single mono signal, and details the user bit handling. Figure 7.9 gives details. Byte 2 determines wordlength as in Figure 7.10. This was made more comprehensive in AES-3–1992. Byte 3 is applicable only to multichannel applications. Byte 4 indicates the suitability of the signal as a sampling rate reference and will be discussed in more detail later in this chapter.

There are two slots of four bytes each which are used for alphanumeric source and destination codes. These can be used for routing. The bytes contain seven-bit ASCII characters (printable characters only) sent LSB first with the eighth bit set to zero acording to AES-3–1992. The destination code can be used to operate an automatic router, and the source code will allow the origin of the audio and other remarks to be displayed at the destination.

Bytes 14–17 convey a 32-bit sample address which increments every channel status frame. It effectively numbers the samples in a relative manner from an arbitrary starting point. Bytes 18–21 convey a similar number, but this is a time-of-day count, which starts from zero at midnight. As many digital audio devices do not have real-time clocks built in, this cannot be relied upon.

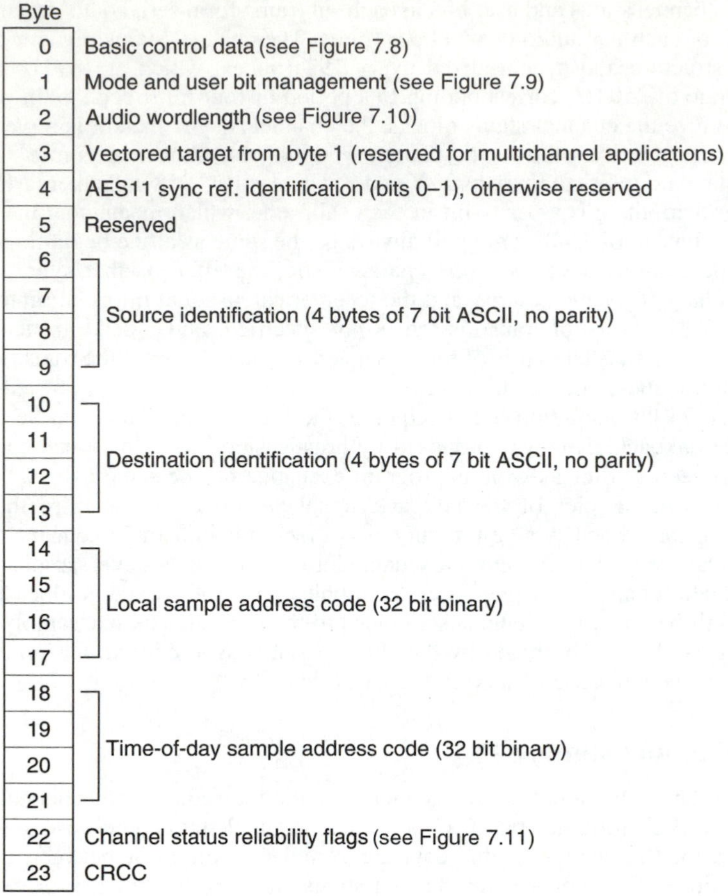

Byte

Byte	
0	Basic control data (see Figure 7.8)
1	Mode and user bit management (see Figure 7.9)
2	Audio wordlength (see Figure 7.10)
3	Vectored target from byte 1 (reserved for multichannel applications)
4	AES11 sync ref. identification (bits 0–1), otherwise reserved
5	Reserved
6	Source identification (4 bytes of 7 bit ASCII, no parity)
7	
8	
9	
10	Destination identification (4 bytes of 7 bit ASCII, no parity)
11	
12	
13	
14	Local sample address code (32 bit binary)
15	
16	
17	
18	Time-of-day sample address code (32 bit binary)
19	
20	
21	
22	Channel status reliability flags (see Figure 7.11)
23	CRCC

Figure 7.7 Overall format of the professional channel status block.

AES-3–92 specified that the time-of-day bytes should convey the real time at which a recording was made, making it rather like timecode. There are enough combinations in 32 bits to allow a sample count over 24 hours at 48 kHz. The sample count has the advantage that it is universal and independent of local supply frequency. In theory if the sampling rate is known, conventional hours, minutes, seconds, frames timecode can be calculated from the sample count, but in practice it is a lengthy computation and users have proposed alternative formats in which the data from EBU or SMPTE timecode are transmitted directly in these bytes. Some of these proposals are in service as *de-facto* standards.

The penultimate byte contains four flags which indicate that certain sections of the channel-status information are unreliable (see Figure 7.11). This allows the transmission of an incomplete channel-status block where the entire structure is not needed or where the information is not available. For example, setting bit 5 to a logical one would mean that no origin or destination data would be interpreted by the receiver, and so it need not be sent.

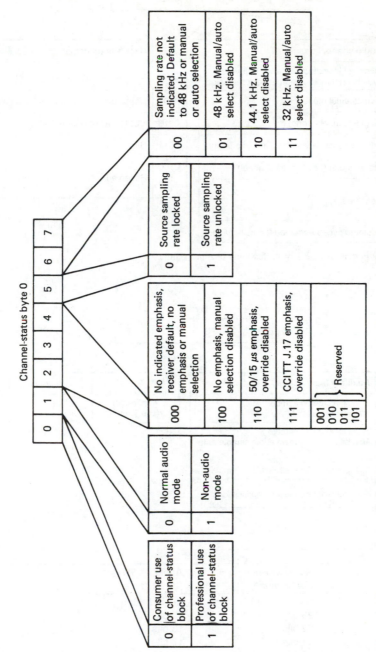

Figure 7.8 The first byte of the channel-status information in the AES/EBU standard deals primarily with emphasis and sampling-rate control.

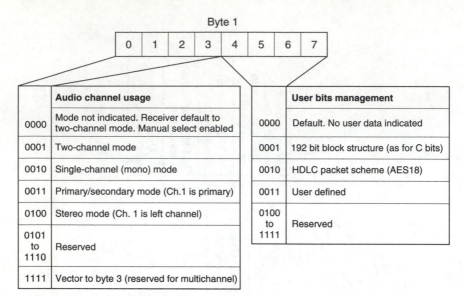

Figure 7.9 Format of byte 1 of professional channel status.

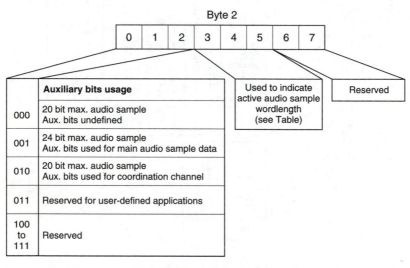

Bits states 3 4 5	Audio wavelength (24 bit mode)	Audio wavelength (20 bit mode)
0 0 0	Not indicated	Not indicated
0 0 1	23 bits	19 bits
0 1 0	22 bits	18 bits
0 1 1	21 bits	17 bits
1 0 0	20 bits	16 bits
1 0 1	24 bits	20 bits

Figure 7.10 Format of byte 2 of professional channel status.

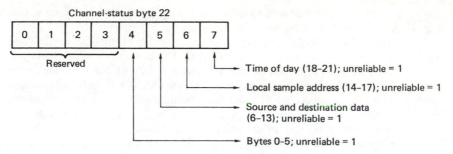

Figure 7.11 Byte 22 of channel status indicates if some of the information in the block is unreliable.

The final byte in the message is a CRCC which converts the entire channel-status block into a codeword (see Chapter 6). The channel status message takes 4 ms at 48 kHz and in this time a router could have switched to another signal source. This would damage the transmission, but will also result in a CRCC failure so the corrupt block is not used. Error correction is not necessary, as the channel status data are either stationary, i.e. they stay the same, or change at a predictable rate, e.g. timecode. Stationary data will only change at the receiver if a good CRCC is obtained.

7.4 User bits

The user channel consists of one bit per audio channel per sample period. Unlike channel status, which only has a 192-bit frame structure, the user channel can have a flexible frame length. Figure 7.9 showed how byte 1 of the channel status frame describes the state of the user channel. Many professional devices do not use the user channel at all and would set the all-zeros code. If the user channel frame has the same length as the channel status frame then code 0001 can be set. One user channel format which is standardized is the data packet scheme of AES18-1992.[9,10] This was developed from proposals to employ the user channel for labelling in an asynchronous format.[11] A computer industry standard protocol known as HDLC (High-level Data Link Control)[12] is employed in order to take advantage of readily available integrated circuits.

The frame length of the user channel can be conveniently made equal to the frame period of an associated device. For example, it may be locked to Film, TV or DAT frames. The frame length may vary in NTSC as there are not an integer number of audio samples in a TV frame.

7.5 MADI – Multi-channel audio digital interface

Whilst the AES/EBU digital interface excels for the interconnection of stereo equipment, it is at a disadvantage when a large number of channels is required. MADI[13] was designed specifically to address the requirement for digital connection between multitrack recorders and mixing consoles by a working group set up jointly by Sony, Mitsubishi, Neve and SSL.

The standard provides for 56 simultaneous digital audio channels which are conveyed point-to-point on a single 75 Ω coaxial cable fitted with BNC connectors (as used for analog video) along with a separate synchronization signal. A distance of at least 50 m can be achieved.

Essentially MADI takes the subframe structure of the AES/EBU interface and multiplexes 56 of these into one sample period rather than the original two. Clearly this will result in a considerable bit rate, and the FM channel code of the AES/EBU standard would require excessive bandwidth. A more efficient code is used for MADI. In the AES/EBU interface the data rate is proportional to the sampling rate in use. Losses will be greater at the higher bit rate of MADI, and the use of a variable bit rate in the channel would make the task of achieving optimum equalization difficult. Instead the data bit rate is made a constant 100 megabits per second, irrespective of sampling rate. At lower sampling rates, the audio data are padded out to maintain the channel rate.

The MADI standard is effectively a superset of the AES/EBU interface in that the subframe data content is identical. This means that a number of separate AES/EBU signals can be fed into a MADI channel and recovered in their entirety on reception. The only caution required with such an application is that all channels must have the same synchronized sampling rate. The primary application of MADI is to multitrack recorders, and in these machines the sampling rates of all tracks are intrinsically synchronous. When the replay speed of such machines is varied, the sampling rate of all channels will change by the same amount, so they will remain synchronous.

At one extreme, MADI will accept a 32 kHz recorder playing $12\frac{1}{2}$ per cent slow, and at the other extreme a 48 kHz recorder playing $12\frac{1}{2}$ per cent fast. This is almost a factor of 2:1. Figure 7.12 shows some typical MADI configurations.

The data transmission of MADI is made using a group code, where groups of four data bits are represented by groups of five channel bits. Four bits have sixteen combinations, whereas five bits have 32 combinations. Clearly only 16 out of these 32 are necessary to convey all possible data. It is then possible to use

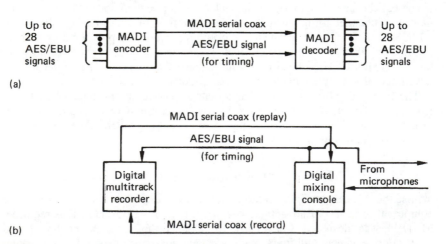

(a)

(b)

Figure 7.12 Some typical MADI applications. In (a) a large number of two-channel digital signals are multiplexed into the MADI cable to achieve economy of cabling. Note the separate timing signal. In (b) a pair of MADI links is necessary to connect a recorder to a mixing console. A third MADI link could be used to feed microphones into the desk from remote convertors.

(a)

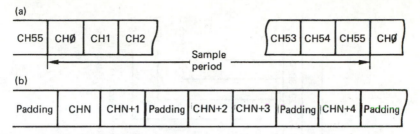

(b)

Figure 7.13 In (a) all 56 channels are sent in numerical order serially during the sample period. For simplicity no padding symbols are shown here. In (b) the use of padding symbols is illustrated. These are necessary to maintain the channel bit rate at 125 M bits/s irrespective of the sample rate in use. Padding can be inserted flexibly, but it must only be placed between the channels.

some of the remaining patterns when it is required to pad out the data rate. The padding symbols will not correspond to a valid data symbol and so they can be recognized and thrown away on reception. A further use of this coding technique is that the 16 patterns of 5 bits which represent real data are chosen to be those which will have the best balance between high and low states, so that DC offsets at the receiver can be minimized. The 4/5 code adopted is the same one used for a computer transmission format known as FDDI, so that existing hardware can be used.

Figure 7.13(a) shows the frame structure of MADI. In one sample period, 56 time slots are available, and these each contain eight 4/5 symbols, corresponding to 32 data bits or 40 channel bits. Depending on the sampling rate in use, more or less padding symbols will need to be inserted in the frame to maintain a constant channel bit rate. Since the receiver does not interpret the padding symbols as data, it is effectively blind to them, and so there is considerable latitude in the allowable positions of the padding. Figure 7.13(b) shows some possibilities. The padding must not be inserted within a channel, only between channels, but the channels need not necessarily be separated by padding. At one extreme, all channels can be butted together, followed by a large padding area, or the channels can be evenly spaced throughout the frame. Although this sounds rather vague, it is intended to allow freedom in the design of associated hardware. Multitrack recorders generally have some form of internal multiplexed data bus, and these have various architectures and protocols. The timing flexibility allows an existing bus timing structure to be connected to a MADI link with the minimum of hardware. Since the channels can be inserted at a variety of places within the frame, the necessity of a separate synchronizing link between transmitter and receiver becomes clear.

Figure 7.14 shows the MADI channel format, which should be compared with the AES/EBU subframe shown in Figure 7.5. The last 28 bits are identical, and differences are only apparent in the synchronizing area. In order to remain transparent to an AES/EBU signal, which can contain two audio channels, MADI must tell the receiver whether a particular channel contains the A leg or B leg, and when the AES/EBU channel status block sync occurs. Bits 2 and 3 perform these functions. As the 56 channels of MADI follow one another in numerical order, it is necessary to identify channel zero so that the channels are not mixed up. This is the function of bit 0, which is set in channel zero and reset in all other

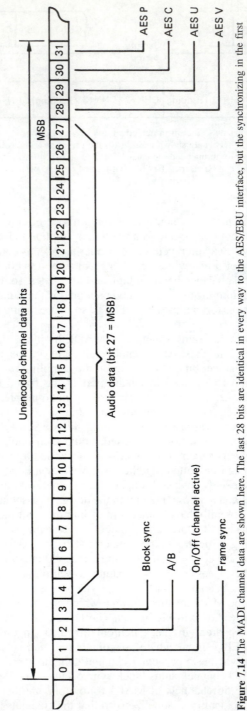

Figure 7.14 The MADI channel data are shown here. The last 28 bits are identical in every way to the AES/EBU interface, but the synchronizing in the first four bits differs. There is a frame sync bit to identify channel 0, and a channel active bit. The A/B leg of a possible AES/EBU input to MADI is conveyed, as is the channel status block sync.

channels. Finally bit 1 is set to indicate an active channel, for the case when less than 56 channels are being fed down the link. Active channels have bit 1 set, and must be consecutive starting at channel zero. Inactive channels have all bits set to zero, and must follow the active channels.

7.6 Fibre-optic interfacing

Whereas a parallel bus is ideal for a distributed multichannel system, for a point-to-point connection, the use of fibre optics is feasible, particularly as distance increases. An optical fibre is simply a glass filament which is encased in such a way that light is constrained to travel along it. Transmission is achieved by modulating the power of an LED or small laser coupled to the fibre. A phototransistor converts the received light back to an electrical signal.

Optical fibres have numerous advantages over electrical cabling. The bandwidth available is staggering. Optical fibres neither generate, nor are prone to, electromagnetic interference and, as they are insulators, ground loops cannot occur.[14] The disadvantage of optical fibres is that the terminations of the fibre where transmitters and receivers are attached suffer optical losses, and while these can be compensated in point-to-point links, the use of a bus structure is not really feasible. Fibre-optic links are already in service in digital audio mixing consoles.[15] The fibre implementation by Toshiba and known as the TOSLink is popular in consumer products, and the protocol is identical to the consumer electrical format.

7.7 Synchronizing

When digital audio signals are to be assembled from a variety of sources, either for mixing down or for transmission through a TDM (time-division multiplexing) system, the samples from each source must be synchronized to one another in both frequency and phase. The source of samples must be fed with a reference sampling rate from some central generator, and will return samples at that rate. The same will be true if digital audio is being used in conjunction with VTRs. As the scanner speed and hence the audio block rate is locked to video, it follows that the audio sampling rate must be locked to video. Such a technique has been used since the earliest days of television in order to allow vision mixing, but now that audio is conveyed in discrete samples, these too must be genlocked to a reference for most production purposes.

AES11–1991[16] documented standards for digital audio synchronization and requires professional equipment to be able to genlock either to a separate reference input or to the sampling rate of an AES/EBU input.

As the interface uses serial transmission, a shift register is required in order to return the samples to parallel format within equipment. The shift register is generally buffered with a parallel loading latch which allows some freedom in the exact time at which the latch is read with respect to the serial input timing. Accordingly the standard defines synchronism as an identical sampling rate, but with no requirement for a precise phase relationship. Figure 7.15 shows the timing tolerances allowed. The beginning of a frame (the frame edge) is defined as the leading edge of the X preamble. A device which is genlocked must correctly decode an input whose frame edges are within ± 25 per cent of the sample period. This is quite a generous margin, and corresponds to the timing

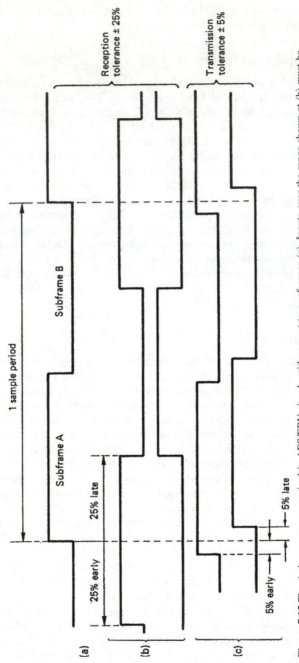

Figure 7.15 The timing accuracy required in AES/EBU signals with respect to a reference (a). Inputs over the range shown at (b) must be accepted, whereas outputs must be closer in timing to the reference as shown at (c).

shift due to putting about a kilometre of cable in series with a signal. In order to prevent tolerance build-up when passing through several devices in series, the output timing must be held within ± 5 per cent of the sample period.

The reference signal may be an AES/EBU signal carrying program material, or it may carry muted audio samples; the so-called digital audio silence signal. Alternatively it may just contain the sync patterns. The accuracy of the reference is specified in bits 0 and 1 of byte 4 of channel status (see Figure 7.7). Two zeros indicates the signal is not reference grade (but some equipment may still be able to lock to it). 01 indicates a Grade 1 reference signal which is ± 1 ppm accurate, whereas 10 indicates a Grade 2 reference signal which is ± 10 ppm accurate. Clearly devices which are intended to lock to one of these references must have an appropriate phase-locked-loop capture range.

In addition to the AES/EBU synchronization approach, some older equipment carries a word clock input which accepts a TTL level square wave at the sampling frequency. This is the reference clock of the old Sony SDIF-2 interface.

Modern digital audio devices may also have a video input for synchronizing purposes. Video syncs (with or without picture) may be input, and a phase-locked loop will multiply the video frequency by an appropriate factor to produce a synchronous audio sampling clock.

7.8 Asynchronous operation

In practical situations, genlocking is not always possible. In a satellite transmission, it is not really practicable to genlock a studio complex half-way around the world to another. Outside broadcasts may be required to generate their own master timing for the same reason. When genlock is not achieved, there will be a slow slippage of sample phase between source and destination due to such factors as drift in timing generators. This phase slippage will be corrected by a synchronizer, which is intended to work with frequencies that are nominally the same. It should be contrasted with the sampling-rate convertor which can work at arbitrary but generally greater frequency relationships. Although a sampling-rate convertor can act as a synchronizer, it is a very expensive way of doing the job. A synchronizer can be thought of as a lower-cost version of a sampling-rate convertor which is constrained in the rate difference it can accept.

In one implementation of a digital audio synchronizer,[17] memory is used as a timebase corrector. Samples are written into the memory with the frequency and phase of the source and, when the memory is half-full, samples are read out with the frequency and phase of the destination. Clearly if there is a net rate difference, the memory will either fill up or empty over a period of time, and in order to recentre the address relationship, it will be necessary to jump the read address. This will cause samples to be omitted or repeated, depending on the relationship of source rate to destination rate, and would be audible on program material. The solution is to detect pauses or low-level passages and permit jumping only at such times. The process is illustrated in Figure 7.16. An alternative to address jumping is to undertake sampling-rate conversion for a short period (Figure 7.17) in order to slip the input/output relationship by one sample.[18] If this is done when the signal level is low, short wordlength logic can be used. However, now that sampling rate convertors are available as a low-cost

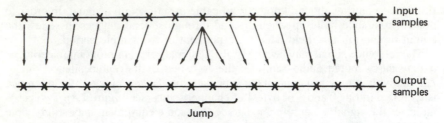

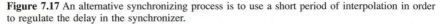

Figure 7.16 In jump synchronizing, input samples are subjected to a varying delay to align them with output timing. Eventually, the sample relationship is forced to jump to prevent delay building up. As shown here, this results in several samples being repeated, and can only be undertaken during program pauses, or at very low audio levels. If the input rate exceeds the output rate, some samples will be lost.

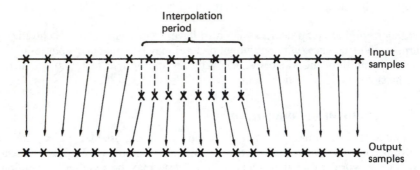

Figure 7.17 An alternative synchronizing process is to use a short period of interpolation in order to regulate the delay in the synchronizer.

single chip, these solutions are found less often in hardware, although they may be used in software-controlled processes.

The difficulty of synchronizing unlocked sources is eased when the frequency difference is small. This is one reason behind the clock accuracy standards for AES/EBU timing generators.[19]

7.9 Routing and networks

Routing is the process of directing audio and video signals betwen a large number of devices so that any one can be connected to any other. The principle of a router is not dissimilar to that of a telephone exchange. In analog routers, there is the potential for quality loss due to the switching element. Digital routers handling AES/EBU audio signals are attractive because they need introduce no loss whatsoever. In addition, the switching is performed on a binary signal and therefore the cost can be lower. Routers can be either cross-point or time-division multiplexed.

In a TDM system, channel reassignment is easy. If the audio channels are transmitted in address sequence, it is only necessary to change the addresses which the receiving channels recognize, and a given input channel will emerge from a different output channel. The only constraint in the use of TDM systems is that all channels must have synchronized sampling rates. Given that the MADI interface uses TDM, it is also possible to perform routing functions using MADI-based hardware.

For asynchronous systems, or where several sampling rates are found simultaneously, a cross-point type of channel-assignment matrix will be necessary, using AES/EBU signals. In such a device, the switching can be performed by logic gates at low cost, and in the digital domain there is, of course, no quality degradation.

Routers are the traditional approach to audio switching. However, given that digital audio is just another kind of data, it is to be expected that computer-based network technology can provide such functions. It is fundamental in a network that any port can communicate with any other port. Figure 7.18 shows a primitive three-port network. Clearly each port must select one or other of the remaining ports in a trivial switching system. However, if it were attempted to redraw Figure 7.18 with one hundred ports, each one would need a 99-way switch and the number of wires needed would be phenomenal. Another approach is needed.

Figure 7.19 shows that the common solution is to have an exchange, also known as a router, hub or switch, which is connected to every port by a single cable. In this case when a port wishes to communicate with another, it instructs

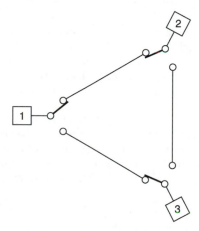

Figure 7.18 Switching is simple with a small number of ports.

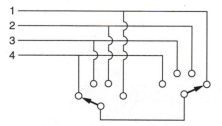

Figure 7.19 An exchange or switch can connect any input to any output, but extra switching is needed to support more than one connection.

the switch to make the connection. The complexity of the switch varies with its performance. The minimal case may be to install a single input selector and a single output selector. This allows any port to communicate with any other, but only one at a time. If more simultaneous communications are needed, further switching is needed. The extreme case is where every possible pair of ports can communicate simultaneously.

The amount of switching logic needed to implement the extreme case is phenomenal and in practice it is unlikely to be needed. One fundamental property of networks is that they are seldom implemented with the extreme case supported. There will be an economic decision made balancing the number of simultaneous communications with the equipment cost. Most of the time the user will be unaware that this limit exists, until there is a statistically abnormal condition which causes more than the usual number of nodes to attempt communication.

The phrase 'the switchboard was jammed' has passed into the language and stayed there despite the fact that manual switchboards are only seen in museums. This is a characteristic of networks. They generally only work up to a certain throughput and then there are problems. This doesn't mean that networks aren't useful, far from it. What it means is that with care, networks can be very useful, but without care they can be a nightmare.

7.10 Networks

There are two key factors to get right in a network. The first is that it must have enough throughput, bandwidth or connectivity to handle the anticipated usage and the second is that a priority system or algorithm is chosen which has appropriate behaviour during overload. These two characteristics are quite different, but often come as a pair in a network corresponding to a particular standard.

Where each device is individually cabled, the result is a radial network shown in Figure 7.20(a). It is not necessary to have one cable per device and several devices can co-exist on a single cable if some form of multiplexing is used. This might be time-division multiplexing (TDM) or frequency division multiplexing (FDM). In TDM, shown in Figure 7.20(b), the time axis is divided into steps

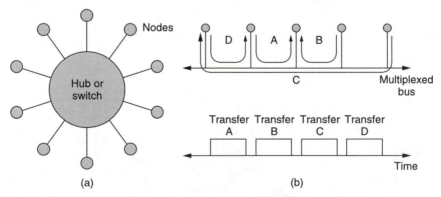

Figure 7.20 (a) Radial installations need a lot of cabling. Time-division multiplexing, where transfers occur during different time frames, reduces this requirement (b).

which may or may not be equal in length. In Ethernet, for example, these are called frames. During each time step or frame a pair of nodes have exclusive use of the cable. At the end of the time step another pair of nodes can communicate. Rapidly switching between steps gives the illusion of simultaneous transfer between several pairs of nodes. In FDM, simultaneous transfer is possible because each message occupies a different band of frequencies in the cable. Each node has to 'tune' to the correct signal. In practice it is possible to combine FDM and TDM. Each frequency band can be time multiplexed in some applications.

Data networks originated to serve the requirements of computers and it is a simple fact that most computer processes don't need to be performed in real time or indeed at a particular time at all. Networks tend to reflect that background as many of them, particularly the older ones, are asynchronous.

Asynchronous means that the time taken to deliver a given quantity of data is unknown. A TDM system may chop the data into several different transfers and each transfer may experience delay according to what other transfers the system is engaged in. Ethernet and most storage system buses are asynchronous. For broadcasting purposes an asynchronous delivery system is no use at all, but for copying an audio data file between two storage devices an asynchronous system is perfectly adequate.

The opposite extreme is the synchronous system in which the network can guarantee a constant delivery rate and a fixed and minor delay. An AES/EBU router is a synchronous network.

In between asynchronous and synchronous networks reside the isochronous approaches. These can be thought of as sloppy synchronous networks or more rigidly controlled asynchronous networks. Both descriptions are valid. In the isochronous network there will be maximum delivery time which is not normally exceeded. The data transmission rate may vary, but if the rate has been low for any reason, it will accelerate to prevent the maximum delay being reached. Isochronous networks can deliver near-real-time performance. If a data buffer is provided at both ends, synchronous data such as AES/EBU audio can be fed through an isochronous network. The magnitude of the maximum delay determines the size of the buffer and the length of the fixed overall delay through the system. This delay is responsible for the term 'near-real time'. ATM is an isochronous network.

These three different approaches are needed for economic reasons. Asynchronous systems are very efficient because as soon as one transfer completes, another can begin. This can only be achieved by making every device wait with its data in a buffer so that transfer can start immediately. Asynchronous sytems also make it possible for low bit rate devices to share a network with high bit rate devices. The low bit rate device will only need a small buffer and will therefore send short data blocks, whereas the high bit rate device will send long blocks. Asynchronous systems have no dificulty in handling blocks of varying size, whereas in a synchronous system this is very difficult.

Isochronous systems try to give the best of both worlds, generally by sacrificing some flexibility in block size. FireWire is an example of a network which is part isochronous and part asynchronous so that the advantages of both are available.

A network is basically a communication resource which is shared for economic reasons. Like any shared resource, decisions have to be made somewhere and somehow about how the resource is to be used. In the absence of such decisions the resultant chaos will be such that the resource might as well not exist.

In communications networks the resource is the ability to convey data from any node or port to any other. On a particular cable, clearly only one transaction of this kind can take place at any one instant even though in practice many nodes will simultaneously be wanting to transmit data. Arbitration is needed to determine which node is allowed to transmit.

There are a number of different arbitration protocols and these have evolved to support the needs of different types of network. In small networks, such as LANs, a single point failure which halts the entire network may be acceptable, whereas in a public transport network owned by a telecommunications company, the network will be redundant so that if a particular link fails data may be sent via an alternative route. A link which has reached its maximum capacity may also be supplanted by transmission over alternative routes.

In physically small networks, arbitration may be carried out in a single location. This is fast and efficient, but if the arbitrator fails it leaves the system completely crippled. The processor buses in computers work in this way. In centrally arbitrated systems the arbitrator needs to know the structure of the system and the status of all the nodes. Following a configuration change, due perhaps to the installation of new equipment, the arbitrator needs to be told what the new configuration is, or have a mechanism which allows it to explore the network and learn the configuration. Central arbitration is only suitable for small networks which change their configuration infrequently.

In other networks the arbitration is distributed so that some decision-making ability exists in every node. This is less efficient but is does allow at least some of the network to continue operating after a component failure. Distributed arbitration also means that each node is self-sufficient and so no changes need to be made if the network is reconfigured by adding or deleting a node. This is the only possible approach in wide area networks where the structure may be very complex and change dynamically in the event of failures or overload.

Ethernet uses distributed arbitration. FireWire is capable of using both types of arbitration. A small amount of decision-making ability is built into every node so that distributed arbitration is possible. However, if one of the nodes happens to be a computer, it can run a centralized arbitration algorithm.

The physical structure of a network is subject to some variation as Figure 7.21 shows. In radial networks, (a), each port has a unique cable connection to a device called a *hub*. The hub must have one connection for every port and this limits the number of ports. However, a cable failure will only result in the loss of one port. In a ring system (b) the nodes are connected like a daisy chain with each node acting as a feedthrough. In this case the arbitration requirement must be distributed. With some protocols, a single cable break doesn't stop the network operating. Depending on the protocol, simultaneous transactions may be possible provided they don't require the same cable. For example, in a storage network a disk drive may be outputting data to an editor while another drive is backing up data to a tape streamer. For the lowest cost, all nodes are physically connected in parallel to the same cable. Figure 7.21(c) shows that a cable break would divide the network into two halves, but it is possible that the impedance mismatch at the break could stop both halves working.

One of the concepts involved in arbitration is priority, which is fundamental to providing an appropriate quality of service. If two processes both want to use a

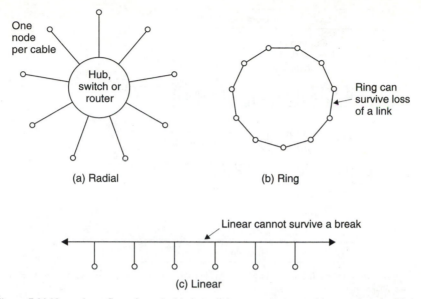

One node per cable

Hub, switch or router

(a) Radial

Ring can survive loss of a link

(b) Ring

Linear cannot survive a break

(c) Linear

Figure 7.21 Network configurations. At (a) the radial system uses one cable to each node. (b) Ring system uses less cable than radial. (c) Linear system is simple but has no redundancy.

network, the one with the highest priority would normally go first. Attributing priority must be done carefully because some of the results are non-intuitive. For example, it may be beneficial to give a high priority to a humble device which has a low data rate for the simple reason that if it is given use of the network it won't need it for long. In a broadcast environment, transactions concerned with on-air processes would have priority over file transfers concerning production and editing.

When a device gains access to the network to perform a transaction, generally no other transaction can take place until it has finished. Consequently it is important to limit the amount of time that a given port can stay on the bus. In this way when the time limit expires, a further arbitration must take place. The result is that the network resource rotates between transactions rather than one transfer hogging the resource and shutting everyone else out.

It follows from the presence of a time (or data quantity) limit that ports must have the means to break large files up into frames or cells and reassemble them on reception. This process is sometimes called *adaptation*. If the data to be sent originally exist at a fixed bit rate, some buffering will be needed so that the data can be time-compressed into the available frames. Each frame must be contiguously numbered and the system must transmit a file size or word count so that the receiving node knows when it has received every frame in the file.

The error-detection system interacts with this process because if any frame is in error on reception, the receiving node can ask for a retransmission of the frame. This is more efficient than retransmitting the whole file. Figure 7.22 shows the flow chart for a receiving node. Breaking files into frames helps to keep down the delay experienced by each process using the network. Figure 7.23 shows that each frame may be stored ready for transmission in a silo memory. It

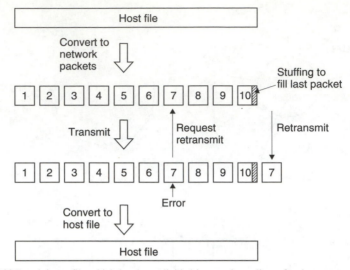

Figure 7.22 Receiving a file which has been divided into packets allows for the retransmission of just the packet in error.

is possible to make the priority a function of the number of frames in the silo, as this is a direct measure of how long a process has been kept waiting. Isochronous systems must do this in order to meet maximum delay specifications. In Figure 7.23 once frame transmission has completed, the arbitrator will determine which process sends a frame next by examining the depth of all the frame buffers. MPEG transport stream multiplexers and networks delivering MPEG data must work in this way because the transfer is isochronous and the amount of buffering in a decoder is limited for economic reasons.

A central arbitrator is relatively simple to implement because when all decisions are taken centrally there can be no timing difficulty (assuming a well-engineered system). In a distributed system, there is an extra difficulty due to the finite time taken for signals to travel down the data paths between nodes.

Figure 7.24 shows the structure of Ethernet which uses a protocol called CSMA/CD (carrier sense multiple access with collision detect) developed by DEC and Xerox. This is a distributed arbitration network where each node follows some simple rules. The first of these is not to transmit if an existing bus signal is detected. The second is not to transmit more than a certain quantity of data before releasing the bus. Devices wanting to use the bus will see bus signals and so will wait until the present bus transaction finishes. This must happen at some point because of the frame size limit. When the frame is completed, signalling on the bus should cease. The first device to sense the bus becoming free and to assert its own signal will prevent any other nodes transmitting according to the first rule. Where numerous devices are present it is possible to give them a priority structure by providing a delay between sensing the bus coming free and beginning a transaction. High-priority devices will have a short delay so they get in first. Lower-priority devices will only be able to start a transaction if the high-priority devices don't need to transfer.

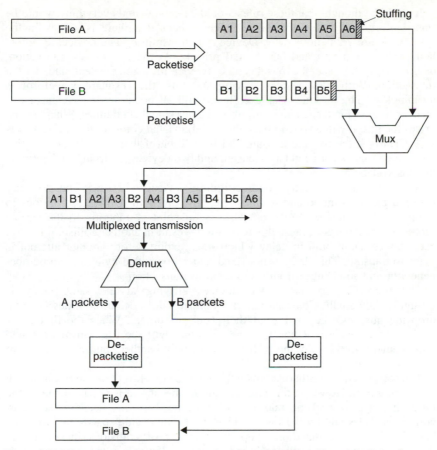

Figure 7.23 Files are broken into frames or packets for multiplexing with packets from other users. Short packets minimize the time between the arrival of successive packets. The priority of the multiplexing must favour isochronous data over asynchronous data.

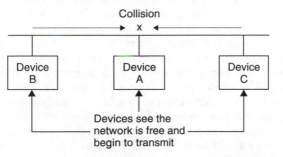
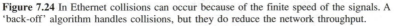

Figure 7.24 In Ethernet collisions can occur because of the finite speed of the signals. A 'back-off' algorithm handles collisions, but they do reduce the network throughput.

It might be thought that these rules would be enough and everything would be fine. Unfortunately the finite signal speed means that there is a flaw in the system. Figure 7.24 shows why. Device A is transmitting and devices B and C both want to transmit and have equal priority. At the end of A's transaction, devices B and C see the bus become free at the same instant and start a transaction. With two devices driving the bus, the resultant waveform is meaningless. This is known as a collision and all nodes must have means to recover from it. First, each node will read the bus signal at all times. When a node drives the bus, it will also read back the bus signal and compare it with what was sent. Clearly if the two are the same all is well, but if there is a difference, this must be because a collision has occurred and two devices are trying to determine the bus voltage at once.

If a collision is detected, both colliding devices will sense the disparity between the transmitted and readback signals, and both will release the bus to terminate the collision. However, there is no point is adhering to the simple protocol to reconnect because this will simply result in another collision. Instead each device has a built-in delay which must expire before another attempt is made to transmit. This delay is not fixed, but is controlled by a random number generator and so changes from transaction to transaction.

The probability of two node devices arriving at the same delay is infinitesimally small. Consequently if a collision does occur, both devices will drop the bus, and they will start their back-off timers. When the first timer expires, that device will transmit and the other will see the transmission and remain silent. In this way the collision is not only handled, but prevented from happening again.

The performance of Ethernet is usually specified in terms of the bit rate at which the cabling runs. However, this rate is academic because it is not available all the time. In a real network bit rate is lost by the need to send headers and error-correction codes and by the loss of time due to interframe spaces and collision handling. As the demand goes up, the number of collisions increases and throughput goes down. Collision-based arbitrators do not handle congestion well.

7.11 FireWire

FireWire[20] is actually an Apple Computers Inc. trade name for the interface which is formally known as IEEE 1394–1995. It was originally intended as a digital audio network, but grew out of recognition. FireWire is more than just an interface as it can be used to form networks and if used with a computer effectively extends the computer's data bus. Devices are simply connected together as any combination of daisy-chain or star network.

Any pair of devices can communicate in either direction, and arbitration ensures that only one device transmits at once. Intermediate devices simply pass on transmissions. This can continue even if the intermediate device is powered down as the FireWire carries power to keep repeater functions active.

Communications are divided into *cycles* which have a period of 125 s. During a cycle, there are 64 time slots. During each time slot, any one node can communicate with any other, but in the next slot, a different pair of nodes may communicate. Thus FireWire is best described as a time-division multiplexed (TDM) system. There will be a new arbitration between the nodes for each cycle.

FireWire is eminently suitable for audio/computer convergent applications because it can simultaneously support asynchronous transfers of non-real-time computer data and isochronous transfers of real-time audio/video data. It can do this because the arbitration process allocates a fixed proportion of slots for isochronous data (about 80 per cent) and these have a higher priority in the arbitration than the asynchronous data. The higher the data rate a given node needs, the more time slots it will be allocated. Thus a given bit rate can be guaranteed throughout a transaction; a prerequisite of real-time A/V data transfer.

It is the sophistication of the arbitration system which makes FireWire remarkable. Some of the arbitration is in hardware at each node, but some is in software which only needs to be at one node. The full functionality requires a computer somewhere in the system which runs the isochronous bus management arbitration. Without this only asynchronous transfers are possible. It is possible to add or remove devices whilst the system is working. When a device is added the system will recognize it through a periodic learning process. Essentially every node on the system transmits in turn so that the structure becomes clear.

The electrical interface of FireWire is shown in Figure 7.25. It consists of two twisted pairs for signalling and a pair of power conductors. The twisted pairs carry differential signals of about 220 mV swinging around a common mode voltage of about 1.9 V with an impedance of 112 W. Figure 7.26 shows how the data are transmitted. The host data are simply serialized and used to modulate

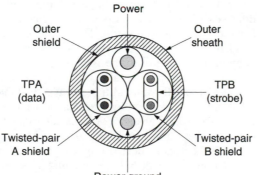

Figure 7.25 Fire Wire uses twin twisted pairs and a power pair.

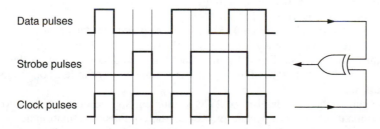

Figure 7.26 The strobe signal is the X-OR of the data and the bit clock. The data and strobe signals together form a self-clocking system.

twisted pair A. The other twisted pair (B) carries a signal called *strobe*, which is the exclusive-OR of the data and the clock. Thus whenever a run of identical bits results in no transitions in the data, the strobe signal will carry transitions. At the receiver another exclusive-OR gate adds data and strobe to recreate the clock.

This signalling technique is subject to skew between the two twisted pairs and this limits cable lengths to about 10 metres between nodes. Thus FireWire is not a long-distance interface technique, instead it is very useful for interconnecting a large number of devices in close proximity. Using a copper interconnect, FireWire can run at 100, 200 or 400 Mbits/s, depending on the specific hardware. It is proposed to create an optical fibre version which would run at gigabit speeds.

7.12 Broadband networks and ATM

Broadband ISDN (B-ISDN) is the successor to N-ISDN and in addition to offering more bandwidth, offers practical solutions to the delivery of any conceivable type of data. The flexibility with which ATM operates means that intermittent or one-off data transactions which only require asynchronous delivery can take place alongside isochronous MPEG video delivery. This is known as *application independence* whereby the sophistication of isochronous delivery does not raise the cost of asynchronous data. In this way, generic data, video, speech and combinations of the above can co-exist.

ATM is multiplexed, but it is not time-division multiplexed. TDM is inefficient because if a transaction does not fill its allotted bandwidth, the capacity is wasted. ATM does not offer fixed blocks of bandwidth, but allows infinitely variable bandwidth to each transaction. This is done by converting all host data into small fixed-size cells at the adaptation layer. The greater the bandwidth needed by a transaction, the more cells per second are allocated to that transaction. This approach is superior to the fixed bandwidth approach, because if the bit rate of a particular transaction falls, the cells released can be used for other transactions so that the full bandwidth is always available.

As all cells are identical in size, a multiplexer can assemble cells from many transactions in an arbitrary order. The exact order is determined by the quality of service required, where the time positioning of isochronous data would be determined first, with asynchronous data filling the gaps.

Figure 7.27 shows how a broadband system might be implemented. The transport network would typically be optical fibre based, using SONET (synchronous optical network) or SDH (synchronous digital hierarchy). These standards differ in minor respects. Figure 7.28 shows the bit rates available in each. Lower bit rates will be used in the access networks which will use different technology such as xDSL.

SONET and SDH assemble ATM cells into a structure known as a *container* in the interests of efficiency. Containers are passed intact between exchanges in the transport network. The cells in a container need not belong to the same transaction, they simply need to be going the same way for at least one transport network leg.

The cell-routing mechanism of ATM is unusual and deserves explanation. In conventional networks, a packet must carry the complete destination address so that at every exchange it can be routed closer to its destination. The exact route by which the packet travels cannot be anticipated and successive packets in the

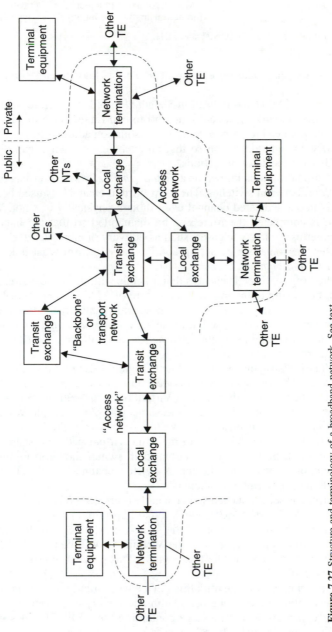

Figure 7.27 Structure and terminology of a broadband network. See text.

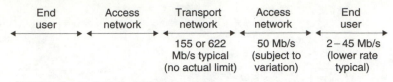

Figure 7.28 Bit rates available in SONET and SDH.

same transaction may take different routes. This is known as a *connectionless* protocol.

In contrast, ATM is a *connection oriented* protocol. Before data can be transferred, the network must set up an end-to-end route. Once this is done, the ATM cells do not need to carry a complete destination address. Instead they only need to carry enough addressing so that an exchange or switch can distinguish between all the expected transactions.

The end-to-end route is known as a *virtual channel* which consists of a series of *virtual links* between switches. The term 'virtual channel' is used because the system acts like a dedicated channel even though physically it is not. When the transaction is completed the route can be dismantled so that the bandwidth is freed for other users. In some cases, such as delivery of a broadcaster's output to a transmitter, the route can be set up continuously to form what is known as a *permanent virtual channel*.

The addressing in the cells ensures that all cells with the same address take the same path, but owing to the multiplexed nature of ATM, at other times and with other cells a completely different routing scheme may exist. Thus the routing structure for a particular transaction always passes cells by the same route, but the next cell may belong to another transaction and will have a different address causing it to routed in another way.

The addressing structure is hierarchical. Figure 7.29(a) shows the ATM cell and its header. The cell address is divided into two fields, the virtual channel identifier and the virtual path identifier. Virtual paths are logical groups of virtual channels which happen to be going the same way. An example would be the output of a video-on-demand server travelling to the first switch. The virtual path concept is useful because all cells in the same virtual path can share the same container in a transport network. A virtual path switch shown in Figure 7.29(b) can operate at the container level whereas a virtual channel switch (c) would need to dismantle and reassemble containers.

When a route is set up, at each switch a table is created. When a cell is received at a switch the VPI and/or VCI code is looked up in the table and used for two purposes. First, the configuration of the switch is obtained, so that this switch will correctly route the cell, second, the VPI and/or VCI codes may be updated so that they correctly control the next switch. This process repeats until the cell arrives at its destination.

In order to set up a path, the initiating device will initially send cells containing an ATM destination address, the bandwidth and quality of service required. The first switch will reply with a message containing the VPI/VCI codes which are to be used for this channel. The message from the initiator will propagate to the destination, creating look-up tables in each switch. At each switch the logic will add the requested bandwidth to the existing bandwidth in use to check that the requested quality of service can be met. If this succeeds for the whole channel,

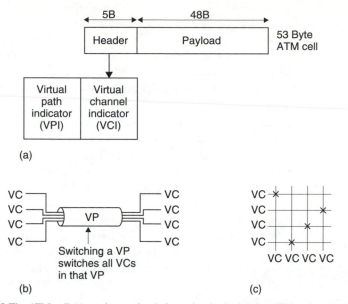

Figure 7.29 The ATM cell (a) carries routing information in the header. ATM paths carrying a group of channels can be switched in a virtual path switch (b). Individual channel switching requires a virtual channel switch which is more complex and causes more delay.

the destination will reply with a connect message which propagates back to the initiating device as confirmation that the channel has been set up. The connect message contains an unique call reference value which identifies this transaction. This is necessary because an initiator such a file server may be initiating many channels and the connect messages will not necessarily return in the same order as the set-up messages were sent.

The last switch will confirm receipt of the connect message to the destination and the initiating device will confirm receipt of the connect message to the first switch.

ATM works by dividing all real data messages into cells of 48 bytes each. At the receiving end, the original message must be recreated. This can take many forms. Figure 7.30 shows some possibilities. The message may be a generic data file having no implied timing structure. The message may be a serial bitstream with a fixed clock frequency, known as UDT (unstructured data transfer). It may be a burst of data bytes from a TDM system.

The application layer in ATM has two sub-layers shown in Figure 7.31. The first is the segmentation and reassembly (SAR) sublayer which must divide the message into cells and rebuild it to get the binary data right. The second is the convergence sublayer (CS) which recovers the timing structure of the original message. It is this feature which makes ATM so appropriate for delivery of audio/ visual material. Conventional networks such as the Internet don't have this ability.

In order to deliver a particular quality of service, the adaptation layer and the ATM layer work together. Effectively the adaptation layer will place constraints on the ATM layer, such as cell delay, and the ATM layer will meet those

Generic data file having no timebase
Constant bit rate serial data stream
Audio/video data requiring a timebase
Compressed A/V data with fixed bit rate
Compressed A/V data with variable bit rate

Figure 7.30 Types of data which may need adapting to ATM.

ATM application layer	Convergence sublayer	Recovers timing of original data
	Segmentation and reassembly	Divides data into cells for transport Reassembles original data format

Figure 7.31 ATM adaption layer has two sublayers, segmentation and convergence.

constraints without needing to know why. Provided the constraints are met, the adaptation layer can rebuild the message. The variety of message types and timing constraints leads to the adaptation layer having a variety of forms.

The adaptation layers which are most relevant to MPEG applications are AAL-1 and AAL-5. AAL-1 is suitable for transmitting MPEG-2 multi-program transport streams at constant bit rate and is standardized for this purpose in ETS 300814 for DVB application. AAL-1 has an integral forward error correction (FEC) scheme. AAL-5 is optimized for single-program transport streams (SPTS) at a variable bit rate and has no FEC.

AAL-1 takes as an input the 188-byte transport stream packets which are created by a standard MPEG-2 multiplexer. The transport stream bit rate must be constant but it does not matter if statistical multiplexing has been used within the transport stream.

The Reed–Solomon FEC of AAL-1 uses a codeword of size 128 so that the codewords consist of 124 bytes of data and 4 bytes of redundancy, making 128 bytes in all. Thirty-one 188-byte TS packets are restructured into this format. The 256-byte codewords are then subject to a block interleave. Figure 7.32 shows that 47 such codewords are assembled in rows in RAM and then columns are read out. These columns are 47 bytes long and, with the addition of an AAL header byte make up a 48-byte ATM packet payload. In this way the interleave block is transmitted in 128 ATM cells.

The result of the FEC and interleave is that the loss of up to four cells in 128 can be corrected, or a random error of up to two bytes can be corrected in each cell. This FEC system allows most errors in the ATM layer to be corrected so that no retransmissions are needed. This is important for isochronous operation. The AAL header has a number of functions. One of these is to identify the first ATM

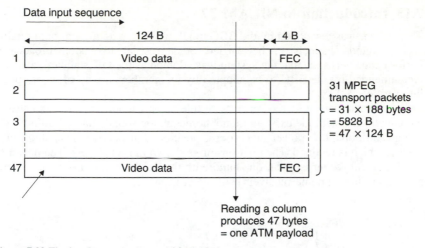

Figure 7.32 The interleave structure used in AAL-1.

cell in the interleave block of 128 cells. Another function is to run a modulo-8 cell counter to detect missing or out-of sequence ATM cells. If a cell simply fails to arrive, the sequence jump can be detected and used to flag the FEC system so that it can correct the missing cell by erasure (see section 6.21). In a manner similar to the use of program clock reference (PCR) in MPEG, AAL-1 embeds a timing code in ATM cell headers. This is called the synchronous residual time stamp (SRTS) and in conjunction with the ATM network clock allows the receiving AAL device to reconstruct the original data bit rate. This is important because in MPEG applications it prevents the PCR jitter specification being exceeded.

In AAL-5 there is no error correction and the adaptation layer simply reformats MPEG TS blocks into ATM cells. Figure 7.33 shows one way in which this can be done. Two TS blocks of 188 bytes are associated with an 8-byte trailer known as CPCS (common part convergence sublayer). The presence of the trailer makes a total of 384 bytes which can be carried in eight ATM cells. AAL-5 does not offer constant delay and external buffering will be required, controlled by reading the MPEG PCRs in order to reconstruct the original time axis.

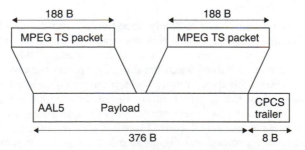

Figure 7.33 The AAL-5 adaptation layer can pack MPEG transport packets in this way.

7.13 Introduction to NICAM 728

This system was developed by the BBC to allow the two additional high-quality digital sound channels to be carried on terrestrial television broadcasts. Performance was such that the system was adopted as the UK standard, and was recommended by the EBU to be adopted by its members, many of whom put it into service.[21]

The introduction of stereo sound with television cannot be at the expense of incompatibility with the existing monophonic analog sound channel. In NICAM 728 an additional low-power subcarrier is positioned just above the analog sound carrier, which is retained. The relationship is shown in Figure 7.34. The power of the digital subcarrier is about one hundredth that of the main vision carrier, and so existing monophonic receivers will reject it.

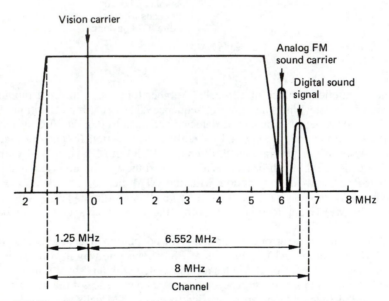

Figure 7.34 The additional carrier needed for digital stereo sound is squeezed in between television channels as shown here. The digital carrier is of much lower power than the analog signals, and is randomized prior to transmission so that it has a broad, low-level spectrum which is less visible on the picture.

Since the digital carrier is effectively shoe-horned into the gap between TV channels, it is necessary to ensure that the spectral width of the intruder is minimized to prevent interference. As a further measure, the power of the existing audio carrier is halved when the digital carrier is present.

Figure 7.35 shows the stages through which the audio must pass. The audio sampling rate used is 32 kHz which offers similar bandwidth to that of an FM stereo radio broadcast. Samples are originally quantized to fourteen-bit resolution in two's complement code. From an analog source this causes no problem, but from a professional digital source having longer wordlength and higher sampling rate it would be necessary to pass through a rate convertor, a

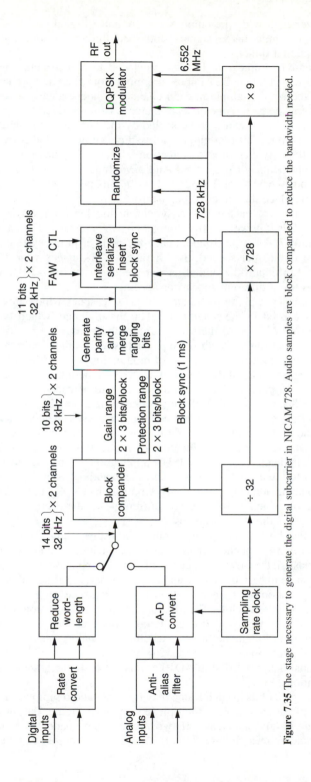

Figure 7.35 The stage necessary to generate the digital subcarrier in NICAM 728. Audio samples are block companded to reduce the bandwidth needed.

digital equalizer to provide pre-emphasis, an optional digital compressor in the case of wide dynamic range signals and then through a truncation circuit incorporating digital dither.

The fourteen-bit samples are block companded to reduce data rate. During each one millisecond block, 32 samples are input from each audio channel. The magnitude of the largest sample in each channel is independently assessed, and used to determine the gain range or scale factor to be used. Every sample in each channel in a given block will then be scaled by the same amount and truncated to ten bits. An eleventh bit present on each sample combines the scale factor of the channel with parity bits for error detection. The encoding process is described as a Near Instantaneously Companded Audio Multiplex, NICAM for short. The resultant data now consists of $2 \times 32 \times 11 = 704$ bits per block. Bit interleaving is employed to reduce the effect of burst errors.

At the beginning of each block a synchronizing byte, known as a Frame Alignment Word, is followed by five control bits and eleven additional data bits, making a total of 728 bits per frame, hence the number in the system name. As there are 1000 frames per second, the bit rate is 728 kbits/s. In the UK this is multiplied by 9 to obtain the digital carrier frequency of 6.552 MHz but some other countries use a different subcarrier spacing.

The digital carrier is phase modulated. It has four states which are 90° apart. Information is carried in the magnitude of a phase change which takes place every 18 cycles, or 2.74 μs. As there are four possible phase changes, two bits are conveyed in every change. The absolute phase has no meaning, only the changes are interpreted by the receiver. This type of modulation is known as differentially encoded quadrature phase shift keying (DQPSK), sometimes called four-phase DPSK. In order to provide consistent timing and to spread the carrier energy throughout the band irrespective of audio content, randomizing is used, except during the frame alignment word. On reception, the FAW is detected and used to synchronize the pseudo-random generator to restore the original data.

Figure 7.36 shows the general structure of a frame. Following the sync pattern or FAW is the application control field. The application control bits determine the significance of following data, which can be stereo audio, two independent mono signals, mono audio and data or data only. Control bits C_1, C_2 and C_3 have eight combinations, of which only four are currently standardized. Receivers are designed to mute audio if C_3 becomes 1.

The frame flag bit C_0 spends eight frames high then eight frames low in an endless sixteen-frame sequence which is used to synchronize changes in channel usage. In the last sixteen-frame sequence of the old application, the application control bits change to herald the new application, whereas the actual data change to the new application on the next sixteen-frame sequence.

The reserve sound switching flag, C_4, is set to 1 if the analog sound being broadcast is derived from the digital stereo. This fact can be stored by the receiver and used to initiate automatic switching to analog sound in the case of loss of the digital channels.

The additional data bits AD0 to AD10 are as yet undefined, and reserved for future applications.

The remaining 704 bits in each frame may be either audio samples or data. The two channels of stereo audio are multiplexed into each frame, but multiplexing does not occur in any other case. If two mono audio channels are sent, they occupy alternate frames. Figure 7.36(a) shows a stereo frame, where the A

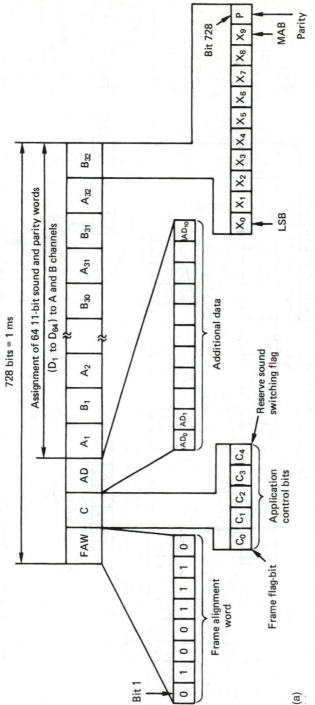

(a)

Figure 7.36 In (a) the block structure of a stereo signal multiplexes samples from both channels (A and B) into one block.

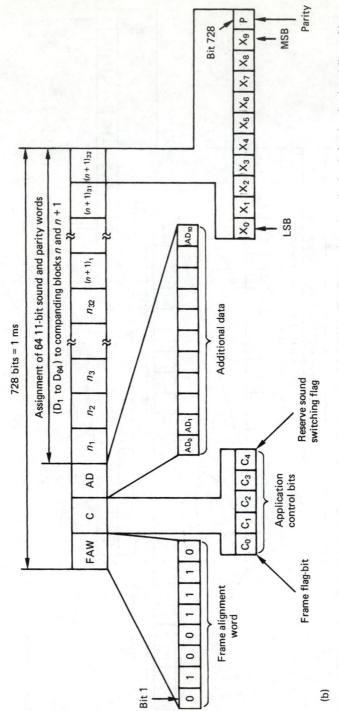

Figure 7.36 In mono, shown in (b), samples from one channel only occupy a given block. The diagrams here show the data before interleaving. Adjacent bits shown here actually appear at least sixteen bits apart in the data stream.

channel is carried in odd-numbered samples, whereas Figure 7.36(b) shows a mono frame, where the M1 channel is carried in odd-numbered frames.

7.14 Audio in digital television broadcasting

Digital television broadcasting relies on the combination of a number of fundamental technologies. These are: compression to reduce the bit rate, multiplexing to combine picture and sound data into a common bitstream, digital modulation schemes to reduce the RF bandwidth needed by a given bit rate and error correction to reduce the error statistics of the channel down to a value acceptable to MPEG data.

Compressed video and audio are both highly sensitive to bit errors, primarily because they confuse the recognition of variable-length codes so that the decoder loses synchronization. However, MPEG is a compression and multiplexing standard and does not specify how error correction should be performed. Consequently a transmission standard must define a system which has to correct essentially all errors such that the delivery mechanism is transparent.

Essentially a transmission standard specifies all the additional steps needed to deliver an MPEG transport stream from one place to another. This transport stream will consist of a number of elementary streams of video and audio, where the audio may be coded according to MPEG audio standard or AC-3. In a system working within its capabilities, the picture and sound quality will be determined only by the performance of the compression system and not by the RF transmission channel. This is the fundamental difference between analog and digital broadcasting.

Whilst in one sense an MPEG transport stream is only data, it differs from generic data in that it must be presented to the viewer at a particular rate. Generic data are usually asynchronous, whereas baseband video and audio are synchronous. However, after compression and multi-plexing audio and video are no longer precisely synchronous and so the term *isochronous* is used. This means a signal which was at one time synchronous and will be displayed synchronously, but which uses buffering at transmitter and receiver to accommodate moderate timing errors in the transmission.

Clearly another mechanism is needed so that the time axis of the original signal can be recreated on reception. The time stamp and program clock reference system of MPEG does this.

Figure 7.37 shows that the concepts involved in digital television broadcasting exist at various levels which have an independence not found in analog technology. In a given configuration a transmitter can radiate a given payload data bit rate. This represents the useful bit rate and does not include the necessary overheads needed by error correction, multiplexing or synchronizing. It is fundamental that the transmission system does not care what this payload bit rate is used for. The entire capacity may be used up by one high-definition channel, or a large number of heavily compressed channels may be carried. The details of this data usage are the domain of the *transport stream*. The multiplexing of transport streams is defined by the MPEG standards, but these do not define any error correction or transmission technique.

At the lowest level in Figure 7.38 the source coding scheme, in this case MPEG compression, results in one or more elementary streams, each of which carries a video or audio channel. Elementary streams are multiplexed into a

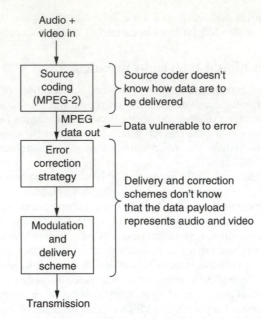

Figure 7.37 Source coder doesn't know delivery mechanism and delivery mechanism doesn't need to know what the data mean.

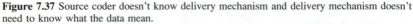

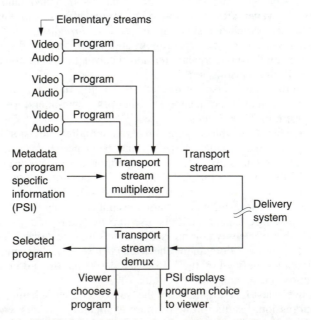

Figure 7.38 Program Specific Information helps the demultiplexer to select the required program.

transport stream. The viewer then selects the desired elementary stream from the transport stream. Metadata in the transport stream ensures that when a video elementary stream is chosen, the appropriate audio elementary stream will automatically be selected.

7.15 Packets and time stamps

The video elementary stream is an endless bitstream representing pictures which take a variable length of time to transmit. Bidirection coding means that pictures are not necessarily in the correct order. Storage and transmission systems prefer discrete blocks of data and so elementary streams are packetized to form a PES (packetized elementary stream). Audio elementary streams are also packetized. A packet is shown in Figure 7.39. It begins with a header containing an unique packet start code and a code which identifies the type of data stream. Optionally the packet header also may contain one or more *time stamps* which are used for synchronizing the video decoder to real time and for obtaining lip-sync.

Figure 7.40 shows that a time stamp is a sample of the state of a counter which is driven by a 90 kHz clock. This is obtained by dividing down the master 27 MHz clock of MPEG-2. This 27 MHz clock must be locked to the video frame rate and the audio sampling rate of the program concerned. There are two types of time stamp: PTS and DTS. These are abbreviations for presentation time stamp and decode time stamp. A presentation time stamp determines when the associated picture should be displayed on the screen, whereas a decode time stamp determines when it should be decoded. In bidirectional coding these times can be quite different.

Audio packets only have presentation time stamps. Clearly if lip-sync is to be obtained, the audio sampling rate of a given program must have been locked to the same master 27 MHz clock as the video and the time stamps must have come from the same counter driven by that clock.

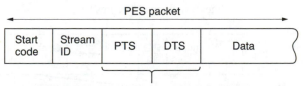

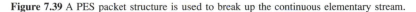

Figure 7.39 A PES packet structure is used to break up the continuous elementary stream.

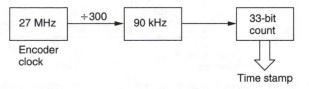

Figure 7.40 Time stamps are the result of sampling a counter driven by the encoder clock.

In practice the time between input pictures is constant and so there is a certain amount of redundancy in the time stamps. Consequently PTS/DTS need not appear in every PES packet. Time stamps can be up to 100 ms apart in transport streams. As each picture type (I, P or B) is flagged in the bitstream, the decoder can infer the PTS/DTS for every picture from the ones actually transmitted.

7.16 MPEG transport streams

The MPEG-2 transport stream is intended to be a multiplex of many TV programs with their associated sound and data channels, although a single program transport stream (SPTS) is possible. The transport stream is based upon packets of constant size so that multiplexing, adding error-correction codes and interleaving in a higher layer is eased. Figure 7.41 shows that these are always 188 bytes long.

Transport stream packets always begin with a header. The remainder of the packet carries data known as the payload. For efficiency, the normal header is relatively small, but for special purposes the header may be extended. In this case

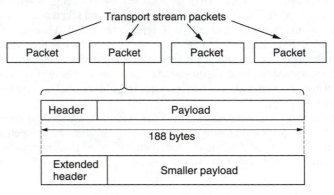

Packet is always 188 bytes long

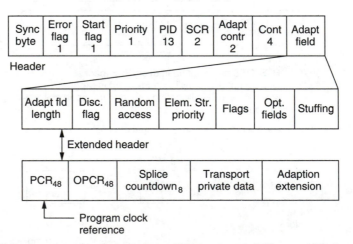

Figure 7.41 Transport stream packets are always 188 bytes long to facilitate multiplexing and error correction.

the payload gets smaller so that the overall size of the packet is unchanged. Transport stream packets should not be confused with PES packets which are larger and which vary in size. PES packets are broken up to form the payload of the transport stream packets.

The header begins with a sync byte which is an unique pattern detected by a demultiplexer. A transport stream may contain many different elementary streams and these are identified by giving each an unique thirteen-bit packet identification code or PID which is included in the header. A multiplexer seeking a particular elementary stream simply checks the PID of every packet and accepts only those which match.

In a multiplex there may be many packets from other programs in between packets of a given PID. To help the demultiplexer, the packet header contains a continuity count. This is a four-bit value which increments at each new packet having a given PID.

This approach allows statistical multiplexing as it does not matter how many or how few packets have a given PID; the demux will still find them. Statistical multiplexing has the problem that it is virtually impossible to make the sum of the input bit rates constant. Instead the multiplexer aims to make the average data bit rate slightly less than the maximum and the overall bit rate is kept constant by adding 'stuffing' or null packets. These packets have no meaning, but simply keep the bit rate constant. Null packets always have a PID of 8191 (all ones) and the demultiplexer discards them.

7.17 Clock references

A transport stream is a multiplex of several TV programs and these may have originated from widely different locations. It is impractical to expect all the programs in a transport stream to be genlocked and so the stream is designed from the outset to allow unlocked programs. A decoder running from a transport stream has to genlock to the encoder and the transport stream has to have a mechanism to allow this to be done independently for each program. The synchronizing mechanism is called program clock reference (PCR).

Figure 7.42 shows how the PCR system works. The goal is to recreate at the decoder a 27 MHz clock which is synchronous with that at the encoder. The encoder clock drives a 48-bit counter which continuously counts up to the maximum value before overflowing and beginning again.

A transport stream multiplexer will periodically sample the counter and place the state of the count in an extended packet header as a PCR. The demultiplexer selects only the PIDs of the required program, and it will extract the PCRs from the packets in which they were inserted.

The PCR codes are used to control a numerically locked loop (NLL) described in section 3.17. The NLL contains a 27 MHz VCXO (voltage-controlled crystal oscillator). This is a variable-frequency oscillator based on a crystal which has a relatively small frequency range.

The VCXO drives a 48-bit counter in the same way as in the encoder. The state of the counter is compared with the contents of the PCR and the difference is used to modify the VCXO frequency. When the loop reaches lock, the decoder counter would arrive at the same value as is contained in the PCR and no change in the VCXO would then occur. In practice the transport stream packets will suffer from transmission jitter and this will create phase noise in the loop. This

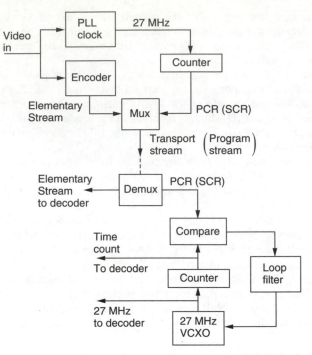

Figure 7.42 Program or System Clock Reference codes regenerate a clock at the decoder. See text for details.

is removed by the loop filter so that the VCXO effectively averages a large number of phase errors.

A heavily damped loop will reject jitter well, but will take a long time to lock. Lock-up time can be reduced when switching to a new program if the decoder counter is jammed to the value of the first PCR received in the new program. The loop filter may also have its time constants shortened during lock-up.

Once a synchronous 27 MHz clock is available at the decoder, this can be divided down to provide the 90 kHz clock which drives the time stamp mechanism.

The entire timebase stability of the decoder is no better than the stability of the clock derived from PCR. MPEG-2 sets standards for the maximum amount of jitter which can be present in PCRs in a real transport stream.

Clearly if the 27 MHz clock in the receiver is locked to one encoder it can only receive elementary streams encoded with that clock. If it is attempted to decode, for example, an audio stream generated from a different clock, the result will be periodic buffer overflows or underflows in the decoder. Thus MPEG defines a program in a manner which relates to timing. A program is a set of elementary streams which have been encoded with the same master clock.

7.18 Program Specific Information (PSI)

In a real transport stream, each elementary stream has a different PID, but the demultiplexer has to be told what these PIDs are and what audio belongs with

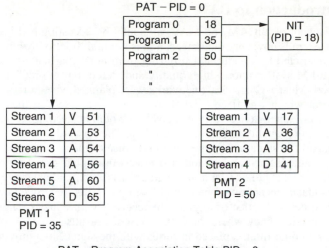

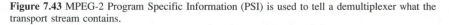

PAT – Program Association Table PID = 0
CAT – Conditional Access Table PID = 1
NIT – Network Information Table
Null packets – PID = 8191

Figure 7.43 MPEG-2 Program Specific Information (PSI) is used to tell a demultiplexer what the transport stream contains.

what video before it can operate. This is the function of PSI which is a form of metadata. Figure 7.43 shows the structure of PSI. When a decoder powers up, it knows nothing about the incoming transport stream except that it must search for all packets with a PID of zero. PID zero is reserved for the Program Association Table (PAT). The PAT is transmitted at regular intervals and contains a list of all the programs in this transport stream. Each program is further described by its own Program Map Table (PMT) and the PIDs of of the PMTs are contained in the PAT.

Figure 7.43 also shows that the PMTs fully describe each program. The PID of the video elementary stream is defined, along with the PID(s) of the associated audio and data streams. Consequently when the viewer selects a particular program, the demultiplexer looks up the program number in the PAT, finds the right PMT and reads the audio, video and data PIDs. It then selects elementary streams having these PIDs from the transport stream and routes them to the decoders.

Program 0 of the PAT contains the PID of the Network Information Table (NIT). This contains information about what other transport streams are available. For example, in the case of a satellite broadcast, the NIT would detail the orbital position, the polarization, carrier frequency and modulation scheme. Using the NIT a set-top box could automatically switch between transport streams.

Apart from 0 and 8191, a PID of 1 is also reserved for the Conditional Access Table (CAT). This is part of the access control mechanism needed to support pay per view or subscription viewing.

7.19 Introduction to DAB

Until the advent of NICAM and MAC, all sound broadcasting had been analog. The AM system is now very old indeed, and is not high fidelity by any standards, having a restricted bandwidth and suffering from noise, particularly at night. In theory, the FM system allows high quality and stereo, but in practice things are not so good. Most FM broadcast networks were planned when a radio set was a sizeable unit which was fixed. Signal strengths were based on the assumption that a fixed FM antenna in an elevated position would be used. If such an antenna is used, reception quality is generally excellent. The forward gain of a directional antenna raises the signal above the front-end noise of the receiver and noise-free stereo is obtained. Such an antenna also rejects unwanted signal reflections.

Reception on car radios is at a disadvantage as directional antennae cannot be used. This makes reception prone to multipath problems. Figure 7.44 shows that when the direct and reflected signals are received with equal strength, nulling occurs at any frequency where the path difference results in a 180° phase shift. Effectively a comb filter is placed in series with the signal. In a moving vehicle, the path lengths change, and the comb response slides up and down the band. When a null passes through the station tuned in, a burst of noise is created. Reflections from aircraft can cause the same problem in fixed receivers.

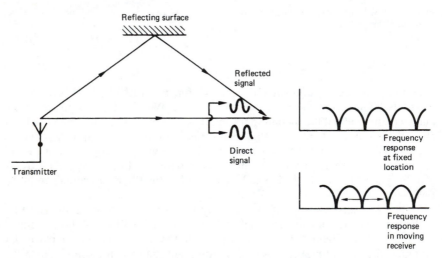

Figure 7.44 Multipath reception. When the direct and reflected signals are received with equal strength, nulling occurs at any frequency where the path difference results in a 180° phase shift.

Digital audio broadcasting (DAB), also known as digital radio, is designed to overcome the problems which beset FM radio, particularly in vehicles. Not only does it do that, it does so using less bandwidth. With increasing pressure for spectrum allocation from other services, a system using less bandwidth to give better quality is likely to be favourably received.

DAB relies on a number of fundamental technologies which are combined into an elegant system. Compression is employed to cut the required bandwidth. Transmission of digital data is inherently robust as the receiver has only to decide between a small number of possible states.

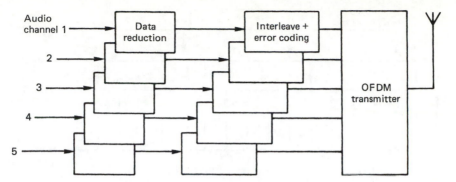

Figure 7.45 Block diagram of a DAB transmitter. See text for details.

Sophisticated modulation techniques help to eliminate multipath reception problems whilst further economizing on bandwidth. Error correction and concealment allow residual data corruption to be handled before conversion to analog at the receiver.

The system can only be realized with extremely complex logic in both transmitter and receiver, but with modern VLSI technology this can be inexpensive and reliable. In DAB, the concept of one-carrier-one-program is not used. Several programs share the same band of frequencies. Receivers will be easier to use since conventional tuning will be unnecessary. 'Tuning' consists of controlling the decoding process to select the desired program. Mobile receivers will automatically switch between transmitters as a journey proceeds.

Figure 7.45 shows the block diagram of a DAB transmitter. Incoming digital audio at 32 kHz is passed into the compression unit which uses the techniques described in Chapter 5 to cut the data rate to some fraction of the original. The compression unit could be at the studio end of the line to cut the cost of the link. The data for each channel are then protected against errors by the addition of redundancy. Convolutional codes described in Chapter 6 are attractive in the broadcast environment. Several such data-reduced sources are interleaved together and fed to the modulator, which may employ techniques such as randomizing which were also introduced in Chapter 6.

Figure 7.46 shows how the multiple carriers in a DAB band are allocated to different program channels on an interleaved basis. Using this technique, it will be evident that when a notch in the received spectrum occurs due to multipath cancellation this will damage a small proportion of all programs rather than a large part of one program. This is the spectral equivalent of physical interleaving on a recording medium. The result is the same in that error bursts are broken up according to the interleave structure into more manageable sizes which can be corrected with less redundancy.

A serial digital waveform has a $\sin x/x$ spectrum and when this waveform is used to phase modulate a carrier the result is a symmetrical $\sin x/x$ spectrum centred on the carrier frequency. Nulls in the spectrum appear at multiples of the phase switching rate away from the carrier. This distance is equal to 90° or one quadrant of $\sin x$. Further carriers can be placed at spacings such that each is centred at the nulls of the others. Owing to the quadrant spacing, these carries are mutually orthogonal, hence the term orthogonal frequency division.[22,23] A

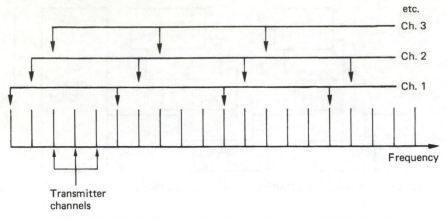

Figure 7.46 Channel interleaving is used in DAB to reduce the effect of multipath notches on a given program.

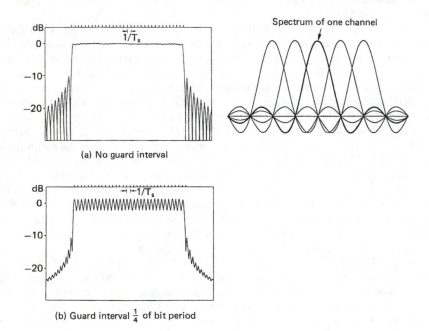

(a) No guard interval

Spectrum of one channel

(b) Guard interval $\frac{1}{4}$ of bit period

Figure 7.47 (a) When mutually orthogonal carriers are stacked in a band, the resultant spectrum is virtually flat. (b) When guard intervals are used, the spectrum contains a peak at each channel centre.

number of such carriers will interleave to produce an overall spectrum which is almost rectangular as shown in Figure 7.47(a). The mathematics describing this process is exactly the same as that of the reconstruction of samples in a low-pass filter. Effectively sampling theory has been transformed into the frequency domain.

In practice, perfect spectral interleaving does not give sufficient immunity from multipath reception. In the time domain, a typical reflective environment turns a transmitted pulse into a pulse train extending over several micro-seconds.[24] If the bit rate is too high, the reflections from a given bit coincide with later bits, destroying the orthogonality between carriers. Reflections are opposed by the use of guard intervals in which the phase of the carrier returns to an unmodulated state for a period which is greater than the period of the reflections. Then the reflections from one transmitted phase decay during the guard interval before the next phase is transmitted.[25] The principle is not dissimilar to the technique of spacing transitions in a recording further apart than the expected jitter. As expected, the use of guard intervals reduces the bit rate of the carrier because for some of the time it is radiating carrier not data. A typical reduction is to around 80 per cent of the capacity without guard intervals. This capacity reduction does, however, improve the error statistics dramatically, such that much less redundancy is required in the error correction system. Thus the effective transmission rate is improved. The use of guard intervals also moves more energy from the sidebands back to the carrier. The frequency spectrum of a set of carriers is no longer perfectly flat but contains a small peak at the centre of each carrier as shown in Figure 7.47(b).

A DAB receiver must receive the set of carriers corresponding to the required program channel. Owing to the close spacing of carriers, it can only do this by performing fast Fourier transforms (FFTs) on the DAB band. If the carriers of a given program are evenly spaced, a partial FFT can be used which only detects energy at spaced frequencies and requires much less computation. This is the DAB equivalent of tuning. The selected carriers are then demodulated and combined into a single bitstream. The error-correction codes will then be de-interleaved so that correction is possible. Corrected data then pass through the expansion part of the data reduction coder, resulting in conventional PCM audio which drives DACs.

It should be noted that in the European DVB standard, COFDM transmission is also used. In some respects, DAB is simply a form of DVB without the picture data.

References

1. Audio Engineering Society, AES recommended practice for digital audio engineering – serial transmission format for linearly represented digital audio data. *J. Audio Eng. Soc.*, **33**, 975–984 (1985)
2. EIAJ CP-340, *A Digital Audio Interface*, Tokyo: EIAJ (1987)
3. EIAJ CP-1201, *Digital Audio Interface (revised)*, Tokyo: EIAJ (1992)
4. EIA RS-422A. Electronic Industries Association, 2001 Eye St NW, Washington, DC 20006, USA
5. Smart, D.L., Transmission performance of digital audio serial interface on audio tie lines. *BBC Designs Dept Technical Memorandum*, 3.296/84
6. European Broadcasting Union, Specification of the digital audio interface. *EBU Doc. Tech.*, 3250
7. Rorden, B. and Graham, M., A proposal for integrating digital audio distribution into TV production. *J. SMPTE*, 606–608 (Sept.1992)
8. Gilchrist, N., Co-ordination signals in the professional digital audio interface. In *Proc. AES/EBU Interface Conf.*, 13–15. Burnham: Audio Engineering Society (1989)
9. AES18–1992, Format for the user data channel of the AES digital audio interface. *J. Audio Eng. Soc.*, **40** 167–183 (1992)

10. Nunn, J.P., Ancillary data in the AES/EBU digital audio interface. In *Proc. 1st NAB Radio Montreux Symp.*, 29–41 (1992)
11. Komly, A and Viallevieille, A., Programme labelling in the user channel. In *Proc. AES/ EBU Interface Conf.*, 28–51. Burnham: Audio Engineering Society (1989)
12. ISO 3309, *Information processing systems – data communications – high level data link frame structure* (1984)
13. AES10–1991, Serial multi-channel audio digital interface (MADI). *J. Audio Eng. Soc.*, **39**, 369–377 (1991)
14. Ajemian, R.G. and Grundy, A.B., Fiber-optics – the new medium for audio: a tutorial. *J.Audio Eng. Soc.*, **38** 160–175 (1990)
15. Lidbetter, P.S. and Douglas, S., A fibre-optic multichannel communication link developed for remote interconnection in a digital audio console. Presented at the 80th Audio Engineering Society Convention (Montreux, 1986), Preprint 2330
16. Dunn, J., Considerations for interfacing digital audio equipment to the standards AES3, AES5 and AES11. In *Proc. AES 10th International Conf.*, 122, New York: Audio Engineering Society (1991)
17. Gilchrist, N.H.C., Digital sound: sampling-rate synchronization by variable delay. *BBC Research Dept Report*, 1979/17
18. Lagadec, R., A new approach to sampling rate synchronisation. Presented at the 76th Audio Engineering Society Convention (New York, 1984), Preprint 2168
19. Shelton, W.T., Progress towards a system of synchronization in a digital studio. Presented at the 82nd Audio Engineering Society Convention (London, 1986), Preprint 2484(K7)
20. Wicklegren, I.J., The facts about FireWire. *IEEE Spectrum*, 19–25 (1997)
21. Anon., NICAM 728: specification for two additional digital sound channels with System I television. BBC Engineering Information Dept (London, 1988)
22. Cimini, L.J., Analysis and simulation of a digital mobile channel using orthogonal frequency division multiplexing. *IEEE Trans. Commun.*, **COM-33**, No.7 (1985)
23. Pommier, D. and Wu, Y., Interleaving or spectrum spreading in digital radio intended for vehicles. *EBU Tech. Review*, No. 217, 128–142 (1986)
24. Cox, D.C., Multipath delay spread and path loss correlation for 910 MHz urban mobile radio propagation. *IEEE Trans. Vehic. Tech.*, **VT-26** (1977)
25. Alard, M. and Lasalle, R., Principles of modulation and channel coding for digital broadcasting for mobile receivers. *EBU Tech. Review*, No. 224, 168–190 (1987)

Digital audio tape recorders

Tape recording is divided into stationary head and rotary-head recorders. In this chapter the two systems will be contrasted and illustrated with examples from actual formats. The reader is referred to Chapter 6 for an explanation of coding and error-correction principles.

8.1 Rotary versus stationary heads

The high bit rate required for digital audio can be recorded in two ways. The head can remain fixed, and the tape can be transported rapidly, or the tape can travel relatively slowly, and the head can be moved. The latter is the principle of the rotary-head recorder.

In helical-scan recorders, shown in Figure 8.1, the tape is wrapped around the drum in such a way that it enters and leaves in two different planes. This causes the rotating heads to record long slanting tracks where the width of the space between tracks is determined by the linear tape speed. The track pitch can easily be made much smaller than in stationary-head recorders.

8.2 PCM adaptors

If digital sample data are encoded to resemble a video waveform, which is known as pseudo-video or composite digital, they can be recorded on a fairly standard video recorder. The device needed to format the samples in this way is called a PCM adaptor. PCM adaptors were popular before high-density digital recording developed. Today they are obsolete.

Figure 8.2 shows a block diagram of a PCM adaptor. The unit has five main sections. Central to operation is the sync and timing generation, which produces sync pulses for control of the video waveform generator and locking the video recorder, in addition to producing sampling-rate clocks and timecode. An ADC allows a conventional analog audio signal to be recorded, but this can be bypassed if a suitable digital input is available. Similarly a DAC is provided to monitor recordings, and this too can be bypassed by using the direct digital output. Also visible in Figure 8.2 are the encoder and decoder stages which convert between digital sample data and the pseudo-video signal.

A typical line of pseudo-video is shown in Figure 8.3. The line is divided into bit cells and, within them, black level represents a binary zero, and about 60 per

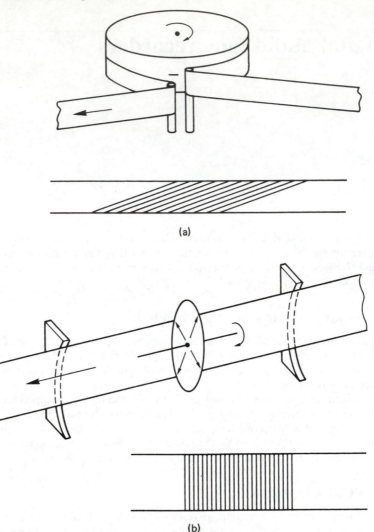

Figure 8.1 Types of rotary-head recorder. (a) Helical scan records long diagonal tracks. (b) Transverse scan records short tracks across the tape.

cent of peak white represents binary one. The use of a two-level input to a frequency modulator means that the recording is essentially frequency-shift keyed (FSK).

As the video recorder is designed to switch heads during the vertical interval, no samples can be recorded there. Time compression is used to squeeze the samples into the active parts of unblanked lines.

8.3 Introduction to DAT

DAT (digital audio tape) was the first digital recorder for consumer use to incorporate a dedicated tape deck. By designing for a specific purpose, the tape

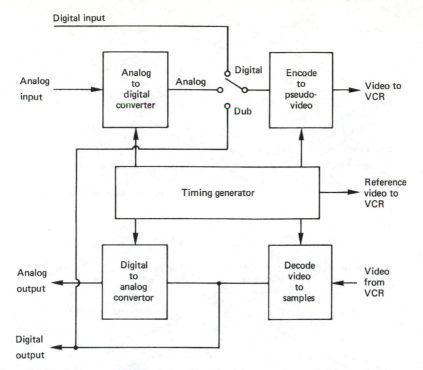

Figure 8.2 Block diagram of PCM adaptor. Note the dub connection needed for producing a digital copy between two VCRs.

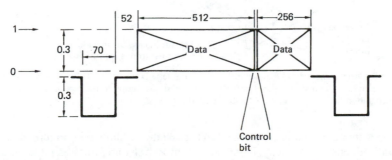

Figure 8.3 Typical line of video from PCM-1610. The control bit conveys the setting of the pre-emphasis switch or the sampling rate depending on position in the frame. The bits are separated using only the timing information in the sync pulses.

consumption can be made very much smaller than that of a converted video machine. In fact the DAT format achieved more bits per square inch than any other form of magnetic recorder at the time of its introduction. The origins of DAT are in an experimental machine built by Sony,[1] but the DAT format has grown out of that through a process of standardization involving some eighty companies.

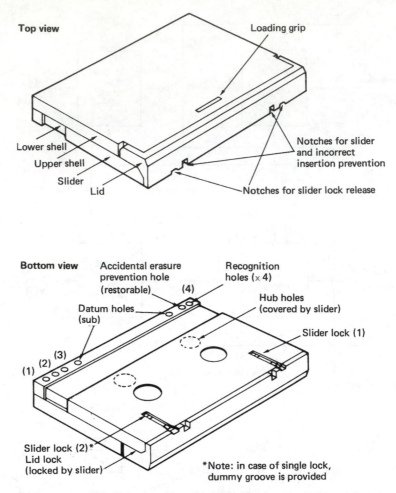

Figure 8.4 Appearance of DAT cassette. Access to the tape is via a hinged lid, and the hub-drive holes are covered by a sliding panel, affording maximum protection to the tape. Further details of the recognition holes are given in Table 8.1. (Courtesy TDK)

The general appearance of the DAT cassette is shown in Figure 8.4. The overall dimensions are only 73 mm × 54 mm × 10.5 mm which is rather smaller than the Compact Cassette. The design of the cassette incorporates some improvements over its analog ancestor,[2] As shown in Figure 8.5, the apertures through which the heads access the tape are closed by a hinged door, and the hub drive openings are covered by a sliding panel which also locks the door when the cassette is not in the transport. The act of closing the door operates brakes that act on the reel hubs. This results in a cassette that is well sealed against contamination due to handling or storage. The short wavelengths used in digital recording make it more sensitive to spacing loss caused by contamination. As in the Compact Cassette, the tape hubs are flangeless, and the edge guidance of the tape pack is achieved by the use of liner sheets. The flangeless approach allows the hub centres to be closer together for a given length of tape.

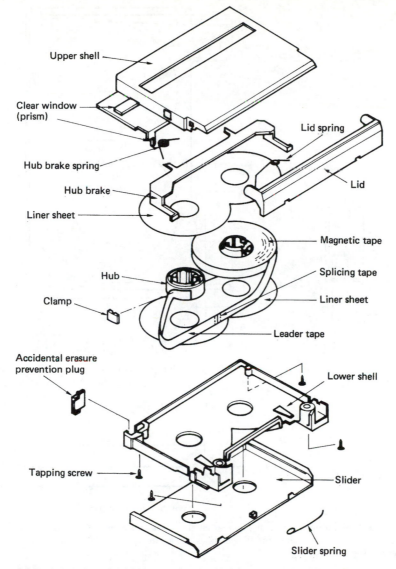

Upper shell

Clear window (prism)

Lid spring

Hub brake spring

Lid

Hub brake

Liner sheet

Magnetic tape

Hub

Splicing tape

Clamp

Liner sheet

Leader tape

Accidental erasure prevention plug

Lower shell

Tapping screw

Slider

Slider spring

Figure 8.5 Exploded view of DAT cassette showing intricate construction. When the lid opens, it pulls the ears on the brake plate, releasing the hubs. Note the EOT/BOT sensor prism moulded into the corners of the clear window. (Courtesy TDK)

The cassette has recognition holes in four standard places so that players can automatically determine what type of cassette has been inserted. In addition there is a write-protect (record-lockout) mechanism which is actuated by a small plastic plug sliding between the cassette halves. The end-of-tape condition is detected optically and the leader tape is transparent. There is some freedom in the design of the EOT sensor. As can be seen in Figure 8.6, transmitted-light sensing can be used across the corner of the cassette, or reflected-light sensing can be used, because the

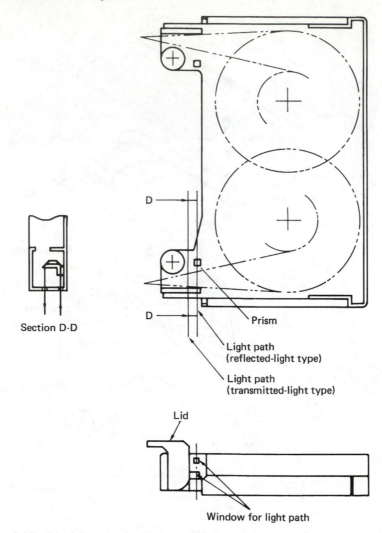

Section D-D

D

Prism

Light path
(reflected-light type)

Light path
(transmitted-light type)

Lid

Window for light path

Figure 8.6 Tape sensing can be either by transmission across the corner of the cassette, or by reflection through an integral prism. In both cases, the apertures are sealed when the lid closes. (Courtesy TDK)

cassette incorporates a prism which reflects light around the back of the tape. Study of Figure 8.6 will reveal that the prisms are moulded integrally with the corners of the transparent insert used for the cassette window. The high coercivity (typically 1480 Oersteds) metal powder tape is 3.81 mm wide, the same width as Compact Cassette tape. The standard overall thickness is 13 μm.

When the cassette is placed in the transport, the slider is moved back as it engages. This releases the lid lock. Continued movement into the transport pushes the slider right back, revealing the hub openings. The cassette is then lowered onto the hub drive spindles and tape guides, and the door is fully opened to allow access to the tape.

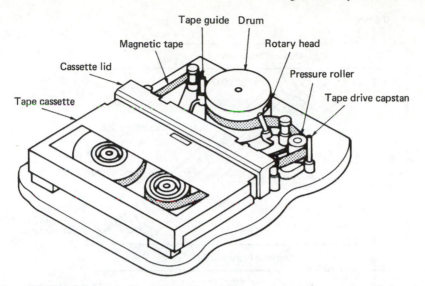

Figure 8.7 The simple mechanism of DAT. The guides and pressure roller move towards the drum and capstan and threading is complete.

As was shown in section 1.6, time compression is used to squeeze continuous samples into an intermittent recording. The angle of wrap of the tape around the drum can be reduced, which makes threading easier. In DAT the wrap angle is only 90° on the commonest drum size. As the heads are 180° apart, this means that for half the time neither head is in contact with the tape. Figure 8.7 shows that the partial-wrap concept allows the threading mechanism to be very simple indeed. As the cassette is lowered into the transport, the pinch roller and several guide pins pass behind the tape. These then simply move towards the capstan and drum and threading is complete. A further advantage of partial wrap is that the friction between the tape and drum is reduced, allowing power saving in portable applications, and allowing the tape to be shuttled at high speed without the partial unthreading needed by videocassettes. In this way the player can read subcode during shuttle to facilitate rapid track access.

The track pattern laid down by the rotary heads is shown in Figure 8.8. The heads rotate at 2000 rev/min in the same direction as tape motion, but because the drum axis is tilted, diagonal tracks 23.5 mm long result, at an angle of just over 6° to the edge. The diameter of the scanner needed is not specified, because it is the track pattern geometry which ensures interchange compatibility. It will be seen from Figure 8.8 that azimuth recording is employed as was described in Chapter 6. This requires no spaces or guard bands between the tracks. The chosen azimuth angle of ±20° reduces crosstalk to the same order as the noise, with a loss of only 1 dB due to the apparent reduction in writing speed.

In addition to the diagonal tracks, there are two linear tracks, one at each edge of the tape, where they act as protection for the diagonal tracks against edge damage. Owing to the low linear tape speed the ·use of these edge tracks is somewhat limited.

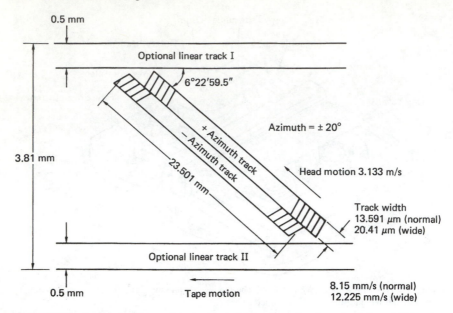

Figure 8.8 The two heads of opposite azimuth angles lay down the above track format. Tape linear speed determines track pitch.

8.4 DAT specification

Several related modes of operation are available, some of which are mandatory whereas the remainder are optional. These are compared in Table 8.1. The most important modes use a sampling rate of 48 kHz or 44.1 kHz, with sixteen-bit two's complement uniform quantization. At a linear tape speed of 8.15 mm/s, the standard cassette offers 120 min unbroken playing time. Initially it was proposed that all DAT machines would be able to record and play at 48 kHz, whereas only professional machines would be able to record at 44.1 kHz. For consumer machines, playback only of prerecorded media was proposed at 44.1 kHz, so that the same software could be released on CD or prerecorded DAT tape. Once SCMS (serial copying management system) was incorporated into consumer machines, they too recorded at 44.1 kHz. For reasons which will be explained later, contact duplicated tapes run at 12.225 mm/s to offer a playing time of 80 min. The above modes are mandatory if a machine is to be considered to meet the format.

Option 1 is identical to 48 kHz mode except that the sampling rate is 32 kHz. Option 2 is an extra-long-play mode. In order to reduce the data rate, the sampling rate is 32 kHz and the samples change to twelve-bit two's complement with non-linear quantizing. Halving the subcode rate allows the overall data rate necessary to be halved. The linear tape speed and the drum speed are both halved to give a playing time of four hours. All the above modes are stereo, but option 3 uses the sampling parameters of option 2 with four audio channels. This doubles the data rate with respect to option 2, so the standard tape speed of 8.15 mm/s is used.

Table 8.1 The significance of the recognition holes on the DAT cassette. Holes 1, 2 and 3 form a coded pattern; whereas hole 4 is independent.

Hole 1	Hole 2	Hole 3	Function
0	0	0	Metal powder tape or equivalent/13 μm thick
0	1	0	MP tape or equivalent/thin tape
0	0	1	1.5 TP/13 μm thick
0	1	1	1.5 TP/thin tape
1	×	×	(Reserved)

Hole 4		1 = Hole present
		0 = Hole blanked off
0	Non-prerecorded tape	
1	Prerecorded tape	

8.5 DAT block diagram

Figure 8.9 shows a block diagram of a typical DAT recorder. In order to make a recording, an analog signal is fed to an input ADC, or a direct digital input is taken from an AES/EBU interface. The incoming samples are subject to interleaving to reduce the effects of error bursts. Reading the memory at a higher rate than it was written performs the necessary time compression. Additional bytes of redundancy computed from the samples are added to the data stream to permit subsequent error correction. Subcode information such as the content of the AES/EBU channel status message is added, and the parallel byte structure is fed to the channel encoder, which combines a bit clock with the data, and produces a recording signal according to the 8/10 code which is free of DC (see Chapter 6). This signal is fed to the heads via a rotary transformer to make the binary recording, which leaves the tape track with a pattern of transitions between the two magnetic states.

On replay, the transitions on the tape track induce pulses in the head, which are used to recreate the record current waveform. This is fed to the 10/8 decoder which converts it to the original data stream and a separate clock. The subcode data are routed to the subcode output, and the audio samples are fed into a de-interleave memory which, in addition to time-expanding the recording, functions to remove any wow or flutter due to head-to-tape speed variations. Error correction is performed partially before and partially after de-interleave. The corrected output samples can be fed to DACs or to a direct digital output.

In order to keep the rotary heads following the very narrow slant tracks, alignment patterns are recorded as well as the data. The automatic track-following system processes the playback signals from these patterns to control the drum and capstan motors. The subcode and ID information can be used by the control logic to drive the tape to any desired location specified by the user.

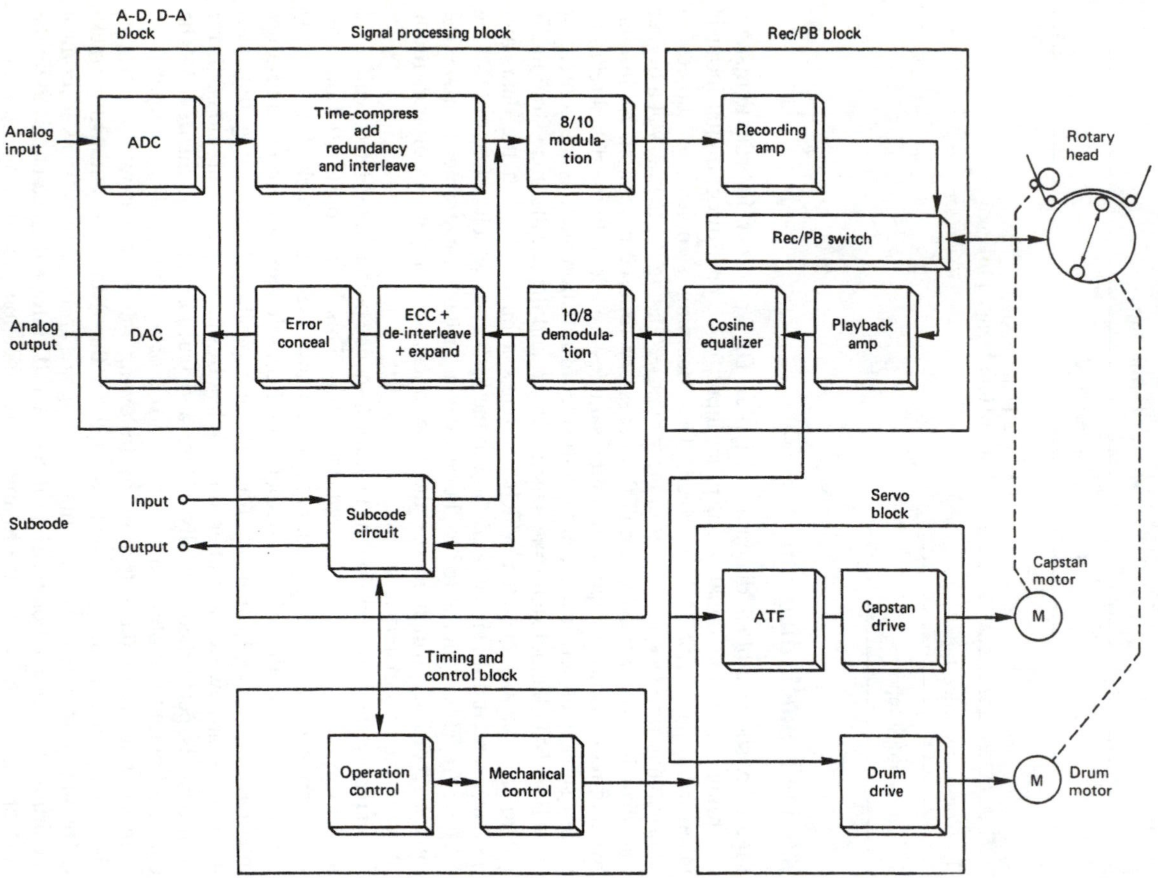

Figure 8.9 Block diagram of DAT.

8.6 Track following in DAT

The high-output metal tape used in DAT allows an adequate signal-to-noise ratio to be obtained with very narrow tracks on the tape. This reduces tape consumption and allows a small cassette, but it becomes necessary actively to control the relative position of the head and the track in order to maximize the replay signal and minimize the error rate. The track width and the coercivity of the tape largely define the signal-to-noise ratio. A track width has been chosen which makes the signal-to-crosstalk ratio dominant in cassettes which are intended for user recording.

Prerecorded tapes are made by contact duplication, and this process only works if the coercivity of the copy is less than that of the master. The output from prerecorded tapes at the track width of $13.59\,\mu m$ would be too low, and would be noise-dominated, which would cause the error rate to rise. The solution to this problem is that in prerecorded tapes the track width is increased to be the same as the head pole. The noise and crosstalk are both reduced in proportion to the reduced output of the medium, and the same error rate is achieved as for normal high-coercivity tape. The 50 per cent increase in track width is achieved by raising the linear tape speed from 8.15 to 12.225 mm/s, and so the playing time of a prerecorded cassette falls to 80 min as opposed to the 120 min of the normal tape. As DAT failed as a consumer product, prerecorded tapes are understandably rare.

The track-following principles are the same for prerecorded and normal cassettes except for dimensional differences. Tracking is achieved in conventional video recorders by the use of a linear control track which contains one pulse for every diagonal track. The phase of the pulses picked up by a fixed head is compared with the phase of pulses generated by the drum, and the error is used to drive the capstan. This method is adequate for the wide tracks of analog video recorders, but errors in the mounting of the fixed head and variations in tape tension rule it out for high-density use. In any case the control-track head adds undesirable mechanical complexity. In DAT, the tracking is achieved by reading special alignment patterns on the tape tracks themselves, and using the information contained in them to control the capstan.

DAT uses a technique called area-divided track following (ATF) in which separate parts of the track are set aside for track-following purposes. Figure 8.10 shows the basic way in which a tracking error is derived. The tracks at each side of the home track have bursts of pilot tone recorded in two different places. The frequency of the pilot tone is 130 kHz, which has been chosen to be relatively low so that it is not affected by azimuth loss. In this way an A head following an A track will be able to detect the pilot tone from the adjacent B tracks.

In Figure 8.11(a) the case of a correctly tracking head is shown. The amount of side-reading pilot tone from the two adjacent B tracks is identical. If the head is off track for some reason, as shown in (b), the amplitude of the pilot tone from one of the adjacent tracks will increase, and the other will decrease. The tracking error is derived by sampling the amplitude of each pilot-tone burst as it occurs, and holding the result so the relative amplitudes can be compared.

There are some practical considerations to be overcome in implementing this simple system, which result in some added complication. The pattern of pilot tones must be such that they occur at different times on each side of every track. To achieve this there must be a burst of pilot tone in every track, although the

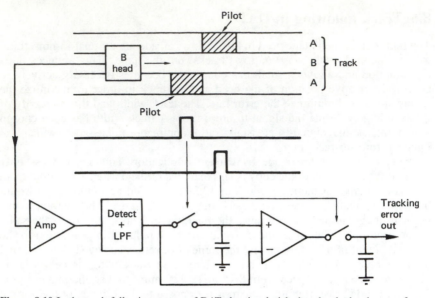

Figure 8.10 In the track-following system of DAT, the signal picked up by the head comes from pilot tones recorded in adjacent tracks at different positions. These pilot tones have low frequency, and are unaffected by azimuth error. The system samples the amplitude of the pilot tones, and subtracts them.

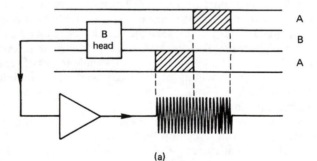

(a)

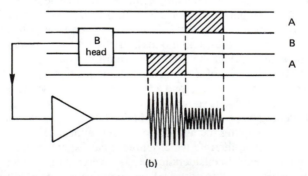

(b)

Figure 8.11 (a) A correctly tracking head produces pilot-tones bursts of identical amplitude. (b) The head is off track, and the first pilot burst becomes larger, whereas the second becomes smaller. This produces the tracking error in the circuit of Figure 8.10.

pilot tone in the home track does not contribute to the development of the tracking error. Additionally there must be some timing signals in the tracks to determine when the samples of pilot tone should be made. The final issue is to prevent the false locking which could occur if the tape happened to run at twice normal speed.

Figure 8.12 shows how the actual track-following pattern of DAT is laid out.[3] The pilot burst is early on A tracks and late on B tracks. Although the pilot bursts have a two-track cycle, the pattern is made to repeat over four tracks by changing the period of the sync patterns which control the pilot sampling. This can be used to prevent false locking. When an A head enters the track, it finds the home pilot-burst first, followed by pilot from the B track above, then pilot from the B track below. The tracking error is derived from the latter two. When a B head enters the track, it sees pilot from the A track above first, A track below next, and finally home pilot. The tracking error in this case is derived from the former two. The machine can easily tell which processing mode to use because the sync signals have a different frequency depending on whether they are in A tracks (522 kHz) or B tracks (784 kHz). The remaining areas are recorded with the interblock gap frequency of 1.56 MHz which serves no purpose except to erase earlier recordings.

Although these pilot and synchronizing frequencies appear strange, they are chosen so that they can be simply obtained by dividing down the master channel-bit-rate clock by simple factors. The channel-bit-rate clock, F_{ch}, is 9.408 MHz; pilot, the two sync frequencies and erase are obtained by dividing it by 72, 18, 12 and 6 respectively. The time at which the pilot amplitude in adjacent tracks should be sampled is determined by the detection of the synchronizing frequencies. As the head sees part of three tracks at all times, the sync detection in the home track has to take place in the presence of unwanted signals. On one side of the home sync signal will be the interblock gap frequency, which is high enough to be attenuated by azimuth. On the other side is pilot, which is unaffected by azimuth. This means that sync detection is easier in the tracking-error direction away from pilot than in the direction towards it. There is an effective working range of about +4 and −5 μm due to this asymmetry, with a dead band of 4 μm between tracks. Since the track-following servo is designed to minimize the tracking error, once lock is achieved the presence of the dead zone becomes academic.

The differential amplitude of the pilot tones produces the tracking error, and so the gain of the servo loop is proportional to the playback gain, which can fluctuate due to head contact variations and head tolerance. This problem is overcome by using AGC in the servo system. In addition to subtracting the pilot amplitudes to develop the tracking error, the circuitry also adds them to develop an AGC voltage. Two sample-and-hold stages are provided which store the AGC parameter for each head separately. The heads can thus be of different sensitivities without upsetting the servo. This condition could arise from manufacturing tolerances, or if one of the heads became contaminated.

8.7 DAT data channel

The channel code used in DAT is designed to function well in the presence of crosstalk, to have zero DC component to allow the use of a rotary transformer, and to have a small ratio of maximum and minimum run lengths to ease

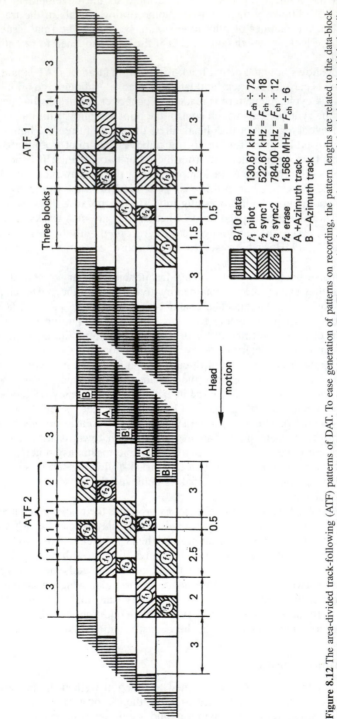

Figure 8.12 The area-divided track-following (ATF) patterns of DAT. To ease generation of patterns on recording, the pattern lengths are related to the data-block dimensions and the frequencies used are obtained by dividing down the channel bit clock F_{ch}. The sync signals are used to control the timing with which the pilot amplitude is sampled.

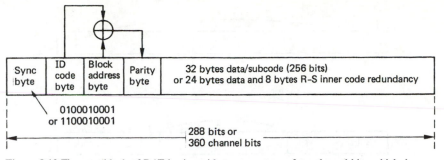

Figure 8.13 The sync block of DAT begins with a sync pattern of ten channel bits, which does not correspond to eight data bits. The header consists of an ID code byte and a block address. Parity is formed on the header bytes. The sync blocks alternate between 32 data (or outer code) bytes and 24 data bytes and 8 bytes of R–S redundancy for the inner codes.

overwrite erasure. The code used is a group code where eight data bits are represented by ten channel bits, hence the name 8/10.

The basic unit of recording is the sync block shown in Figure 8.13. This consists of the sync pattern, a three-byte header and 32 bytes of data, making 36 bytes in total, or 360 channel bits. The subcode areas each consist of eight of these blocks, and the PCM audio area consists of 128 of them. Note that a preamble is only necessary at the beginning of each area to allow the data separator to phase-lock before the first sync block arrives. Synchronism should be maintained throughout the area, but the sync pattern is repeated at the beginning of each sync block in case sync is lost due to dropout.

The first byte of the header contains an ID code which in the PCM audio blocks specifies the sampling rate in use, the number of audio channels, and whether there is a copy-prohibit in the recording. The second byte of the header specifies whether the block is subcode or PCM audio with the first bit. If set, the least significant four bits specify the subcode block address in the track, whereas if it is reset, the remaining seven bits specify the PCM audio block address in the track. The final header byte is a parity check and is the exclusive-OR sum of header bytes one and two.

The data format within the tracks can now be explained. The information on the track has three main purposes, PCM audio, subcode data and ATF patterns. It is necessary to be able to record subcode at a different time from PCM audio in professional machines in order to update or post-stripe the timecode. The subcode is placed in separate areas at the beginning and end of the tracks. When subcode is recorded on a tape with an existing PCM audio recording, the heads have to go into record at just the right time to drop a new subcode area onto the track. This timing is subject to some tolerance, and so some leeway is provided by the margin area which precedes the subcode area and the interblock gap (IBG) which follows. Each area has its own preamble and sync pattern so the data separator can lock to each area individually even though they were recorded at different times or on different machines.

The track-following system will control the capstan so that the heads pass precisely through the centre of the ATF area. Figure 8.14 shows that, in the presence of track curvature, the tracking error will be smaller overall if the

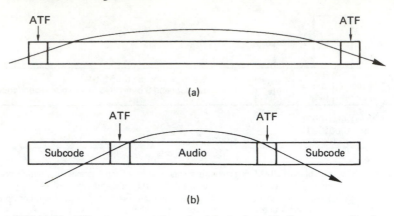

Figure 8.14 (a) The ATF patterns are at the ends of the track, and in the presence of track curvature the tracking error is exaggerated. (b) The ATF patterns are part-way down the track, minimizing mistracking due to curvature, and allowing a neat separation between subcode and audio blocks.

ATF pattern is placed part-way down the tracks. This explains why the ATF patterns are between the subcode areas and the central PCM audio area.

The data interleave is block-structured. One pair of tape tracks (one + azimuth and one – azimuth) corresponding to one drum revolution, make up an interleave block. Since the drum turns at 2000 rev/min, one revolution takes 30 ms and, in this time, 1440 samples must be stored for each channel for 48 kHz working.

The first interleave performed is to separate both left- and right-channel samples into odd and even. The right-channel odd samples followed by the left even samples are recorded in the + azimuth track, and the left odd samples followed by the right even samples are recorded in the – azimuth track. Figure 8.15 shows that this interleave allows uncorrectable errors to be concealed by interpolation. At (b) a head becomes clogged and results in every other track having severe errors. The split between right and left samples means that half of the samples in each channel are destroyed instead of every sample in one channel. The missing right even samples can be interpolated from the right odd samples, and the missing left odd samples are interpolated from the left even samples. Figure 8.15(c) shows the effect of a longitudinal tape scratch. A large error burst occurs at the same place in each head sweep. As the positions of left- and right-channel samples are reversed from one track to the next, the errors are again spread between the two channels and interpolation can be used in this case also. The error-correction system of DAT uses product codes which were treated in Chapter 6.

8.8 Multi-channel rotary-head recorders

Several audio channels can be accommodated on a single tape track by the use of multiplexing. In DAT, there are two channels, but these cannot be recorded singly. However, in other formats, if suitable edit gaps and block identification codes are present along the track, it is possible to record only those channels which are required, leaving the remainder intact. The head switches between play and record dynamically at the right place in each track. In this way a practical

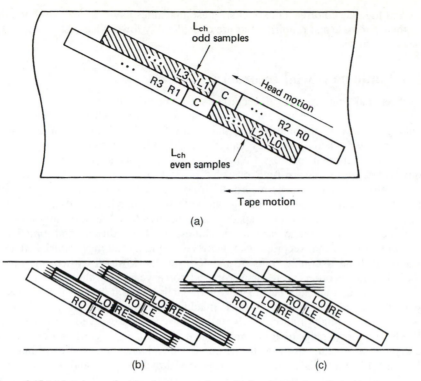

Figure 8.15 (a) Interleave of odd and even samples and left and right channels to permit concealment in case of gross errors. (b) Clogged head loses every other track. Half of the samples of each channel are still available, and interpolation is possible. (c) A linear tape scratch destroys odd samples in both channels. Interpolation is again possible.

multi-track digital audio recorder can be made using rotary heads. In other respects such recorders work on the same principles as the DAT format described above.

Following work which suggests that a helical-scan machine can accept spliced tape, Kudelski[4] proposed a format for 1/4-inch tape using a rotary head which became that of the NAGRA D. This machine offers four independently recordable channels of up to twenty-bit wordlength and timecode faciliies. The block structure is basically that of the audio channels of the D-1 DVTR. The format is restricted to low-density recording because of the potential for contamination with open reels. Whilst the recording density is not as great as in DAT, it is still competitive with professional analog machines and as the NAGRA D is a professional-only product, tape consumption is of less consequence than reliability. Manual splicing of a helical scan tape causes a serious tracking and data loss problem at the splice. The principle of jump editing is used so that the area of the splice is not played.

A number of manufacturers have developed low-cost digital multitrack recorders for the home studio market. These are based on either VHS or Video-8 rotary head cassette tape decks and generally offer eight channels of audio. Some models have timecode and include synchronizers so that several machines can be locked together to offer more tracks. These machines have become very popular as their purchase and running costs are considerably lower than that of stationary head machines.

8.9 Stationary-head recorders

Professional stationary-head recorders are specifically designed for record production and mastering, and have to be able to offer all the features of an analog multitrack. It could be said that many digital multitracks mimic analog machines so exactly that they can be installed in otherwise analog studios with the minimum of fuss. When the stationary-head formats were first developed, the necessary functions of a professional machine were: independent control of which tracks record and play, synchronous recording, punch-in/punch-out editing, tape-cut editing, variable-speed playback, offtape monitoring in record, various tape speeds and bandwidths, autolocation and the facilities to synchronize several machines. In both theory and practice a modern rotary-head recorder can achieve a higher storage density than a stationary-head recorder, thus using less tape. However, when multitrack digital audio recorders were first proposed some years ago, the adaptation of a video-recorder transport had to be ruled out because it lacked the necessary bandwidth. For example, a 24-track machine requires about 20 megabits per second. Accordingly, multitrack digital audio recorders evolved with stationary heads and open reels to look and behave like analog recorders even to the extent of supporting splicing. In the context of advances in data recording and the use of hard disks, the open reel multitrack digital audio recorder is unlikely to develop further.

A stationary-head recorder is basically quite simple, as the block diagram of Figure 8.16 shows. The transport is not dissimilar to that of an analog recorder. The tape is very thin, rather like videotape, to allow it to conform closely to the heads for short-wavelength working. Control of the capstan is more like that of a video recorder. The capstan turns at constant speed when a virgin tape is being recorded, but for replay, it will be controlled to run at whatever speed is necessary to make the offtape sample rate equal to the reference rate. In this way, several machines can be kept in exact synchronism by feeding them with a common reference. Variable-speed replay can be achieved by changing the reference frequency. It should be emphasized that, when variable speed is used, the output sampling rate changes. This may not be of any consequence if the samples are returned to the analog domain, but it prevents direct connection to a digital mixer, since these usually have fixed sampling rates.

The major items in the block diagram have been discussed in the relevant chapters. Samples are interleaved, redundancy is added, and the bits are converted into a suitable channel code. In stationary-head recorders, the frequencies in each head are low, and complex coding is not difficult. The lack of the rotary transformer of the rotary-head machine means that DC content is a less important issue. The codes used generally try to emphasize density ratio, which keeps down the linear tape speed, and the jitter window, since this helps to reject the inevitable crosstalk between the closely spaced heads. On replay there are the usual data separators, timebase correctors and error-correction circuits.

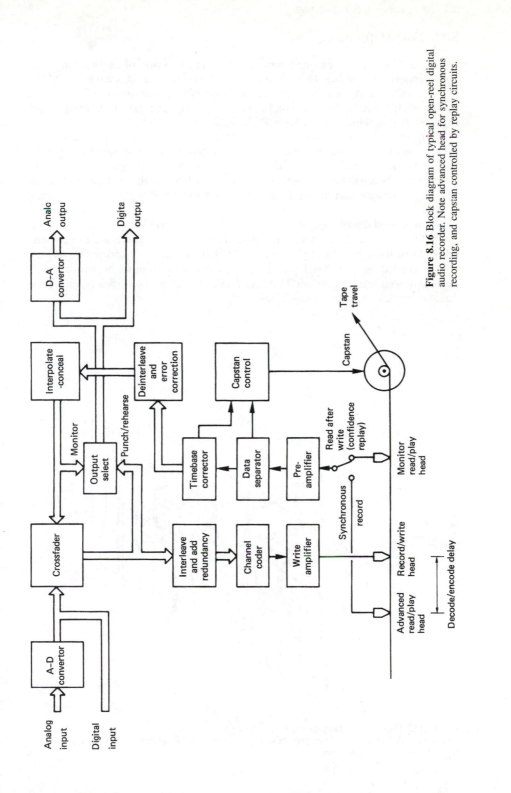

Figure 8.16 Block diagram of typical open-reel digital audio recorder. Note advanced head for synchronous recording, and capstan controlled by replay circuits.

8.10 DASH format

The DASH[5] format is not one format as such, but a family of like formats, and thus supports a number of different track layouts. The quarter-inch DASH formats are obsolete and not considered here. With ferrite-head technology, it was possible to record 24 tracks on half-inch tape (H). The most frequently found member of this family is the Sony PCM-3324.

Using thin-film heads, the magnetic circuits and windings are produced by deposition on a substrate at right angles to the tape plane, and as seen in Figure 8.17 they can be made very accurately at small track spacings. Perhaps more importantly, because the magnetic circuits do not have such large parallel areas, mutual inductance and crosstalk are smaller allowing a higher practical track density.

The so-called double-density version, known as DASH II, uses such thin-film heads to obtain 48 digital tracks on quarter-inch in tape. The track dimensions allow 24 of the replay head gaps on a DASH II machine to align with and play tapes recorded on a DASH I machine. PCM–3348 machines can play 24-track tapes and even record a further 24 tracks on them, but such 48-track tapes cannot then be played on 24-track machines.

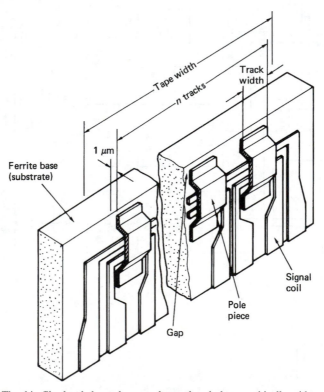

Figure 8.17 The thin-film head shown here can be produced photographically with very small dimensions. Flat structure reduces crosstalk. This type of head is suitable for DASH II which has twice as many tracks as DASH I.

The DASH format supports three sampling rates and the tape speed is normalized to 30 in/s at the highest rate. The three rates are 32 kHz, 44.1 kHz and 48 kHz. In fact most stationary-head recorders will record at any reasonable sampling rate just by supplying them with an external reference, or word clock, at the appropriate frequency. Under these conditions, the sampling-rate switch on the machine only controls the status bits in the recording which set the default playback rate.

The error-correction strategy of DASH forms codewords which are confined to single-tape tracks. In all practical recorders measures have to be taken for the rare cases when the error correction is overwhelmed by gross corruption. In open-reel stationary-head recorders, one obvious mechanism is the act of splicing the tape and the resultant contamination due to fingerprints. The use of interleaving is essential to handle burst errors; unfortunately it conflicts with the requirements of tape-cut editing. Figure 8.18 shows that a splice in cross-interleave destroys codewords for the entire constraint length of the interleave. The longer the constraint length, the greater the resistance to burst errors, but the more damage is done by a splice.

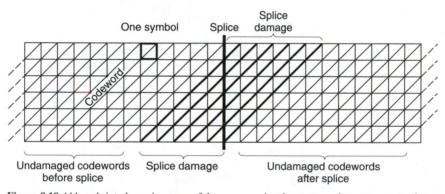

Figure 8.18 Although interleave is a powerful weapon against burst errors, it causes greater data loss when tape is spliced because many codewords are replayed in two unrelated halves.

In order to handle dropouts or splices, samples from the convertor or direct digital input are first sorted into odd and even. The odd/even distance has to be greater than the crossinterleave constraint length. In DASH, the constraint length is 119 blocks, or 1428 samples, and the odd/even delay is 204 blocks, or 2448 samples. In the case of a severe dropout, after the replay de-interleave process, the effect will be to cause two separate error bursts, first in the odd samples, then in the even samples. The odd samples can be interpolated from the even and vice versa in order to conceal the dropout. In the case of a splice, samples are destroyed for the constraint length, but Figure 8.19 shows that this occurs at different times for the odd and even samples.

Using interpolation, it is possible to obtain simultaneously the end of the old recording and the beginning of the new one. A digital crossfade is made between the old and new recordings. The interpolation during concealment and splices causes a momentary reduction in frequency response which may result in aliasing if there is significant audio energy above one quarter of the sampling rate.

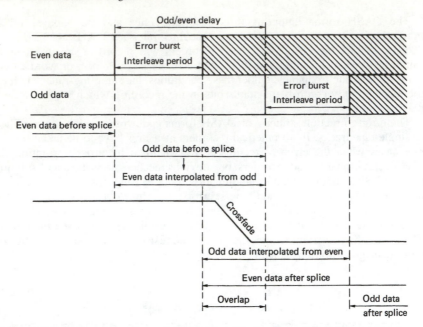

Figure 8.19 Following de-interleave, the effect of a splice is to cause odd and even data to be lost at different times. Interpolation is used to provide the missing samples, and a crossfade is made when both recordings are available in the central overlap.

8.11 DCC – Digital Compact Cassette

DCC is a stationary-head format in which the tape transport is designed to play existing analog Compact Cassettes in addition to making and playing digital recordings. This backward compatibility means that an existing Compact Cassette collection can still be enjoyed whilst newly made or purchased recordings will be digital.[6] To achieve this compatibility, DCC tape is the same width as analog Compact Cassette tape (3.81 mm) and travels at the same speed ($1\frac{7}{8}$ in/s or 4.76 cm/s). The formulation of the DCC tape is different; it resembles conventional chrome video tape, but the principle of playing one 'side' of the tape in one direction and then playing the other side in the opposite direction is retained.

Although the DCC cassette has similar dimensions to the Compact Cassette so that both can be loaded in the same transport, the DCC cassette is of radically different construction. The DCC cassette only fits in the machine one way, it cannot be physically turned over as it only has hub drive apertures on one side. The head access bulge has gone and the cassette has a uniform rectangular cross-section, taking up less space in storage. The transparent windows have also been deleted as the amount of tape remaining is displayed on the panel of the player. This approach has the advantage that labelling artwork can cover almost the entire top surface. The same approach has been used in pre-recorded MiniDiscs. As the cassette cannot be turned over, all transports must be capable of playing in both directions. Thus DCC is an auto-reverse format. In addition to a record lockout plug, the cassette body carries identification holes. Combinations of these specify six different playing times from 45 min to 120 min as in Table 8.1.

The apertures for hub drive, capstans, pinch rollers and heads are covered by a sliding cover formed from metal plate. The cover plate is automatically slid aside when the cassette enters the transport. The cover plate also operates hub brakes when it closes and so the cassette can be left out of its container. The container fits the cassette like a sleeve and has space for an information booklet.

DCC uses a form of compression which Philips call Precision Adaptive Sub-band Coding (PASC). This is quite similar to the MPEG Layer I coding described in Chapter 5 and its use allows the recorded data rate to be about one quarter that of the original PCM audio. Conventional chromium tape may then be used with a minimum wavelength of about one micrometre instead of the more expensive high-coercivity tapes normally required for use with shorter wavelengths. Linear tracks were chosen so that tape duplication could be carried out at high speed. Even with compression the only way in which the bit rate can be accommodated is to use many tracks in parallel.

Figure 8.20 shows that in DCC audio data are distributed over eight parallel tracks along with a subcode track which together occupy half the width of the tape. At the end of the tape the head rotates about an axis perpendicular to the tape and plays the remaining tracks in reverse. The other half of the head is fitted with magnetic circuits sized for analog tracks and so the head rotation can also select the head type which is in use for a given tape direction.

Compression followed by distribution over eight tracks means that each track runs at only *96 kbits/s*. The linear tape speed is incredibly low by stationary-head digital standards in order to obtain the desired playing time. The rate of change of flux in the replay head is very small due to the low tape speed, and conventional inductive heads are at a severe disadvantage because their self-noise drowns the signal. Magnetoresistive heads are necessary because they do

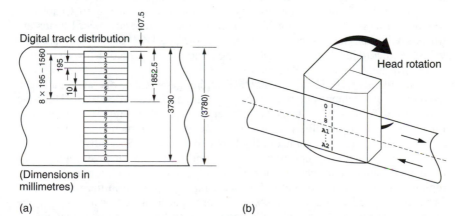

(a)   (b)

Figure 8.20 In DCC audio and auxiliary data are recorded on nine parallel tracks along each side of the tape as shown in (a). The replay head shown in (b) carries magnetic poles which register with one set of nine tracks. At the end of the tape, the replay head rotates 180° and plays a further nine tracks on the other side of the tape. The replay head also contains a pair of analog audio magnetic circuits which will be swung into place if an analog cassette is to be played.

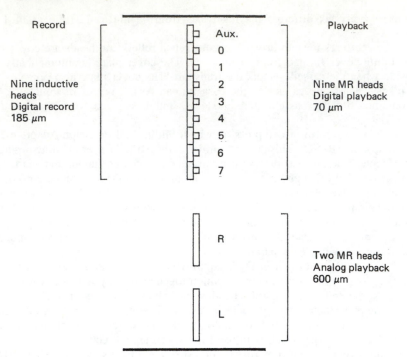

Figure 8.21 The head arrangement used in DCC. There are nine record heads which leave tracks wider than the MR replay heads to allow for misregistration. Two MR analog heads allow compact cassette replay.

not have a derivative action, and so the signal is independent of speed. A magnetoresistive head uses an element whose resistance is influenced by the strength of flux from the tape and its operation was discussed in Chapter 6. Magneto-resistive heads are unable to record, and so separate record heads are necessary. Figure 8.21 shows a schematic outline of a DCC head. There are nine inductive record heads for the digital tracks, and these are recorded with a width of 185 μm and a pitch of 195 μm.

Alongside the record head are nine MR replay gaps. These operate on a 70μm band of the tape which is nominally in the centre of the recorded track. There are two reasons for this large disparity between the record and replay track widths. Firstly, replay signal quality is unaffected by a lateral alignment error of ± 57 μm and this ensures tracking compatibility between machines. Secondly, the loss due to incorrect azimuth is proportional to track width and the narrower replay track is thus less sensitive to the state of azimuth adjustment. In addition to the digital replay gaps, a further two analog MR head gaps are present in the replay stack. These are aligned with the two tracks of a stereo pair in a Compact Cassette. The twenty-gap head could not be made economically by conventional techniques. Instead it is made lithographically using thin film technology.

Tape guidance is achieved by a combination of guides on the head block and pins in the cassette. Figure 8.22 shows that at each side of the head is fitted a C-shaped tape guide. This guide is slightly narrower than the nominal tape width.

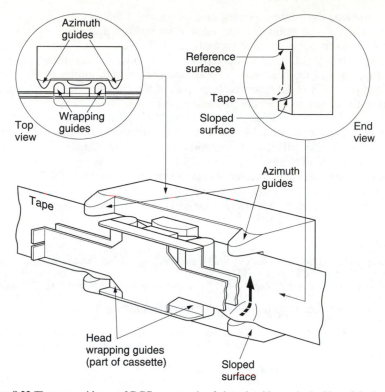

Figure 8.22 The tape guidance of DCC uses a pair of shaped guides on both sides of the head. See text for details.

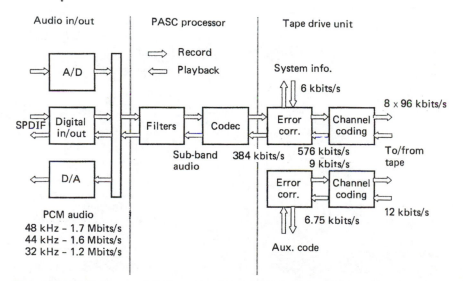

Figure 8.23 Block diagram of DCC machine. This is basically similar to any stationary-head recorder except for the compression (PASC) unit between the convertors and the transport.

The reference edge of the tape runs against a surface that is at right angles to the guide, whereas the non-reference edge runs against a sloping surface. Tape tension tends to force the tape towards the reference edge. As there is such a guide at both sides of the head, the tape cannot wander in the azimuth plane. The tape wrap around the head stack and around the azimuth guides is achieved by a pair of pins behind the tape which are part of the cassette. Between the pins is a conventional sprung pressure pad and screen.

Figure 8.23 shows a block diagram of a DCC machine. The audio interface contains convertors which allow use in analog systems. The digital interface may be used as an alternative. DCC supports 48, 44.1 and 32 kHz sampling rates, offering audio bandwidths of 22, 20 and 14.5 kHz respectively with eighteen-bit dynamic range. Between the interface and the tape subsystem is the PASC coder. The tape subsystem requires error correction and channel coding systems not only for the audio data, but also for the auxiliary data on the ninth track.

References

1. Nakajima, H. and Odaka, K., A rotary-head high-density digital audio tape recorder. *IEEE Trans. Consum. Electron.*, **CE–29**, 430–437 (1983)
2. Itoh, F., Shiba, H., Hayama, M. and Satoh, T., Magnetic tape and cartridge of R-DAT. *IEEE Trans. Consum. Electron*, **CE-32**, 442–452 (1986)
3. Hitomi, A. and Taki, T., Servo technology of R-DAT. *IEEE Trans. Consum. Electron.*, **CE–32**, 425–432 (1986)
4. Kudelski, S., *et al.*, Digital audio recording format offering extensive editing capabilities. Presented at the 82nd Audio Engineering Society Convention (London, 1987), Preprint 2481(H-7)
5. Doi, T.T., Tsuchiya, Y., Tanaka, M. and Watanabe, N., A format of stationary-head digital audio recorder covering wide range of applications. Presented at the 67th Audio Engineering Society Convention (New York, 1980), Preprint 1677(H6)
6. Lokhoff, G.C.P., DCC: Digital compact cassette. *IEEE Trans. Consum. Electron.*, **CE-37**, 702–706 (1991)

9

Magnetic disk drives

Disk drives came into being as random-access file-storage devices for digital computers. However, the explosion in personal computers has fuelled demand for low-cost high-density magnetic disk drives and the rapid access offered is increasingly finding applications in digital audio to the detriment of tape-based storage.

9.1 Types of disk drive

The disk drive was developed specifically to offer rapid random access to stored data. Figure 9.1 shows that, in a magnetic disk drive, the data are recorded on a circular track. In floppy disks, the magnetic medium is flexible, and the head touches it. This restricts the rotational speed. In hard-disk drives, the disk rotates at several thousand rev/min so that the head-to-disk speed is of the order of one hundred miles per hour. At this speed no contact can be tolerated, and the head flies on a boundary layer of air turning with the disk at a height measured in microinches. The longest time it is necessary to wait to access a given data block is a few milliseconds. To increase the storage capacity of the drive without a proportional increase in cost, many concentric tracks are recorded on the disk surface, and the head is mounted on a positioner that can rapidly bring the head to any desired track. Such a machine is termed a moving-head disk drive. The positioner was usually designed so that it could remove the heads away from the

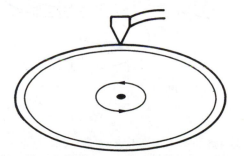

Figure 9.1 The rotating store concept. Data on the rotating circular track are repeatedly presented to the head.

325

disk completely, which could thus be exchanged. The exchangeable-pack moving-head disk drive became the standard for mainframe and minicomputers for a long time, and usually at least two were furnished so that important data could be 'backed up' or copied to a second disk for safe keeping.

Later came the so-called Winchester technology disks, where the disk and positioner formed a sealed unit that allowed increased storage capacity but precluded exchange of the disk pack.

Disk drive development has been phenomenally rapid. The first flying head disks were about 3 feet across, whereas now they are small enough to fit in laptop computers. Despite the reduction in size, the storage capacity is not compromised because the recording density has increased and continues to increase. In fact there is an advantage in making a drive smaller because the moving parts are then lighter and travel a shorter distance, improving access time.

Figure 9.2 shows a typical multi-platter disk pack in conceptual form. Given a particular set of coordinates (cylinder, head, sector), known as a disk physical

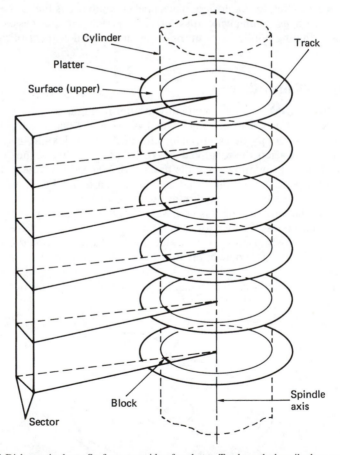

Figure 9.2 Disk terminology. Surface: one side of a platter. Track: path described on a surface by a fixed head. Cylinder: imaginary shape intersecting all surfaces at tracks of the same radius. Sector: angular subdivision of pack. Block: that part of a track within one sector. Each block has a unique cylinder, head and sector address.

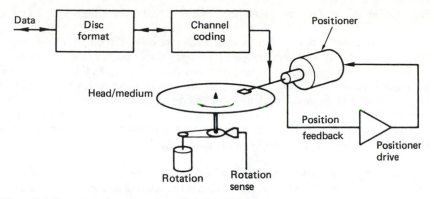

Figure 9.3 The main subsystems of a typical disk drive.

address, one unique data block is defined. The subdivision into sectors is sometimes omitted for special applications. Figure 9.3 introduces the essential subsystems of a disk drive that will be discussed.

9.2 Structure of disk

Rigid or 'hard' disks are made from aluminium alloy. Magnetic-oxide types use an aluminium oxide substrate, or undercoat, giving a flat surface to which the oxide binder can adhere. Later metallic disks are electroplated with the magnetic medium. In both cases the surface finish must be extremely good owing to the very small flying height of the head. As the head-to-disk speed and recording density are functions of track radius, the data are confined to the outer areas of the disks to minimize the change in these parameters. As a result, the centre of the pack is often an empty well. In fixed (i.e. non-interchangeable) disks the drive motor is often installed in the centre well.

9.3 Principle of flying head

Disk drives permanently sacrifice storage density in order to offer rapid access. The use of a flying head with a deliberate air gap between it and the medium is necessary because of the high medium speed, but this causes a severe separation loss that restricts the linear density available. The air gap must be accurately maintained, and consequently the head is of low mass and is mounted flexibly.

The aerohydrodynamic part of the head is known as the slipper; it is designed to provide lift from the boundary layer that changes rapidly with changes in flying height. It is not initially obvious that the difficulty with disk heads is not making them fly, but making them fly close enough to the disk surface. The boundary layer travelling at the disk surface has the same speed as the disk, but as height increases, it slows down due to drag from the surrounding air. As the lift is a function of relative air speed, the closer the slipper comes to the disk, the greater the lift will be. The slipper is therefore mounted at the end of a rigid cantilever sprung towards the medium. The force with which the head is pressed towards the disk by the spring is equal to the lift at the designed flying height.

Because of the spring, the head may rise and fall over small warps in the disk.

It would be virtually impossible to manufacture disks flat enough to dispense with this feature. As the slipper negotiates a warp it will pitch and roll in addition to rising and falling, but it must be prevented from yawing, as this would cause an azimuth error. Downthrust is applied to the aerodynamic centre by a spherical thrust button, and the required degrees of freedom are supplied by a thin flexible gimbal. The slipper has to bleed away surplus air in order to approach close enough to the disk, and holes or grooves are usually provided for this purpose in the same way that pinch rollers on some tape decks have grooves to prevent tape slip.

In exchangeable-pack drives, there will be a ramp on the side of the cantilever that engages a fixed block when the heads are retracted in order to lift them away from the disk surface.

9.4 Reading and writing

Figure 9.4 shows how disk heads are made. The magnetic circuit of disk heads was originally assembled from discrete magnetic elements. As the gap and flying height became smaller to increase linear recording density, the slipper was made from ferrite, and became part of the magnetic circuit. This was completed by a small C-shaped ferrite piece that carried the coil. In thin-film heads, the magnetic circuit and coil are both formed by deposition on a substrate that becomes the rear of the slipper.

In a moving-head device it is difficult to position separate erase, record and playback heads accurately. Usually, erase is by overwriting, and reading and writing are both carried out by the same head. The presence of the air film causes severe separation loss, and peak shift distortion is a major problem. The flying height of the head varies with the radius of the disk track, and it is difficult to provide accurate equalization of the replay channel because of this. The write current is often controlled as a function of track radius so that the changing reluctance of the air gap does not change the resulting record flux. Automatic gain control (AGC) is used on replay to compensate for changes in signal amplitude from the head.

Equalization may be used on recording in the form of precompensation, which moves recorded transitions in such a way as to oppose the effects of peak shift in addition to any replay equalization used. This was discussed in Chapter 6, which also introduced digital channel coding. Early disks used FM coding, which was easy to decode, but had a poor density ratio. Later disks use group codes.

Typical drives have several heads, but with the exception of special-purpose parallel-transfer machines for digital video or instrumentation work, only one head will be active at any one time, which means that the read and write circuitry can be shared between the heads. Figure 9.5 shows that in one approach the centre-tapped heads are isolated by connecting the centre tap to a negative voltage, which reverse-biases the matrix diodes. The centre tap of the selected head is made positive. When reading, a small current flows through both halves of the head winding, as the diodes are forward-biased. Opposing currents in the head cancel, but read signals due to transitions on the medium can pass through the forward-biased diodes to become differential signals on the matrix bus. During writing, the current from the write generator passes alternately through

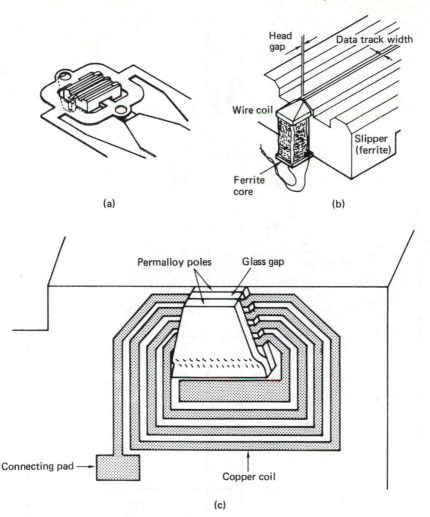

(a)

Head gap

Data track width

Wire coil

Slipper (ferrite)

Ferrite core

(b)

Permalloy poles   Glass gap

Connecting pad →

Copper coil

(c)

Figure 9.4 (a) Winchester head construction showing large air bleed grooves. (b) Close-up of slipper showing magnetic circuit on trailing edge. (c) Thin-film head is fabricated on the end of the slipper using microcircuit technology.

the two halves of the head coil. Further isolation is necessary to prevent the write-current-induced voltages from destroying the read preamplifier input. Alternatively, FET analog switches may be used for head selection.

The read channel usually incorporates AGC, which will be overridden by the control logic between data blocks in order to search for address marks, which are short unmodulated areas of track. As a block preamble is entered, the AGC will be enabled to allow a rapid gain adjustment.

The high bit rates of disk drives, due to the speed of the medium, mean that peak detection in the replay channel is usually by differentiation. The detected peaks are then fed to the data separator.

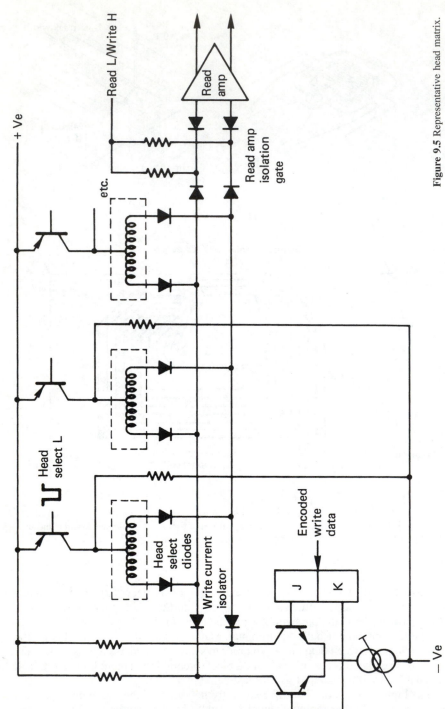

Figure 9.5 Representative head matrix.

9.5 Moving the heads

The servo system required to move the heads rapidly between tracks, and yet to hold them in place accurately for data transfer, is a fascinating and complex piece of engineering. In exchangeable pack drives, the disk positioner moves on a straight axis that passes through the spindle. Motive power is generally by moving-coil drive, because of the small moving mass that this technique permits.

When a drive is track following, it is said to be detented, in fine mode or in linear mode depending on the manufacturer. When a drive is seeking from one track to another, it can be described as being in coarse mode or velocity mode. These are the two major operating modes of the servo. Moving-coil actuators do not naturally detent and require power to stay on track. The servo system needs positional feedback of some kind. The purpose of the feedback will be one or more of the following:

1 To count the number of cylinders crossed during a seek
2 To generate a signal proportional to carriage velocity
3 To generate a position error proportional to the distance from the centre of the desired track

Magnetic and optical drives obtain these feedback signals in different ways. Many drives incorporate a tacho that may be a magnetic moving-coil type or its complementary equivalent the moving-magnet type. Both generate a voltage proportional to velocity, and can give no positional information.

A seek is a process where the positioner moves from one cylinder to another. The speed with which a seek can be completed is a major factor in determining the access time of the drive. The main parameter controlling the carriage during a seek is the cylinder difference, which is obtained by subtracting the current cylinder address from the desired cylinder address. The cylinder difference will be a signed binary number representing the number of cylinders to be crossed to reach the target, direction being indicated by the sign. The cylinder difference is loaded into a counter that is decremented each time a cylinder is crossed. The counter drives a DAC that generates an analog voltage proportional to the cylinder difference. As Figure 9.6 shows, this voltage, known as the scheduled velocity is compared with the output of the carriage-velocity tacho. The difference between the two is a velocity error driving the carriage in such a way that the error is cancelled. As the carriage approaches the target cylinder, the cylinder difference becomes smaller, with the result that the run-in to the target is critically damped to eliminate overshoot.

Figure 9.7(a) shows graphs of scheduled velocity, actual velocity and motor current with respect to cylinder difference during a seek. In the first half of the seek, the actual velocity is less than the scheduled velocity, causing a large velocity error which saturates the amplifier and provides maximum carriage acceleration. In the second half of the graphs, the scheduled velocity is falling below the actual velocity, generating a negative velocity error that drives a reverse current through the motor to slow the carriage down. The scheduled deceleration slope can clearly not be steeper than the saturated acceleration slope. Areas A and B on the graph will be about equal, as the kinetic energy put into the carriage has to be taken out. The current through the motor is continuous, and would result in a heating problem, so to counter that, the DAC is made non-linear

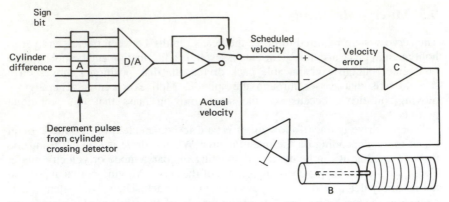

Figure 9.6 Control of carriage velocity by cylinder difference. The cylinder difference is loaded into the difference counter A. A digital-to-analog convertor generates an analog voltage from the cylinder difference, known as the scheduled velocity. This is compared with the actual velocity from the transducer B in order to generate the velocity error which drives the servo amplifier C.

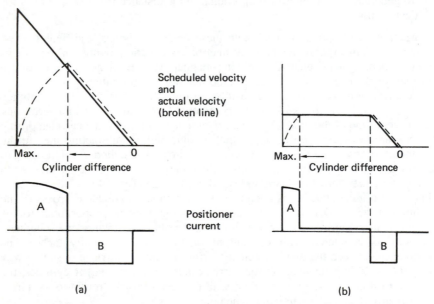

Figure 9.7 In the simple arrangement in (a) the dissipation in the positioner is continuous, causing a heating problem. The effect of limiting the scheduled velocity above a certain cylinder difference is apparent in (b) where heavy positioner current only flows during acceleration and deceleration. During the plateau of the velocity profile, only enough current to overcome friction is necessary. The curvature of the acceleration slope is due to the back EMF of the positioner motor.

so that above a certain cylinder difference no increase in scheduled velocity will occur. This results in the graph of Figure 9.7(b). The actual velocity graph is called a velocity profile. It consists of three regions: acceleration, where the system is saturated; a constant velocity plateau, where the only power needed is to overcome friction; and the scheduled run-in to the desired cylinder. Dissipation is only significant in the first and last regions.

9.6 Rotation

The rotation subsystems of disk drives will now be covered. The track-following accuracy of a drive positioner will be impaired if there is bearing run-out, and so the spindle bearings are made to a high degree of precision. Most modern drives incorporate brushless DC motors with integral speed control. In exchangeable-pack drives, some form of braking is usually provided to slow down the pack for convenient removal.

In order to control reading and writing, the drive control circuitry needs to know which cylinder the heads are on, and which sector is currently under the head. Sector information used to be obtained from a sensor that detects holes or slots cut in the hub of the disk. Modern drives will obtain this information from the disk surface as will be seen. The result is that a sector counter in the control logic remains in step with the physical rotation of the disk. The desired sector address is loaded into a register, which is compared with the sector counter. When the two match, the desired sector has been found. This process is referred to as a search, and usually takes place after a seek. Having found the correct physical place on the disk, the next step is to read the header associated with the data block to confirm that the disk address contained there is the same as the desired address.

9.7 Servo-surface disks

One of the major problems to be overcome in the development of high-density disk drives was that of keeping the heads on track despite changes of temperature. The very narrow tracks used in digital recording have similar dimensions to the amount a disk will expand as it warms up. The cantilevers and the drive base all expand and contract, conspiring with thermal drift in the cylinder transducer to limit track pitch. The breakthrough in disk density came with the introduction of the servo-surface drive. The position error in a servo-surface drive is derived from a head reading the disk itself. This virtually eliminates thermal effects on head positioning and allows great increases in storage density.

In a multi-platter drive, one surface of the pack holds servo information that is read by the servo head. In a ten-platter pack this means that 5 per cent of the medium area is lost, but this is unimportant since the increase in density allowed is enormous. Using one side of a single-platter cartridge for servo information would be unacceptable as it represents 50 per cent of the medium area, so in this case the servo information can be interleaved with sectors on the data surfaces. This is known as an embedded-servo technique. These two approaches are contrasted in Figure 9.8. The servo surface is written at the time of disk pack manufacture, and the disk drive can only read it. Writing the servo surface has nothing to do with disk formatting, which affects the data storage areas only.

9.8 Soft sectoring

It has been seen that a position error and a cylinder count can be derived from the servo surface, eliminating the cylinder transducer. As there is exactly the same number of pulses on every track on the servo surface, it is possible to describe the rotational position of the disk simply by counting them. All that is needed is

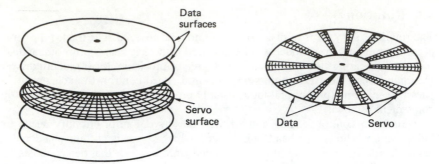

Figure 9.8 In a multiplatter disk pack, one surface is dedicated to servo information. In a single platter, the servo information is embedded in the data on the same surfaces.

an unique pattern of missing pulses once per revolution to act as an index point, and the sector transducer can also be eliminated.

The advantage of deriving the sector count from the servo surface is that the number of sectors on the disk can be varied. Any number of sectors may be accommodated if the pulse signal is fed through a programmable divider. The same disk and drive may then be used in numerous different applications.

9.9 Winchester technology

In order to offer extremely high capacity per spindle, which reduces the cost per bit, a disk drive must have very narrow tracks placed close together, and must use very short recorded wavelengths, which implies that the flying height of the heads must be small. The so-called Winchester technology is one approach to high storage density. The technology was first developed by IBM, and the name came about because the model number of the development drive was the same as that of the famous rifle.

Reduction in flying height magnifies the problem of providing a contaminant-free environment. A conventional disk is well protected whilst inside the drive, but outside the drive the effects of contamination become intolerable.

In exchangeable-pack drives, there is a real limit to the track pitch that can be achieved because of the difficulty or cost of engineering head-alignment mechanisms to make the necessary minute adjustments to give interchange compatibility.

The essence of Winchester technology is that each disk pack has its own set of read/write and servo heads, with an integral positioner. The whole is protected by a dust-free enclosure, and the unit is referred to as a head disk assembly, or HDA.

As the HDA contains its own heads, compatibility problems do not exist, and no head alignment is necessary or provided for. It is thus possible to reduce track pitch considerably compared with exchangeable pack drives. The sealed environment ensures complete cleanliness that permits a reduction in flying height without loss of reliability, and hence leads to an increased linear density. If the rotational speed is maintained, this can also result in an increase in data transfer rate.

The HDA is completely sealed, but some have a small filtered port to equalize pressure. Into this sealed volume of air, the drive motor delivers the majority of its power output. The heat that results is dissipated with the help of fins on the HDA casing. Some HDAs are filled with helium that significantly reduces drag and heat build-up.

An exchangeable-pack drive must retract the heads to facilitate pack removal. With Winchester technology this is not necessary. An area of the disk surface is reserved as a landing strip for the heads. The disk surface is lubricated, and the heads are designed to withstand landing and take-off without damage. Winchester heads have very large air-bleed grooves to allow low flying height with a much smaller down-thrust from the cantilever, and so they exert less force on the disk surface during contact. When the term parking is used in the context of Winchester technology, it refers to the positioning of the heads over the landing area. Disk rotation must be started and stopped quickly to minimize the length of time the heads slide over the medium.

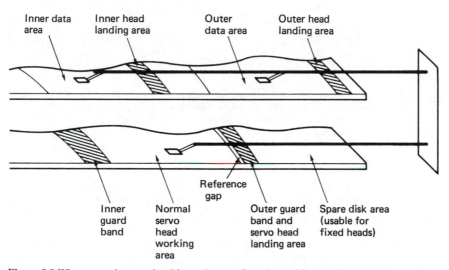

Figure 9.9 When more than one head is used per surface, the positioner still only requires one servo head. This is often arranged to be equidistant from the read/write heads for thermal stability.

A major advantage of contact start/stop is that more than one head can be used on each surface if retraction is not needed. This leads to two gains: first, the travel of the positioner is reduced in proportion to the number of heads per surface, reducing access time; and, second, more data can be transferred at a given detented carriage position before a seek to the next cylinder becomes necessary. This increases the speed of long transfers. Figure 9.9 illustrates the relationships of the heads in such a system.

9.10 Rotary positioners

Figure 9.10 shows that rotary positioners are feasible in Winchester drives; they cannot be used in exchangeable-pack drives because of interchange problems. There are some advantages to a rotary positioner. It can be placed in the corner

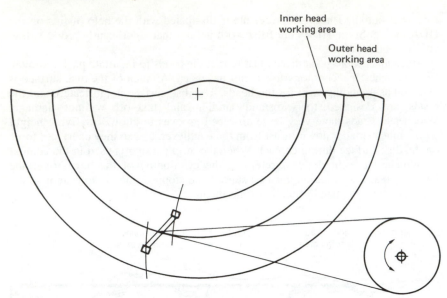

Inner head
working area

Outer head
working area

Figure 9.10 A rotary positioner with two heads per surface. The tolerances involved in the spacing between the heads and the axis of rotation mean that each arm records data in a unique position. Those data can only be read back by the same heads, which rules out the use of a rotary positioner in exchangeable-pack drives. In a head disk assembly the problem of compatibility does not arise.

of a compact HDA allowing smaller overall size. The manufacturing cost will be less than a linear positioner because fewer bearings and precision bars are needed. Significantly, a rotary positioner can be made faster since its inertia is smaller. With a linear positioner all parts move at the same speed. In a rotary positioner, only the heads move at full speed, as the parts closer to the shaft must move more slowly. The principle of many rotary positioners is exactly that of a moving-coil ammeter, where current is converted directly into torque.

One disadvantage of rotary positioners is that there is a component of windage on the heads that tends to pull the positioner in towards the spindle. In linear positioners windage is at right angles to motion and can be neglected. Windage can be overcome in rotary positioners by feeding the current cylinder address to a ROM that sends a code to a DAC. This produces an offset voltage that is fed to the positioner driver to generate a torque to balance the windage whatever the position of the heads.

When extremely small track spacing is contemplated, it cannot be assumed that all the heads will track the servo head due to temperature gradients. In this case the embedded-servo approach must be used, where each head has its own alignment patterns. The servo surface is often retained in such drives to allow coarse positioning, velocity feedback and index and write-clock generation, in addition to locating the guard bands for landing the heads.

Winchester drives have been made with massive capacity, but the problem of backup is then magnified, and the general trend has been for the physical size of the drive to come down as the storage density increases in order to improve access time. Very small Winchester disk drives are now available which plug into standard integrated circuit sockets. These are competing with RAM for memory applications where non-volatility is important.

9.11 The disk controller

A disk controller is a unit interposed between the drives and the rest of the system. It consists of two main parts; that which issues control signals to and obtains status from the drives, and that which handles the data to be stored and retrieved. Operation of the two parts is synchronized by the sequencer. The essentials of a disk controller are determined by the characteristics of drives and the functions needed, and so they do not vary greatly. It is desirable for economic reasons to use a commercially available disk controller intended for computers. Such controllers are adequate for still-store applications, but cannot support the data rate required for real-time moving video unless data reduction is employed. Disk drives are generally built to interface to a standard controller interface, such as the SCSI bus. The disk controller will then be a unit that interfaces the drive bus to the host computer system.

The execution of a function by a disk subsystem requires a complex series of steps, and decisions must be made between the steps to decide what the next will be. There is a parallel with computation, where the function is the equivalent of an instruction, and the sequencer steps needed are the equivalent of the microinstructions needed to execute the instruction. The major failing in this analogy is that the sequence in a disk drive must be accurately synchronized to the rotation of the disk.

Most disk controllers use direct memory access, which means that they have the ability to transfer disk data in and out of the associated memory without the assistance of the processor. In order to cause a file transfer, the disk controller must be told the physical disk address (cylinder, sector, track), the physical memory address where the file begins, the size of the file and the direction of transfer (read or write). The controller will then position the disk heads, address the memory, and transfer the samples. One disk transfer may consist of many contiguous disk blocks, and the controller will automatically increment the disk-address registers as each block is completed. As the disk turns, the sector address increases until the end of the track is reached. The track or head address will then be incremented and the sector address reset so that transfer continues at the beginning of the next track. This process continues until all of the heads have been used in turn. In this case both the head address and sector address will be reset, and the cylinder address will be incremented, which causes a seek. A seek which takes place because of a data transfer is called an implied seek, because it is not necessary formally to instruct the system to perform it. As disk drives are block-structured devices, and the error correction is codeword-based, the controller will always complete a block even if the size of the file is less than a whole number of blocks. This is done by packing the last block with zeros.

The status system allows the controller to find out about the operation of the drive, both as a feedback mechanism for the control process, and to handle any errors. Upon completion of a function, it is the status system that interrupts the control processor to tell it that another function can be undertaken.

In a system where there are several drives connected to the controller via a common bus, it is possible for non data-transfer functions such as seeks to take place in some drives simultaneously with a data transfer in another.

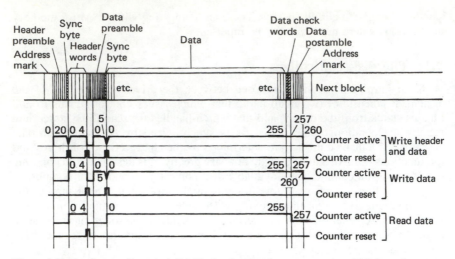

Figure 9.11 The format of a typical disk block related to the count process which is used to establish where in the block the head is at any time. During a read the count is derived from the actual data read, but during a write, the count is derived from the write clock.

Before a data transfer can take place, the selected drive must physically access the desired block, and confirm this by reading the block header. Following a seek to the required cylinder, the positioner will confirm that the heads are on track and settled. The desired head will be selected, and then a search for the correct sector begins. This is done by comparing the number of the desired sector with the contents of the current sector register that is typically incremented by dividing down servo-surface pulses. When the two counts are equal, the head is about to enter the desired block. Figure 9.11 shows the structure of a typical magnetic disk track. In between blocks are placed address marks, which are areas without transitions that the read circuits can detect. Following detection of the address mark, the sequencer is roughly synchronized to begin handling the block. As the block is entered, the data separator locks to the preamble, and in due course the sync pattern will be found. This sets to zero a counter which divides the data-bit rate by eight, allowing the serial recording to be correctly assembled into bytes, and also allowing the sequencer to count the position of the head through the block in order to perform all the necessary steps at the right time.

The first header word is usually the cylinder address, and this is compared with the contents of the desired cylinder register. The second header word will contain the sector and track address of the block, and these will also be compared with the desired addresses. There may also be bad-block flags and/or defect-skipping information. At the end of the header is a CRCC that will be used to ensure that the header was read correctly. Figure 9.12 shows a flowchart of the position verification, after which a data transfer can proceed. The header reading is completely automatic. The only time it is necessary formally to command a header to be read is when checking that a disk has been formatted correctly.

During the read of a data block, the sequencer is employed again. The sync pattern at the beginning of the data is detected as before, following which the actual data arrive. These bits are converted to byte or sample parallel, and sent

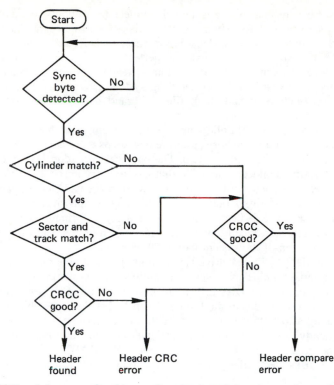

Figure 9.12 The vital process of position confirmation is carried out in accordance with the above flowchart. The appropriate words from the header are compared in turn with the contents of the disk-address registers in the subsystem. Only if the correct header has been found and read properly will the data transfer take place.

to the memory by DMA. When the sequencer has counted the last data-byte off the track, the redundancy for the error-correction system will be following.

During a write function, the header-check function will also take place as it is perhaps even more important not to write in the wrong place on a disk. Once the header has been checked and found to be correct, the write process for the associated data block can begin. The preambles, sync pattern, data block, redundancy and postamble have all to be written contiguously. This is taken care of by the sequencer, which is obtaining timing information from the servo surface to lock the block structure to the angular position of the disk. This should be contrasted with the read function, where the timing comes directly from the data.

9.12 Defect handling

The protection of data recorded on disks differs considerably from the approach used on other media in digital audio. This has much to do with the intolerance of data processors to errors when compared with video data. In particular, it is not possible to interpolate to conceal errors in a computer program or a data file.

In the same way that magnetic tape is subject to dropouts, magnetic disks suffer from surface defects whose effect is to corrupt data. The shorter wavelengths employed as disk densities increase are affected more by a given size of defect. Attempting to make a perfect disk is subject to a law of diminishing returns, and eventually a state is reached where it becomes more cost-effective to invest in a defect-handling system.

In the construction of bad-block files, a brand new disk is tested by the operating system. Known patterns are written everywhere on the disk, and these are read back and verified. Following this the system gives the disk a volume name, and creates on it a directory structure that keeps records of the position and size of every file subsequently written. The physical disk address of every block that fails to verify is allocated to a file that has an entry in the disk directory. In this way, when genuine data files come to be written, the bad blocks appear to the system to be in use storing a fictitious file, and no attempt will be made to write there. Some disks have dedicated tracks where defect information can be written during manufacture or by subsequent verification programs, and these permit a speedy construction of the system bad-block file.

Whilst the MTBF of a disk drive is very high, it is a simple matter of statistics that when a large number of drives is assembled in a system the time between failures becomes shorter. Disk drives are sealed units and the disks cannot be removed if there is an electronic failure. Even if this were possible the system cannot usually afford downtime whilst such a data recovery takes place.

Consequently any system in which the data are valuable must take steps to ensure data integrity. This is commonly done using RAID (redundant array of inexpensive disks) technology. Figure 9.13 shows that in a RAID array data blocks are spread across a number of drives.

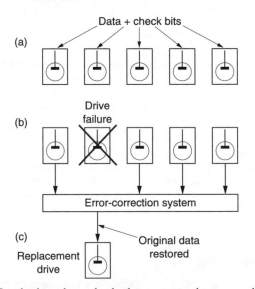

Figure 9.13 In RAID technology, data and redundancy are spread over a number of drives (a). In the case of a drive failure (b) the error-correction system can correct for the loss and continue operation. When the drive is replaced (c) the data can be rewritten so that the system can then survive a further failure.

An error-correcting check symbol (typically Reed–Solomon) is stored on a redundant drive. The error correction is powerful enough to fully correct any error in the block due to a single failed drive. In RAID arrays the drives are designed to be hot-plugged (replaced without removing power) so if a drive fails it is simply physically replaced with a new one. The error-correction system will rewrite the new drive with the data that were lost with the failed unit.

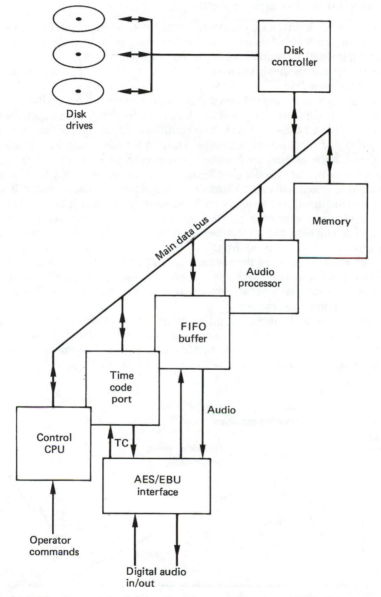

Figure 9.14 The main parts of a digital audio disk system. Memory and FIFO allow continuous audio despite the movement of disk heads between blocks.

When a large number of disk drives are arrayed together, it is necessary and desirable to spread files across all the drives in a RAID array. Whilst this ensures data integrity, it also can also mean that the data transfer rate may be multiplied by the number of drives sharing the data. The data transfer rate can usefully be increased and new approaches are necessary to move the data in and out of the disk system.

9.13 Digital audio disk system

In order to use disk drives for the storage of audio samples, a system like the one shown in Figure 9.14 is needed. The control computer determines where and when samples will be stored and retrieved, and sends instructions to the disk controller that causes the drives to read or write, and transfers samples between them and the memory.

When audio samples are fed into a disk-based system, from a digital interface or from an A/D converter, they will be placed in a memory, from which the disk controller will read them by DMA. The continuous-input sample stream will be split up into disk blocks for disk storage. The disk transfers must by definition be intermittent, because there are headers between contiguous sectors. Once all the sectors on a particular cylinder have been used, it will be necessary to seek to the next cylinder, which will cause a further interruption to the data transfer. If a bad block is encountered, the sequence will be interrupted until it has passed. The instantaneous data rate of a parallel transfer drive is made higher than the continuous audio data rate, so that there is time for the positioner to move whilst the audio output is supplied from the buffer memory. In replay, the drive controller attempts to keep the memory as full as possible by issuing a read command as soon as one block space appears in the memory. This allows the maximum time for a seek to take place before reading must resume. Figure 9.15 shows the action of the memory.

Whilst recording, the drive controller attempts to keep the buffer as empty as possible by issuing write commands as soon as a block of data is present. In this way the amount of time available to seek is maximized in the presence of a continuous audio sample input.

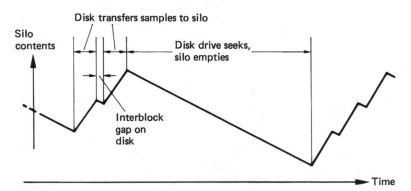

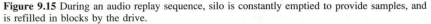

Figure 9.15 During an audio replay sequence, silo is constantly emptied to provide samples, and is refilled in blocks by the drive.

9.14 Arranging the audio data on disk

When playing a tape recording or a disk having a spiral track, it is only necessary to start in the right place, and the data are automatically retrieved in the right order. Such media are also driven at a speed that is proportional to the sampling rate. In contrast, a hard disk has a discontinuous recording and acts more like a RAM in that it must be addressed before data can be retrieved. The rotational speed of the disk is constant and not locked to anything. A vital step in converting a disk drive into an audio recorder is to establish a link between the time through the recording and the location of the data on the disk.

When audio samples are fed into a disk-based system, from an AES/EBU interface or from a convertor, they will be placed initially in RAM, from which the disk controller will read them by DMA. The continuous-input sample stream will be split up into disk blocks for disk storage. The AES/EBU interface carries a timecode in the channel status data, and this timecode, or that from a local generator, will be used to assemble a table that contains a conversion from real time in the recording to the physical disk address of the corresponding audio files. As an alternative, an interface may be supplied which allows conventional SMPTE or EBU timecode to be input. Wherever possible, the disk controller will allocate incoming audio samples to contiguous disk addresses, since this eases the conversion from timecode to physical address.[1] This is not, however, always possible in the presence of defective blocks, or if the disk has become chequerboarded from repeated rerecording.

The table of disk addresses will also be made into a named disk file and stored in an index that will be in a different area of the disk from the audio files. Several recordings may be fed into the system in this way, and each will have an entry in the index.

If it is desired to play back one or more of the recordings, then it is only necessary to specify the starting timecode and the filename. The system will look up the index file in order to locate the physical address of the first and subsequent sample blocks in the desired recording, and will begin to read them from disk and write them into the RAM. Once the RAM is full, the real time replay can begin by sending samples from RAM to the output or to local convertors. The sampling rate clock increments the RAM address and the timecode counter. Whenever a new timecode frame is reached, the corresponding disk address can be obtained from the index table, and the disk drive will read a block in order to keep the RAM topped up.

The disk transfers must by definition take varying times to complete because of the rotational latency of the disk. Once all the sectors on a particular cylinder have been read, it will be necessary to seek to the next cylinder, which will cause a further extension of the reading sequence. If a bad block is encountered, the sequence will be interrupted until it has passed. The RAM buffering is sufficient to absorb all of these access time variations. Thus the RAM acts as a delay between the disk transfers and the sound that is heard. A corresponding advance is arranged in timecodes fed to the disk controller. In effect the actual timecode has a constant added to it so that the disk is told to obtain blocks of samples in advance of real time. The disk takes a varying time to obtain the samples, and the RAM then further delays them to the correct timing. Effectively the disk/RAM subsystem is a timecode-controlled memory. One need only put in the time, and out comes the audio corresponding to that time. This is the characteristic of an audio synchronizer. In most audio equipment the synchronizer is extra; the hard disk needs one to work at all, and so every hard disk comes with a free synchronizer. This makes disk-based systems very flexible as they can be made to lock to almost any reference and care little what sampling rate is used or if it varies. They perform well locked to videotape or film via timecode because no matter how the pictures are shuttled or edited, the timecode link always produces the correct sound to go with the pictures.

A multi-track recording can be stored on a single disk and, for replay, the drive will access the files for each track faster than real time so that they all become present in the memory simultaneously. It is not, however, compulsory to play back the tracks in their original time relationship. For the purpose of synchronization,[2] or other effects, the tracks can be played with any time relationship desired, a feature not possible with multi-track tape drives.

In order to edit the raw audio files fed into the system, it is necessary to listen to them in order to locate the edit points. This can be done by playback of the whole file at normal speed if time is no object, but this neglects the random access capability of a disk-based system. If an event list has been made at the time of the recordings, it can be used to access any part of them within a few tens of milliseconds, which is the time taken for the heads to traverse the entire disk surface. This is far superior to the slow spooling speed of tape recorders.

9.15 Spooling files

If an event list is not available, it will be necessary to run through the recording at a raised speed in order rapidly to locate the area of the desired edit. If the disk can access fast enough, an increase of up to ten times normal speed can be achieved simply by raising the sampling-rate clock, so that the timecode advances more rapidly, and new data blocks are requested from the disk more rapidly. If a constant sampling-rate output is needed, then rate reduction via a digital filter will be necessary.[3,4] Some systems have sophisticated signal processors which allow pitch changing, so that files can be played at non-standard speed but with normal pitch or vice versa.[5] If higher speeds are required, an alternative approach to processing on playback only is to record spooling files[6] at the same time as an audio file is made. A spooling file block contains a sampling-rate-reduced version of several contiguous audio blocks. When played at standard sampling rate, it will sound as if it is playing faster by the factor of rate reduction employed. The spooling files can be accessed less often for a given playback speed, or higher speed is possible within a given access-rate constraint.

Once the rough area of the edit has been located by spooling, the audio files from that area can be played to locate the edit point more accurately. It is often not sufficiently accurate to mark edit points on the fly by listening to the sound at normal speed. In order to simulate the rock-and-roll action of edit-point location in an analog tape recorder, audio blocks in the area of the edit point can be transferred to memory and accessed at variable speed and in either direction by deriving the memory addresses from a hand-turned rotor.

9.16 Broadcast applications

In a radio broadcast environment it is possible to contain all the commercials and jingles in daily use on a disk system thus eliminating the doubtful quality of analog cartridge machines.[7] Disk files may be cued almost instantly by specifying the file name of the wanted piece, and once they are RAM-resident play instantly they are required. Adding extra output modules means that several audio files can be played back simultaneously if a station broadcasts on more than one channel. If a commercial break contains several different spots, these can be chosen at short notice just by producing a new edit list.

9.17 Sampling rate and playing time

The bit rate of a digital audio system is such that high-density recording is mandatory for long playing time. A disk drive can never reach the density of a rotary-head tape machine because it is optimized for fast random access, and so it would be unwise to expect too much of a disk-based system in terms of playing time. In practice, the editing power of a disk-based system far outweighs this restriction.

One high-quality digital audio channel requires nearly a megabit per second, which means that a megabyte of storage (the usual unit for disk measurement) offers about ten seconds of monophonic audio. Using this factor, the playing time can soon be obtained from the drive capacity. There is, however, no compulsion to devote the whole disk to one audio channel, and so two channels could be recorded for half as long, or four channels for one quarter as long and so on. Where compression is used, these playing times can be extended by multiplication by the compression factor.

For broadcast applications, where an audio bandwidth of 15 kHz is imposed by the FM stereo transmission standard, the alternative sampling rate of 32 kHz can be used, which allows about an hour of uncompressed monophonic digital audio from 300 megabytes. Where only speech is required, an even lower rate can be employed. A server, possibly using a RAID array, will be necessary in most musical post-production applications if stereo or multi-track working is contemplated. In practice, multi-track working with disks is better than these calculations would indicate, because on a typical multi-track master tape, all tracks are not recorded continuously. Some tracks will contain only short recordings in a much longer overall session. A tape machine has no option but to leave these tracks unrecorded either side of the wanted recording, whereas a disk system will only store the actual wanted samples. The playing time in a real application will thus be greater than expected.

A further consideration is that hard disks systems do not need to edit the actual data files on disk. The editing is performed in the memory of the control system and is repeated dynamically under the control of an EDL (edit decision list) each time the edited work is required. Thus a lengthy editing session on a hard disk system does not result in the disk becoming fuller as only a few bytes of EDL are generated.

References

1. McNally, G.W., Gaskell, P.S. and Stirling, A.J., Digital audio editing. *BBC Research Dept Report*, RD 1985/10
2. McNally, G.W., Bloom, P.J. and Rose, N.J., A digital signal processing system for automatic dialogue post-synchronisation. Presented at the 82nd Audio Engineering Society Convention (London, 1987), Preprint 2476(K-6)

3. McNally, G.W., Varispeed replay of digital audio with constant output sampling rate. Presented at the 76th Audio Engineering Society Convention (New York, 1984), Preprint 2137(A-9)

4. Gaskell, P.S., A hybrid approach to the variable speed replay of audio. Presented at the 77th Audio Engineering Society Convention (Hamburg, 1985), Preprint 2202(B-1)

5. Gray, E., The Synclavier digital audio system: recent developments in audio post production. *Int. Broadcast Eng.*, **18**, 55 (March 1987)

6. McNally, G.W., Fast edit-point location and cueing in disk-based digital audio editing. Presented at the 78th Audio Engineering Society Convention (Anaheim, 1985), Preprint 2232(D-10)

7. Itoh, T., Ohta, T. and Sohma, Y., Real time transmission system of commercial messages in radio broadcasting. Presented at the 67th Audio Engineering Society Convention (New York, 1980), Preprint 1682(H-1)

Digital audio editing

Digital audio editing takes advantage of the freedom to store data in any suitable medium and the signal processing techniques developed in computation. This chapter shows how the edit process is achieved using combinations of storage media, processing and control systems.

10.1 Introduction

Editing ranges from a punch-in on a multi-track recorder, or the removal of 'ums and ers' from an interview to the assembly of myriad sound effects and mixing them with timecode-locked dialogue in order to create a film soundtrack. Mastering is a form of editing where various tracks are put together to make a master recording that will be duplicated for general sale. The duration of each musical piece, the length of any pauses between pieces and the relative levels of the pieces on the disk have to be determined at the time of mastering. The master recording will be compiled from source media that may each contain only some of the pieces required on the final CD, in any order. The recordings will vary in level, and may contain several retakes of a passage.

The purpose of the digital mastering editor is to take each piece, and insert sections from retakes to correct errors, and then to assemble the pieces in the correct order, with appropriate pauses between and with the correct relative levels to create the master tape.

Digital audio editors work in two basic ways, by assembling or by inserting sections of audio waveform to build the finished waveform. Both terms have the same meaning as in the context of video recording. Assembly begins with a blank master file or recording. The beginning of the work is copied from the source, and new material is successively appended to the end of the previous material. Figure 10.1 shows how a master recording is made from source recordings by the process of assembly. Insert editing begins with an existing recording in which a section is replaced by the edit process. Punch-in in multi-track recorders is a form of insert-editing.

10.2 Editing with random access media

In all types of audio editing the goal is the appropriate sequence of sounds at the appropriate times. In analog audio equipment, editing was almost always

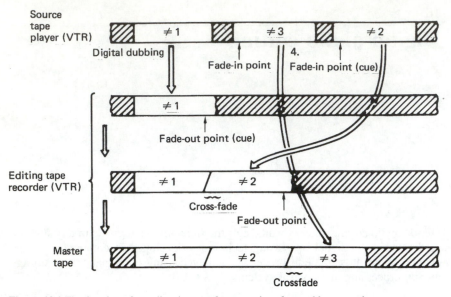

Figure 10.1 The function of an editor is to perform a series of assembles to produce a master tape from source tapes.

performed using tape or magnetically striped film. These media have the characteristic that the time through the recording is proportional to the distance along the track. Editing consisted of physically cutting and splicing the medium, in order to mechanically assemble the finished work, or of copying lengths of source medium to the master.

Whilst open-reel digital audio tape formats support splice editing, in all other digital audio editing samples from various sources are brought from the storage media to various pages of RAM. The edit is performed by crossfading between sample streams retrieved from RAM and by subsequently rewriting on the output medium. Thus the nature of the storage medium does not affect the form of the edit in any way except the amount of time needed to execute it.

Tapes only allow serial access to data, whereas disks and RAM allow random access and so can be much faster. Editing using random access storage devices is very powerful as the shuttling of tape reels is avoided. The technique is often called non-linear editing.

10.3 Editing on recording media

All digital recording media use error correction which requires an interleave, or reordering, of samples to reduce the impact of large errors, and the assembling of many samples into an error correcting codeword. Codewords are recorded in constant-sized blocks on the medium. Audio editing requires the modification of source material in the correct real-time sequence to sample accuracy. This contradicts the interleaved block based codes of real media.

Editing to sample accuracy simply cannot be performed directly on real media. Even if an individual sample could be located in a block, replacing the samples after it would destroy the codeword structure and render the block uncorrectable.

The only solution is to ensure that the medium itself is only edited at block boundaries so that entire error correction codewords are written down. In order to obtain greater editing accuracy, blocks must be read from the medium and de-interleaved into RAM, modified there and re-interleaved for writing back on the medium, the so called *read-modify-write* process.

In disks, blocks are often associated into clusters consisting of a fixed number of blocks in order to increase data throughput. When clustering is used, editing on the disk can only take place by rewriting entire clusters.

10.4 The structure of an editor

The digital audio editor consists of three main areas. First, the various contributory recordings must enter the processing stage at the right time with respect to the master recording. This will be achieved using a combination of timecode, transport synchronization and RAM timebase correction. The synchronizer will take control of the various transports during an edit so that one section reaches its out-point just as another reaches its in-point. Second, the audio

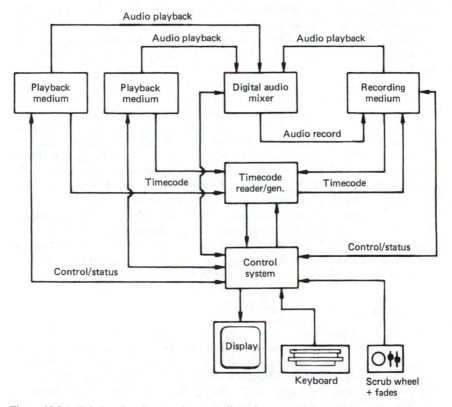

Figure 10.2 A digital audio editor requires an audio path to process the samples, and a timing and synchronizing section to control the time alignment of signals from the various sources. A supervisory control system acts as the interface between the operator and the hardware.

signal path of the editor must take the appropriate action, such as a crossfade, at the edit point. This requires some digital processing circuitry. Third, the editing operation must be supervised by a control system which coordinates the operation of the transports and the signal processing to achieve the desired result.

Figure 10.2 shows a simple block diagram of an editor. Each source device, be it disk, tape or some other medium must produce timecode locked to the audio samples. The synchronizer section of the control system uses the timecode to determine the relative timing of sources and sends remote control signals to the transport to make the timing correct. The master recorder is also fed with timecode in such a way that it can make a contiguous timecode track when performing assembly edits. The control system also generates a master sampling rate clock to which contributing devices must lock in order to feed samples into the edit process. The audio signal processor takes contributing sources and mixes them as instructed by the control system. The mix is then routed to the recorder.

10.5 Timecode

Synchronization between timecode and the sampling rate is essential, otherwise there will be a conflict between the need to lock the various sampling rates in the system with the need to lock the timecodes. This can only be resolved with synchronous timecode. The EBU timecode format relates easily to digital audio sampling rates of 48 kHz, 44.1 kHz and 32 kHz, but it is not so easy with the drop-frame SMPTE timecode necessary for NTSC recording due to the 0.1 per cent slip between the actual field rate and 60 Hz.

The timecode used in the SMPTE standard for 525/60 is shown in Figure 10.3. PAL VTRs use EBU timecode that is basically similar to SMPTE. These store hours, minutes, seconds and frames as binary-coded decimal (BCD) numbers, which are serially encoded along with user bits into an FM channel code (see Chapter 6) which is recorded on one of the linear audio tracks of the tape. Disks also use timecode for audio synchronization, but the timecode forms part of the access mechanism so that samples may be retrieved by specifying the required timecode. This mechanism was detailed in Chapter 9.

A further problem with the use of video-based timecode is that the accuracy to which the edit must be made in audio is much greater than the frame boundary accuracy needed in video. When the exact edit point is chosen in an audio editor, it will be described to great accuracy and is stored as hours, minutes, seconds, frames and the number of the sample within the frame.

10.6 Locating the edit point

Digital audio editors must simulate the 'rock and roll' process of edit-point location in analog tape recorders where the tape reels are moved to and fro by hand. The solution is to transfer the recording in the area of the edit point to RAM in the editor. RAM access can take place at any speed or direction and the precise edit point can then be conveniently found by monitoring audio from the RAM.

Figure 10.4 shows how the area of the edit point is transferred to the memory. The source device is commanded to play, and the operator listens to replay

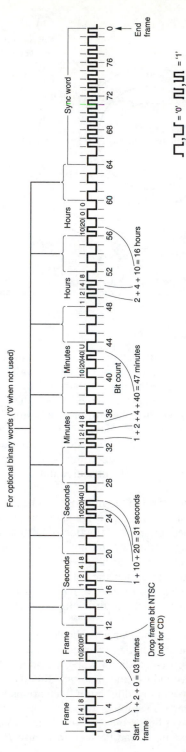

For optional binary words ('0' when not used)

Frame Frame Seconds Seconds Minutes Minutes Hours Hours Sync word

1|2|4|8   10|20|0|FI   1|2|4|8   10|20|40|U   1|2|4|8   10|20|40|U   1|2|4|8   10|20|0|0

0   4   8   12   16   20   24   28   32   36   40   44   48   52   56   60   64   68   72   76   0

Start   End
frame   frame

1 + 2 + 0 = 03 frames

Drop frame bit NTSC
(not for CD)

1 + 10 + 20 = 31 seconds

1 + 2 + 4 + 40 = 47 minutes

Bit count

2 + 4 + 10 = 16 hours

Π,⎸⎹ = '0'   ⊓⎹,⎸⊓ = '1'

Figure 10.3 In SMPTE standard timecode, the frame number and time are stored as eight BCD symbols. There is also space for 32 user-defined bits. The code repeats every frame. Note the asymmetrical sync word which allows the direction of tape movement to be determined.

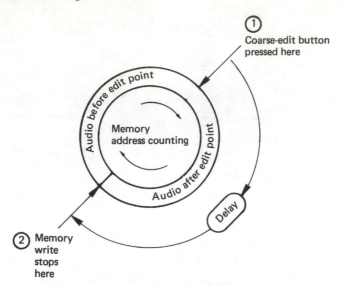

Figure 10.4 The use of a ring memory which overwrites allows storage of samples before and after the coarse edit point.

samples via a DAC in the monitoring system. The same samples are continuously written into a memory within the editor. This memory is addressed by a counter which repeatedly overflows to give the memory a ring-like structure rather like that of a timebase corrector, but somewhat larger. When the operator hears the rough area in which the edit is required, he will press a button. This action stops the memory writing, not immediately, but one half of the memory contents later. The effect is then that the memory contains an equal number of samples before and after the rough edit point. Once the recording is in the memory, it can be accessed at leisure, and the constraints of the source device play no further part in the edit-point location.

There are a number of ways in which the memory can be read. If the memory address is supplied from a counter clocked at the appropriate rate, the edit area can be replayed at normal speed, or at some fraction of normal speed repeatedly. In order to simulate the analog method of finding an edit point, the operator is provided with a *scrub wheel* or rotor, and the memory address will change at a rate proportional to the speed with which the rotor is turned, and in the same direction. Thus the sound can be heard forward or backward at any speed, and the effect is exactly that of manually rocking an analog tape past the heads of an ATR.

The operation of a scrub wheel encoder was shown in Chapter 3. Although a simple device, there are some difficulties to overcome. There are not enough pulses per revolution to create a clock directly and the human hand cannot turn the rotor smoothly enough to address the memory directly without flutter. A phase-locked loop is generally employed to damp fluctuations in rotor speed and multiply the frequency. A standard sampling rate must be recreated to feed the monitor DAC and a rate convertor, or interpolator, is necessary to restore the sampling rate to normal. These items can be seen in Figure 10.5.

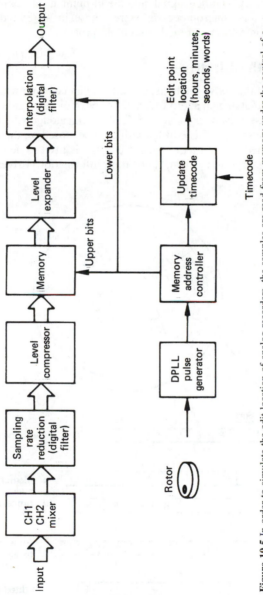

Figure 10.5 In order to simulate the edit location of analog recorders, the samples are read from memory under the control of a hand-operated rotor.

The act of pressing the coarse edit-point button stores the timecode of the source at that point, which is frame-accurate. As the rotor is turned, the memory address is monitored, and used to update the timecode to sample accuracy. Before assembly can be performed, two edit points must be determined, the out-point at the end of the previously recorded signal, and the in-point at the beginning of the new signal. The editor's microprocessor stores these in an edit decision list (EDL) in order to control the automatic assemble process.

10.7 Editing with disk drives

Using one or other of the above methods, an edit list can be made which contains an in-point, an out-point and an audio filename for each of the segments of audio which need to be assembled to make the final work, along with a crossfade period and a gain parameter. This edit list will also be stored on the disk. When a preview of the edited work is required, the edit list is used to determine what files will be necessary and when, and this information drives the disk controller.

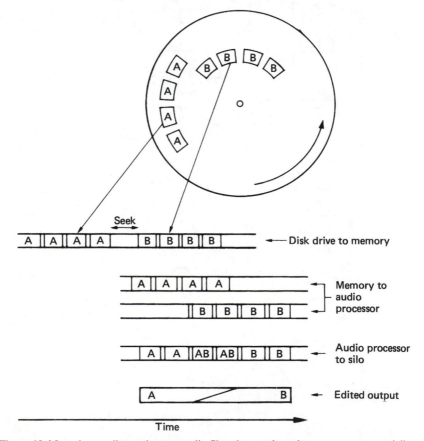

Figure 10.6 In order to edit together two audio files, they are brought to memory sequentially. The audio processor accesses file pages from both together, and performs a crossfade between them. The silo produces the final output at constant steady-sampling rate.

Figure 10.6 shows the events during an edit between two files. The edit list causes the relevant audio blocks from the first file to be transferred from disk to memory. These blocks will be accessed by the signal processor in order to produce the preview output. As the edit point approaches, the disk controller will also place blocks from the incoming file into the memory. It can do this because the rapid data-transfer rate of the drive allows blocks to be transferred to memory much faster than real time, leaving time for the positioner to seek from one file to another. In different areas of the memory there will be simultaneously the end of the outgoing recording and the beginning of the incoming recording. The signal processor will use the fine edit-point parameters to work out the relationship between the actual edit points and the cluster boundaries. The relationship between the cluster on disk and the RAM address to which it was transferred is known, and this allows the memory address to be computed in order to obtain samples with the correct timing.

Before the edit point, only samples from the outgoing recording are accessed, but as the crossfade begins, samples from the incoming recording are also accessed, multiplied by the gain parameter and then mixed with samples from the outgoing recording according to the crossfade period required. The output of the signal processor becomes the edited preview material, which can be checked for the required subjective effect. If necessary the in- or out-points can be trimmed, or the crossfade period changed, simply by modifying the edit-list file. The preview can be repeated as often as needed, until the desired effect is obtained. At this stage the edited work does not exist as a file, but is recreated each time by a further execution of the EDL. Thus a lengthy editing session need not fill up the disk.

It is important to realize that at no time during the edit process were the original audio files modified in any way. The edit process was performed solely by reading the audio files. The power of this approach is that if an edit list is created wrongly, the original recording is not damaged, and the problem can be put right simply by correcting the edit list. The advantage of a disk-based system for such work is that location of edit points, previews and reviews are all performed almost instantaneously, because of the random access of the disk. This can reduce the time taken to edit a program to a quarter of that needed with a tape machine.[1]

During an edit, the disk drive has to provide audio files from two different places on the disk simultaneously, and so it has to work much harder than for a simple playback. If there are many close-spaced edits, the drive may be hard-pressed to keep ahead of real time, especially if there are long crossfades, because during a crossfade the source data rate is twice as great as during replay. A large buffer memory helps this situation because the drive can fill the memory with files before the edit actually begins, and thus the instantaneous sample rate can be met by the memory's emptying during disk-intensive periods. In practice crossfades measured in seconds can be achieved in a disk-based system, a figure not matched by tape systems.

Once the editing is finished, it will generally be necessary to transfer the edited material to form a contiguous recording so that the source files can make way for new work. If the source files already exist on tape the disk files can simply be erased. If the disks hold original recordings they will need to be backed up to tape if they will be required again. In large broadcast systems, the edited work can be broadcast directly from the disk file. In smaller systems it will be necessary to output to some removable medium, since the Winchester drives in the editor have fixed media. It is only necessary to connect the AES/EBU output of the signal processor to any type of digital recorder, and then the edit list is executed once more. The edit sequence will be performed again, exactly as it was during the last preview, and the results will be recorded on the external device.

10.8 Editing in DAT

In order to edit a DAT tape, many of the constraints of video editing apply. Editing can only take place at the beginning of an interleave block, known as a frame, which is contained in two diagonal tracks. The transport would need to perform a preroll, starting before the edit point, so that the drum and capstan servos would be synchronized to the tape tracks before the edit was reached. Fortunately, the very small drum means that mechanical inertia is minute by the standards of video recorders, and lock-up can be very rapid.

Although editing can be done on a DAT machine that can only record or play, a better solution, used in professional machines, is to fit two sets of heads in the drum. The standard permits the drum size to be increased and the wrap angle to be reduced provided that the tape tracks are recorded to the same dimensions. In normal recording, the first heads to reach the tape tracks would make the recording, and the second set of heads would be able to replay the recording immediately afterwards for confidence monitoring. For editing, the situation would be reversed. The first heads to meet a given tape track would play back the existing recording, and this would be de-interleaved and corrected, and presented as a sample stream to the record circuitry. The record circuitry would then interleave the samples ready for recording. If the heads are mounted a suitable distance apart in the scanner along the axis of rotation, the time taken for tape to travel from the first set of heads to the second will be equal to the decode/encode delay. If this process goes on for a few blocks, the signal going to the record head will be exactly the same as the pattern already on the tape, so the record head can be switched on at the beginning of an interleave block. Once this has been done, new material can be crossfaded into the sample stream from the advanced replay head, and an edit will be performed.

If insert editing is contemplated, following the above process, it will be necessary to crossfade back to the advanced replay samples before ceasing

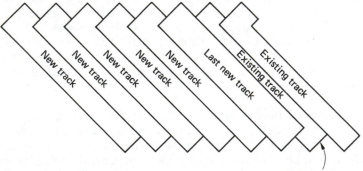

Half-width track

Figure 10.7 When editing a small track-pitch recording, the last track written will be 1.5 times the normal track width, since that is the width of the head. This erases half of the next track of the existing recording.

rerecording at an interleave block boundary. The use of overwrite to produce narrow tracks causes a problem at the end of such an insert. Figure 10.7 shows that this produces a track half the width it should be. Normally the error-correction system would take care of the consequences, but if a series of inserts were made at the same point in an attempt to make fine changes to an edit, the result could be an extremely weak signal for the duration of one track. One solution is to incorporate an algorithm into the editor so that the points at which the tape begins and ends recording change on every attempt. This does not affect the audible result as this is governed by the times at which the crossfader operates.

10.9 Editing in open-reel digital recorders

On many occasions in studio recording it is necessary to replace a short section of a long recording, because a wrong note was played or something fell over and made a noise. The tape is played back to the musicians before the bad section, and they play along with it. At a musically acceptable point prior to the error, the tape machine passes into record, a process known as punch-in, and the offending section is rerecorded. At another suitable time, the machine ceases recording at the punch-out point, and the musicians can subsequently stop playing.

Once more, a read-modify-write approach is necessary, using a record head positioned *after* the replay head. The mechanism necessary is shown in Figure 10.8 Prior to the punch-in point, the replay-head signal is de-interleaved, and this signal is fed to the record channel. The record channel re-interleaves the samples, and after some time will produce a signal identical to what is already on the tape. At a block boundary the record current can be turned on, when the existing recording will be rerecorded. At the punch-in point, the samples fed to the record encoder will be crossfaded to samples from the ADC. The crossfade takes place in the non-interleaved domain. The new recording is made to replace the unsatisfactory section, and at the end, punch-out is performed by crossfading to the samples from the replay head. After some time, the record head will once more be rerecording what is already on the tape, and at a block boundary the record current can be switched off. The crossfade duration can be chosen according to the nature of the recorded material. It is possible to rehearse the punch-in process and monitor what it would sound like by feeding headphones from the crossfader, and doing everything described except that the record head is disabled. The punch-in and punch-out points can then be moved to give the best subjective result. The machine can learn the sector addresses at which the punches take place, so the final punch is fully automatic.

Assemble editing, where parts of one or more source tapes are dubbed from one machine to another to produce a continuous recording, is performed in the same way as a punch-in, except that the punch-out never comes. After the new recording from the source machine is faded in, the two machines continue to dub until one of them is stopped. This will be done some time after the next assembly point is reached.

10.10 Jump editing

Conventional splice handling in stationary head recorders was detailed in Chapter 8. In an extension to the principle, suggested by Lagadec,[2] the samples

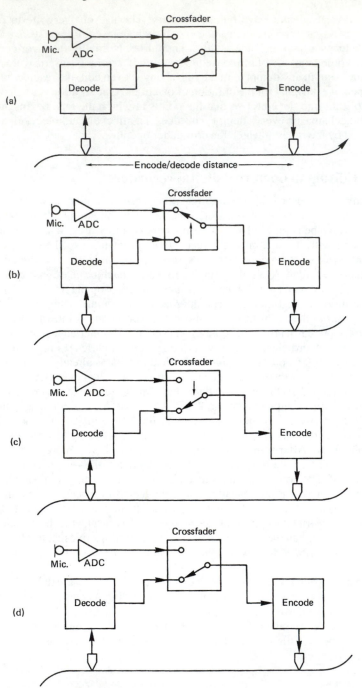

Figure 10.8 The four stages of an insert (punch-in/out) with interleaving: (a) rerecord existing samples for at least one constraint length; (b) crossfade to incoming samples (punch-in point); (c) crossfade to existing replay samples (punch-out point); (d) rerecord existing samples for at least one constraint length. An assemble edit consists of steps (a) and (b) only.

from the area of the splice are not heard. Instead an electronic edit is made between the samples before the splice and those after.

In this system, a tape splice is made physically with excess tape adjacent to the intended edit points. The timebase corrector has two read-address generators that can access the memory independently. It will be seen in Figure 10.9 that when the machine plays the tape, the capstan is phase-advanced so that the timebase corrector is causing a long delay to compensate. As the splice is detected, the corruption due to the splice enters the TBC memory and travels towards the output. As the splice nears the end of the memory, the machine output crossfades to a signal from the second TBC output that has been delayed much less. The data in the area of the tape splice are thus omitted. The capstan will now be effectively lagging because the delay has been shortened, and it will speed up slightly for a short period until the lead condition is re-established. This can be done without

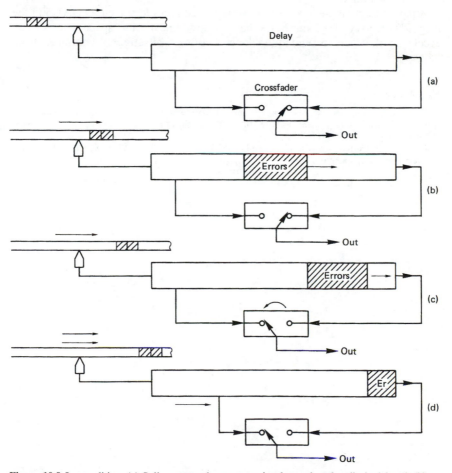

Figure 10.9 Jump editing. (a) Splice approaches, capstan is advanced, and audio is delayed. (b) Splice passes head, and error burst travels down delay. (c) Crossfader fades to signal after splice. (d) Capstan accelerates, and delay increases. When the delay tap reaches the end, the crossfader can switch back ready for the next splice.

ill effect since the sample rate from the memory remains constant throughout. Although the splice is an irrevocable mechanical act, the precise edit timing can be changed at will by controlling the sector address at which the TBC jumps, which determines the out-point, and the address difference, which determines the length of tape omitted, and thus controls the in-point. The size of the jump is limited by the available memory.

If only a short section of audio is to be removed, no splice is necessary at all as a memory jump can be used to omit a short length of the recording. Such a system would be excellent for news broadcasts where it is often necessary to remove many short sections of tape to eliminate hesitations and unwanted pauses from interviews. Control of the jumping could be by programming a CPU to recognize timecode or sector addresses and insert the commands, or by inserting the jump distance in the reference track prior to the splice. In either case machines not equipped to jump would handle any splices with mechanically determined timing.

Jump editing can also be used in rotary head recorders such as DAT and the Nagra-D. Rotary head machines have a low linear tape speed and so can accelerate the tape to omit quite long sections whilst replay continues from memory.

References

1. Todoroki, S., *et al.*, New PCM editing system and configuration of total professional digital audio system in near future. Presented at the 80th Audio Engineering Society Convention (Montreux, 1986), Preprint 2319(A8)
2. Lagadec, R., Current status in digital audio. Presented at the IERE Video and Data Recording Conference (Southampton, 1984)

Optical disks in digital audio

Optical disks are particularly important to digital audio, not least because of the success of the Compact Disc, followed by the MiniDisc, DVD and magneto-optical production recorders.

11.1 Types of optical disk

There are numerous types of optical disk, which have different characteristics.[1] There are, however, three broad groups, shown in Figure 11.1, which can be usefully compared.

1 The Compact Disc and the prerecorded MiniDisc and DVD are read-only laser disks, which are designed for mass duplication by stamping. They cannot be recorded.
2 Some laser disks can be recorded, but once a recording has been made, it cannot be changed or erased. These are usually referred to as write-once-read-many (WORM) disks. Recordable CDs (CD-R) and DVDs (DVD-R) work on this principle.
3 Erasable optical disks have essentially the same characteristic as magnetic disks, in that new and different recordings can be made in the same track indefinitely. Recordable MiniDisc and CD-RW are in this category. Sometimes a separate erase process is necessary before rewriting.

The Compact Disc, generally abbreviated to CD, is a consumer digital audio recording which is intended for mass replication. Philips' approach was to invent an optical medium having the same characteristics as the vinyl disk in that it could be mass replicated by moulding or stamping. The information on it is carried in the shape of flat-topped physical deformities in a layer of plastic. Such relief structures lack contrast and must be read with a technique called phase-contrast microscopy that allows an apparent contrast to be obtained using optical interference.

Figure 11.2(a) shows that the information layer of CD and the prerecorded MiniDisc is an optically flat mirror upon which microscopic bumps are raised. A thin coating of aluminium renders the layer reflective. When a small spot of light is focused on the information layer, the presence of the bumps affects the way in which the light is reflected back, and variations in the reflected light are detected in order to read the disk. Figure 11.2 also illustrates the very small dimensions

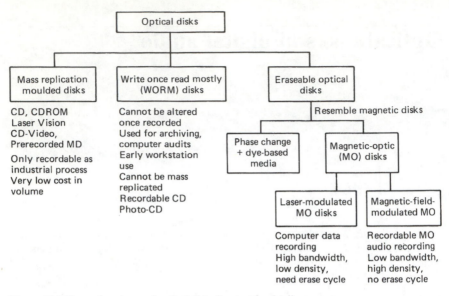

Figure 11.1 The various types of optical disk. See text for details.

common to both disks. For comparison, some sixty CD/MD tracks can be accommodated in the groove pitch of a vinyl LP. These dimensions demand the utmost cleanliness in manufacture.

Figure 11.2(b) shows two types of WORM disk. In the first, the disk contains a thin layer of metal; on recording, a powerful laser melts spots on the layer. Surface tension causes a hole to form in the metal, with a thickened rim around the hole. Subsequently a low-power laser can read the disk because the metal reflects light, but the hole passes it through. Computer WORM disks work on this principle. In the second, the layer of metal is extremely thin, and the heat from the laser heats the material below it to the point of decomposition. This causes gassing which raises a blister or bubble in the metal layer. In a further type, the disk surface is coated with a special dye whose chemical composition is changed by the heat of the recording laser. This changes the opacity of the dye. Clearly once such recordings have been made, they are permanent.

Rerecordable or eraseable optical disks rely on magneto-optics,[2] also known more fully as thermomagneto-optics. Writing in such a device makes use of a thermomagnetic property possessed by all magnetic materials, which is that above a certain temperature, known as the Curie temperature, their coercive force becomes zero. This means that they become magnetically very soft, and take on the flux direction of any externally applied field. On cooling, this field orientation will be frozen in the material, and the coercivity will oppose attempts to change it. Although many materials possess this property, there are relatively few which have a suitably low Curie temperature. Compounds of terbium and gadolinium have been used, and one of the major problems to be overcome is that almost all suitable materials from a magnetic viewpoint corrode very quickly in air.

There are two ways in which magneto-optic (MO) disks can be written. Figure 11.2(c) shows the first system, in which the intensity of laser is modulated with

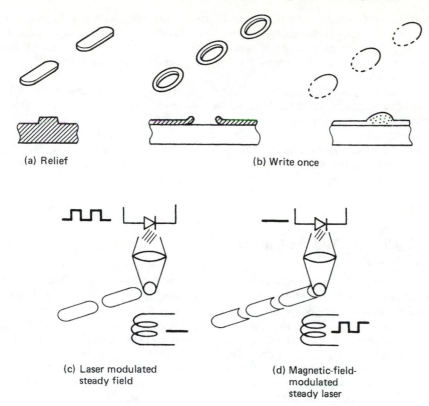

(a) Relief

(b) Write once

(c) Laser modulated
steady field

(d) Magnetic-field-
modulated
steady laser

Figure 11.2 (a) The information layer of CD is reflective and uses interference. (b) Write-once disks may burn holes or raise blisters in the information layer. (c) High data rate MO disks modulate the laser and use a constant magnetic field. (d) At low data rates the laser can run continuously and the magnetic field is modulated.

the waveform to be recorded. If the disk is considered to be initially magnetized along its axis of rotation with the north pole upwards, it is rotated in a field of the opposite sense, produced by a steady current flowing in a coil which is weaker than the room-temperature coercivity of the medium. The field will therefore have no effect. A laser beam is focused on the medium as it turns, and a pulse from the laser will momentarily heat a very small area of the medium past its Curie temperature, whereby it will take on a reversed flux due to the presence of the field coils. This reversed-flux direction will be retained indefinitely as the medium cools.

Alternatively the waveform to be recorded modulates the magnetic field from the coils as shown in Figure 11.2(d). In this approach, the laser is operating continuously in order to raise the track beneath the beam above the Curie temperature, but the recorded magnetic field is determined by the current in the coil at the instant the track cools. Magnetic field modulation is used in the recordable MiniDisc.

In both of these cases, the storage medium is clearly magnetic, but the writing mechanism is the heat produced by light from a laser; hence the term thermomagneto-optics. The advantage of this writing mechanism is that there is no physical contact between the writing head and the medium. The distance can be several millimetres, some of which is taken up with a protective layer to prevent corrosion. In prototypes, this layer is glass, but commercially available disks use plastics.

The laser beam will supply a relatively high power for writing, since it is supplying heat energy. For reading, the laser power is reduced, such that it cannot heat the medium past the Curie temperature, and it is left on continuously. Readout depends on the so-called Kerr effect, which describes a rotation of the plane of polarization of light due to a magnetic field. The magnetic areas written on the disk will rotate the plane of polarization of incident polarized light to two different planes, and it is possible to detect the change in rotation with a suitable pickup.

11.2 CD and MD contrasted

CD and MD have a great deal in common. Both use a laser of the same wavelength that creates a spot of the same size on the disk. The track pitch and speed are the same and both offer the same playing time. The channel code and error correction strategy are the same.

CD carries 44.1 kHz sixteen-bit PCM audio and is recorded in a continuous spiral like a vinyl disk. The CD process, from cutting, through pressing and reading, produces no musical degradation whatsoever, since it simply conveys a series of numbers which are exactly those recorded on the master tape. The only part of a CD player that can cause subjective differences in sound quality in normal operation is the DAC, although in the presence of gross errors some players will correct and/or conceal better than others.

MD begins with the same PCM data, but uses a form of compression known as ATRAC having a compression factor of 0.2. After the addition of subcode and housekeeping data MD has an average data rate which is 0.225 that of CD. However, MD has the same recording density and track speed as CD, so the data rate from the disk greatly exceeds that needed by the audio decoders. The difference is absorbed in RAM as for a hard-drive based machine. The RAM in a typical player is capable of buffering about 3 seconds of audio. When the RAM is full, the disk drive stops transferring data but keeps turning. As the RAM empties into the decoders, the disk drive will top it up in bursts. As the drive need not transfer data for over three quarters of the time, it can reposition between transfers and so it is capable of editing in the same way as a magnetic hard disk. A further advantage of the RAM buffer is that if the pickup is knocked off track by an external shock the RAM continues to provide data to the audio decoders and provided the pickup can get back to the correct track before the RAM is exhausted there will be no audible effect.

When recording an MO disk, the MiniDisc drive also uses the RAM buffer to allow repositioning so that a continuous recording can be made on a disk which has become chequerboarded through selective erasing. The full total playing time is then always available irrespective of how the disk is divided into different recordings.The sound quality of MiniDisc is a function of the performance of the convertors and of the data reduction system, with the latter typically being responsible for most of the degradation.

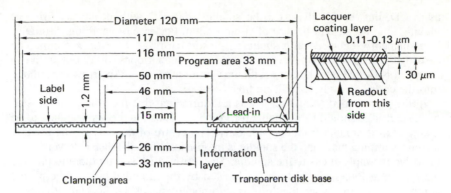

Figure 11.3 Mechanical specification of CD. Between diameters of 46 and 117 mm is a spiral track 5.7 km long.

11.3 CD and MD – disk construction

Figure 11.3 shows the mechanical specification of CD. Within an overall diameter of 120 mm the program area occupies a 33 mm-wide band between the diameters of 50 and 116 mm. Lead-in and lead-out areas increase the width of this band to 35.5 mm. As the track pitch is a constant 1.6 μm, there will be

$$\frac{35.6 \times 1000}{1.6} = 22\,188$$

tracks crossing a radius of the disk. As the track is a continuous spiral, the track length will be given by the above figure multiplied by the average circumference.

$$\text{Length} = 2 \times \pi \times \frac{58.5 + 23}{2} \times 22\,188 = 5.7\,\text{km}$$

Figure 11.4 shows the mechanical specification of prerecorded MiniDisc. Within an overall diameter of 64 mm the lead-in area begins at a diameter of 29 mm and the program area begins at 32 mm. The track pitch is exactly the same

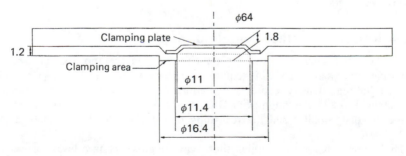

Figure 11.4 The mechanical dimensions of MiniDisc.

as in CD, but the MiniDisc can be smaller than CD without any sacrifice of playing time because of the use of compression. For ease of handling, MiniDisc is permanently enclosed in a shuttered plastic cartridge $72 \times 68 \times 5$ mm. The cartridge resembles a smaller version of a $3\frac{1}{2}$-inch floppy disk, but unlike a floppy, it is slotted into the drive with the shutter at the side. An arrow is moulded into the cartridge body to indicate this.

In the prerecorded MiniDisc, it was a requirement that the whole of one side of the cartridge should be available for graphics. Thus the disk is designed to be secured to the spindle from one side only. The centre of the disk is fitted with a ferrous clamping plate and the spindle is magnetic. When the disk is lowered into the drive it simply sticks to the spindle. The ferrous disk is only there to provide the clamping force. The disk is still located by the moulded hole in the plastic component. In this way the ferrous component needs no special alignment accuracy when it is fitted in manufacture. The back of the cartridge has a centre opening for the hub and a sliding shutter to allow access by the optical pickup.

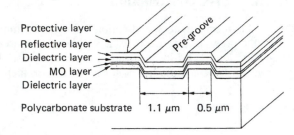

Figure 11.5 The construction of the MO recordable MiniDisc.

The recordable MiniDisc and cartridge has the same dimensions as the prerecorded MiniDisc, but access to both sides of the disk is needed for recording. Thus the recordable MiniDisc has a shutter which opens on both sides of the cartridge, rather like a double-sided floppy disk. The opening on the front allows access by the magnetic head needed for MO recording, leaving a smaller label area. Figure 11.5 shows the construction of the MO MiniDisc. The 1.1 μm wide tracks are separated by grooves which can optically be tracked. Once again the track pitch is the same as in CD. The MO layer is sandwiched between protective layers.

11.4 Rejecting surface contamination

A fundamental goal of consumer optical disks is that no special working environment or handling skill is required. The bandwidth required by PCM audio is such that high-density recording is mandatory if reasonable playing time is to be obtained in CD. Although MiniDisc uses compression, it does so in order to make the disk smaller and the recording density is actually the same as for CD.

High-density recording implies short wavelengths. Using a laser focused on the disk from a distance allows short-wavelength recordings to be played back

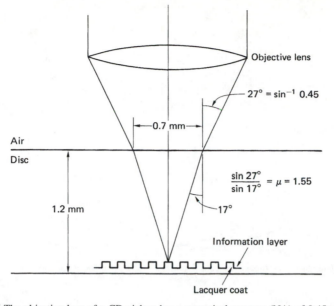

Figure 11.6 The objective lens of a CD pickup has a numerical aperture (NA) of 0.45; thus the outermost rays will be inclined at approximately 27° to the normal. Refraction at the air/disk interface changes this to approximately 17° within the disk. Thus light focused to a spot on the information layer has entered the disk through a 0.7 mm diameter circle, giving good resistance to surface contamination.

without physical contact, whereas conventional magnetic recording requires intimate contact and implies a wear mechanism, the need for periodic cleaning, and susceptibility to contamination. The information layer of CD and MD is read through the thickness of the disk. Figure 11.6 shows that this approach causes the readout beam to enter and leave the disk surface through the largest possible area. The actual dimensions involved are shown in the figure. Despite the minute spot size of about 1.2 μm diameter, light enters and leaves through a 0.7 mm-diameter circle. As a result, surface debris has to be three orders of magnitude larger than the readout spot before the beam is obscured. This approach has the further advantage in MO drives that the magnetic head, on the opposite side to the laser pickup, is then closer to the magnetic layer in the disk.

The bending of light at the disk surface is due to refraction of the wavefronts arriving from the objective lens. Wave theory of light suggests that a wavefront advances because an infinite number of point sources can be considered to emit spherical waves that will only add when they are all in the same phase. This can only occur in the plane of the wavefront. Figure 11.7 shows that at all other angles, interference between spherical waves is destructive. When such a wavefront arrives at an interface with a denser medium, such as the surface of an optical disk, the velocity of propagation is reduced; therefore the wavelength in the medium becomes shorter, causing the wavefront to leave the interface at a different angle (Figure 11.8). This is known as refraction. The ratio of velocity *in vacuo* to velocity in the medium is known as the refractive index of that medium;

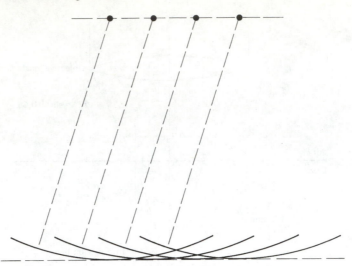

Figure 11.7 Plane-wave propagation considered as infinite numbers of spherical waves.

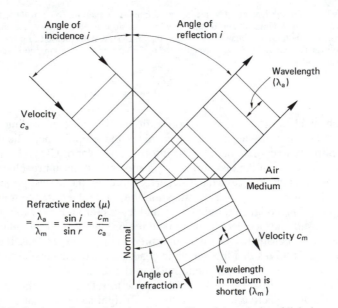

Figure 11.8 Reflection and refraction, showing the effect of the velocity of light in a medium.

it determines the relationship between the angles of the incident and refracted wavefronts.

The size of the entry circle in Figure 11.6 is a function of the refractive index of the disk material, the numerical aperture of the objective lens and the thickness of the disk. MiniDiscs are permanently enclosed in a cartridge, and scratching is unlikely. This is not so for CD, but fortunately the method of readout through the disk thickness tolerates surface scratches very well. In extreme cases of damage, a scratch can often be successfully removed with metal polish. By way of contrast, the label side is actually more vulnerable than the readout side, since the lacquer coating is only 30 μm thick. For this reason, writing on the label side of CD is not recommended.

The base material is in fact a polycarbonate plastic produced by (among others) Bayer under the trade name of Makrolon. It has excellent mechanical and optical stability over a wide temperature range, and lends itself to precision moulding and metallization. It is often used for automotive indicator clusters for the same reasons. An alternative material is polymethyl methacrylate (PMMA), one of the first optical plastics, known by such trade names as Perspex and Plexiglas. Polycarbonate is preferred by some manufacturers since it is less hygroscopic than PMMA. The differential change in dimensions of the lacquer coat and the base material can cause warping in a hygroscopic material. Audio disks are too small for this to be a problem, but the larger video disks are actually two disks glued together back-to-back to prevent this warpage.

11.5 Playing optical disks

A typical laser disk drive resembles a magnetic drive in that it has a spindle drive mechanism to revolve the disk, and a positioner to give radial access across the disk surface. The positioner has to carry a collection of lasers, lenses, prisms, gratings and so on, and cannot be accelerated as fast as a magnetic-drive positioner. A penalty of the very small track pitch possible in laser disks, which gives the enormous storage capacity, is that very accurate track following is needed, and it takes some time to lock on to a track. For this reason tracks on laser disks are usually made as a continuous spiral, rather than the concentric rings of magnetic disks. In this way, a continuous data transfer involves no more than track following once the beginning of the file is located.

In order to record MO disks or replay any optical disk, a source of monochromatic light is required. The light source must have low noise otherwise the variations in intensity due to the noise of the source will mask the variations due to reading the disk. The requirement for a low noise monochromatic light source is economically met using a semiconductor laser that is a relative of the light-emitting diode (LED). Both operate by raising the energy of electrons to move them from one valence band to another conduction band. Electrons falling back to the valence band emit a quantum of energy as a photon whose frequency is proportional to the energy difference between the bands. The process is described by Planck's Law:

Energy difference $E = H \times f$

where H = Planck's Constant

$= 6.6262 \times 10^{-34}$ Joules/Hertz

For gallium arsenide, the energy difference is about 1.6 eV, where 1 eV is 1.6×10^{-19} Joules. Using Planck's Law, the frequency of emission will be:-

$$f = \frac{1.6 \times 1.6 \times 10^{-19}}{6.6262 \times 10^{-34}} \text{Hz}$$

The wavelength will be c/f where c = the velocity of light = 3×10^8 m/s.

$$\text{Wavelength} = \frac{3 \times 10^8 \times 6.6262 \times 10^{-34}}{2.56 \times 10^{-19}} \text{ m} = 780 \text{ nanometres}$$

In the LED, electrons fall back to the valence band randomly, and the light produced is incoherent. In the laser, the ends of the semiconductor are optically flat mirrors, which produce an optically resonant cavity. One photon can bounce to and fro, exciting others in synchronism, to produce coherent light. This is known as Light Amplification by Stimulated Emission of Radiation, mercifully abbreviated to LASER, and can result in a runaway condition, where all available energy is used up in one flash. In injection lasers, equilibrium is reached between energy input and light output, allowing continuous operation. The equilibrium is delicate, and such devices are usually fed from a current source. To avoid runaway when temperature change disturbs the equilibrium, a photosensor is often fed back to the current source. Such lasers have a finite life, and become steadily less efficient.

Some of the light reflected back from the disk re-enters the aperture of objective lens. The pickup must be capable of separating the reflected light from the incident light. Figure 11.9 shows two systems. In (a) an intensity beam-splitter consisting of a semi-silvered mirror is inserted in the optical path and reflects some of the returning light into the photosensor. This is not very efficient, as half of the replay signal is lost by transmission straight on. In the example at (b) separation is by polarization.

Rotation of the plane of polarization is a useful method of separating incident and reflected light in a laser pickup. Using a quarter-wave plate, the plane of polarization of light leaving the pickup will have been turned 45°, and on return it will be rotated a further 45°, so that it is now at right angles to the plane of polarization of light from the source. The two can easily be separated by a polarizing prism, which acts as a transparent block to light in one plane, but as

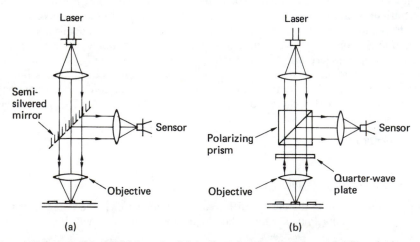

(a)   (b)

Figure 11.9 (a) Reflected light from the disk is directed to the sensor by a semisilvered mirror. (b) A combination of polarizing prism and quarter-wave plate separates incident and reflected light.

a prism to light in the other plane, such that reflected light is directed towards the sensor.

In a CD player, the sensor is concerned only with the intensity of the light falling on it. When playing MO disks, the intensity does not change, but the magnetic recording on the disk rotates the plane of polarization one way or the other depending on the direction of the vertical magnetization. MO disks cannot be read with circular polarized light. Light incident on the medium must be plane polarized and so the quarter-wave plate of the CD pickup cannot be used. Figure 11.10(a) shows that a polarizing prism is still required to linearly polarize the

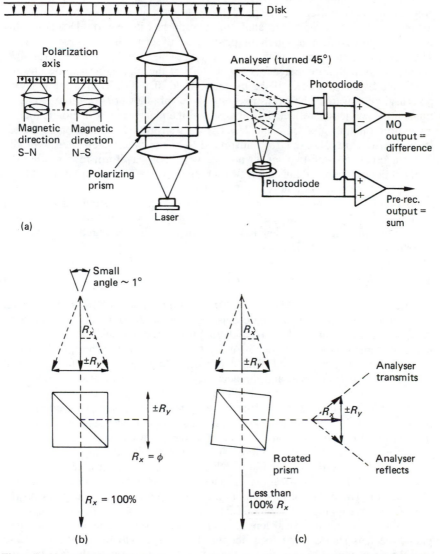

Figure 11.10 A pickup suitable for the replay of magneto-optic disks must respond to very small rotations of the plane of polarization.

light from the laser on its way to the disk. Light returning from the disk has had its plane of polarization rotated by approximately ±1 degree. This is an extremely small rotation. Figure 11.10(b) shows that the returning rotated light can be considered to be comprised of two orthogonal components. R_x is the component which is in the same plane as the illumination and is called the *ordinary* component and R_y is the component due to the Kerr effect rotation and is known as the *magneto-optic* component.

A polarizing beam splitter mounted squarely would reflect the magneto-optic component R_y very well because it is at right angles to the transmission plane of the prism, but the ordinary component would pass straight on in the direction of the laser. By rotating the prism slightly a small amount of the ordinary component is also reflected. Figure 11.10(c) shows that when combined with the magneto-optic component, the angle of rotation has increased.[3] Detecting this rotation requires a further polarizing prism or analyser as shown in Figure 11.10. The prism is twisted such that the transmission plane is at 45° to the planes of R_x and R_y. Thus with an unmagnetized disk, half of the light is transmitted by the prism and half is reflected. If the magnetic field of the disk turns the plane of polarization towards the transmission plane of the prism, more light is transmitted and less is reflected. Conversely if the plane of polarization is rotated away from the transmission plane, less light is transmitted and more is reflected. If two sensors are used, one for transmitted light and one for reflected light, the difference between the two sensor outputs will be a waveform representing the angle of polarization and thus the recording on the disk. This differential analyser eliminates common-mode noise in the reflected beam.[4]

As Figure 11.10 shows, the output of the two sensors is summed as well as subtracted in a MiniDisc player. When playing MO disks, the difference signal is used. When playing prerecorded disks, the sum signal is used and the effect of the second polarizing prism is disabled.

11.6 Focus and tracking systems

The frequency response of the laser pickup and the amount of crosstalk are both a function of the spot size and care must be taken to keep the beam focused on the information layer. Disk warp and thickness irregularities will cause focal-plane movement beyond the depth of focus of the optical system, and a focus servo system will be needed. The depth of field is related to the numerical aperture, which is defined, and the accuracy of the servo must be sufficient to keep the focal plane within that depth, which is typically ±1 μm.

The focus servo moves a lens along the optical axis in order to keep the spot in focus. Since dynamic focus-changes are largely due to warps, the focus system must have a frequency response in excess of the rotational speed. A focus-error system is necessary to drive the lens. There are a number of ways in which this can be derived, the most common of which will be described here.

In Figure 11.11 a cylindrical lens is installed between the beam splitter and the photosensor. The effect of this lens is that the beam has no focal point on the sensor. In one plane, the cylindrical lens appears parallel-sided, and has negligible effect on the focal length of the main system, whereas in the other plane, the lens shortens the focal length. The image will be an ellipse whose aspect ratio changes as a function of the state of focus. Between the two foci, the image will be circular. The aspect ratio of the ellipse, and hence the focus error,

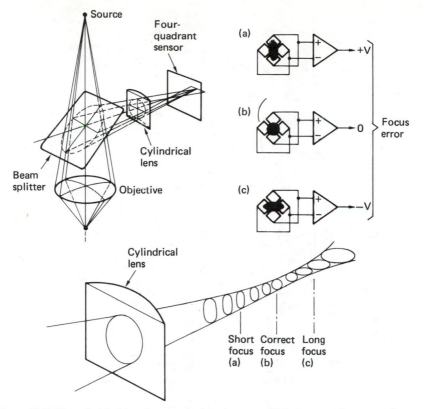

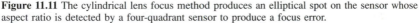

Figure 11.11 The cylindrical lens focus method produces an elliptical spot on the sensor whose aspect ratio is detected by a four-quadrant sensor to produce a focus error.

can be found by dividing the sensor into quadrants. When these are connected as shown, the focus-error signal is generated. The data readout signal is the sum of the quadrant outputs.

Figure 11.12 shows the knife-edge method of determining focus. A split sensor is also required. At (a) the focal point is coincident with the knife-edge, so it has little effect on the beam. At (b) the focal point is to the right of the knife-edge, and rising rays are interrupted, reducing the output of the upper sensor. At (c) the focal point is to the left of the knife-edge, and descending rays are interrupted, reducing the output of the lower sensor. The focus error is derived by comparison of the outputs of the two halves of the sensor. A drawback of the knife-edge system is that the lateral position of the knife-edge is critical, and adjustment is necessary. To overcome this problem, the knife-edge can be replaced by a pair of prisms, as shown in Figure 11.12(d)–(f). Mechanical tolerances then only affect the sensitivity, without causing a focus offset.

The cylindrical lens method has a smaller capture range than the knife-edge/prism method and a focus-search mechanism will be required, which moves the focus servo over its entire travel, looking for a zero crossing. At this time the feedback loop will be completed, and the sensor will remain on the linear part of its characteristic. The spiral track of CD and MiniDisc starts at the inside and

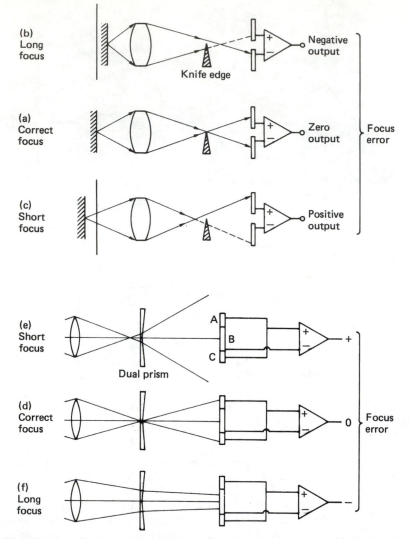

Figure 11.12 (a)–(c) Knife-edge focus method requires only two sensors, but is critically dependent on knife-edge position. (d)–(f) Twin-prism method requires three sensors (A, B, C), where focus error is (A + C) – B. Prism alignment reduces sensitivity without causing focus offset.

works outwards. This was deliberately arranged because there is less vertical run-out near the hub, and initial focusing will be easier.

The track pitch is only 1.6 μm, and this is much smaller than the accuracy to which the player chuck or the disk centre hole can be made; on a typical player, run-out will swing several tracks past a fixed pickup. A track-following servo is necessary to keep the spot centralized on the track. There are several ways in which a tracking error can be derived.

In the three-spot method, two additional light beams are focussed on the disk track, one offset to each side of the track centre-line. Figure 11.13 shows that, as

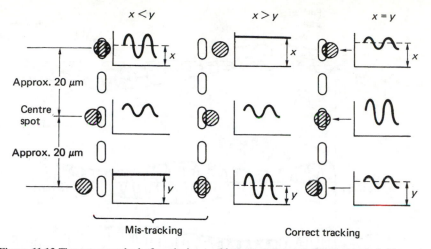

Figure 11.13 Three-spot method of producing tracking error compares average level of side-spot signals. Side spots are produced by a diffraction grating and require their own sensors.

one side spot moves away from the track into the mirror area, there is less destructive interference and more reflection. This causes the average amplitude of the side spots to change differentially with tracking error. The laser head contains a diffraction grating that produces the side spots, and two extra photosensors onto which the reflections of the side spots will fall. The side spots feed a differential amplifier, which has a low-pass filter to reject the channel-code information and retain the average brightness difference. Some players use a delay line in one of the side-spot signals whose period is equal to the time taken for the disk to travel between the side spots. This helps the differential amplifier to cancel the channel code.

The alternative approach to tracking-error detection is to analyse the diffraction pattern of the reflected beam. The effect of an off-centre spot is to rotate the radial diffraction pattern about an axis along the track. Figure 11.14 shows that, if a split sensor is used, one half will see greater modulation than the

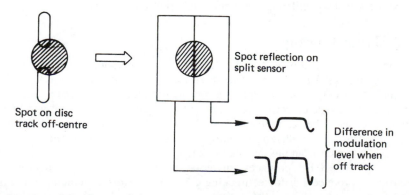

Figure 11.14 Split-sensor method of producing tracking error focuses image of spot onto sensor. One side of spot will have more modulation when off-track.

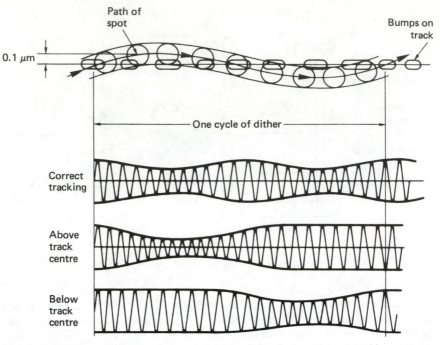

Figure 11.15 Dither applied to readout spot modulates the readout envelope. A tracking error can be derived.

other when off-track. Such a system may be prone to develop an offset due either to drift or to contamination of the optics, although the capture range is large. A further tracking mechanism is often added to obviate the need for periodic adjustment. Figure 11.15 shows that in a dither-based system, a sinusoidal drive is fed to the tracking servo, causing a radial oscillation of spot position of about ±50 nm. This results in modulation of the envelope of the readout signal, which can be synchronously detected to obtain the sense of the error. The dither can be produced by vibrating a mirror in the light path, which enables a high frequency to be used, or by oscillating the whole pickup at a lower frequency.

11.7 Typical pickups

It is interesting to compare different designs of laser pickup. Figure 11.16 shows an early Philips laser head.[5] The dual-prism focus method is used, which combines the output of two split sensors to produce a focus error. The focus amplifier drives the objective lens mounted on a parallel motion formed by two flexural arms. The capture range of the focus system is sufficient to accommodate normal tolerances without assistance. A radial differential tracking signal is extracted from the sensors as shown in the figure. Additionally, a dither frequency of 600 Hz produces envelope modulation that is synchronously rectified to produce a drift-free tracking error. Both errors are combined to drive the tracking system. As only a single spot is used, the pickup is relatively insensitive to angular errors, and a rotary positioner can be used, driven by a

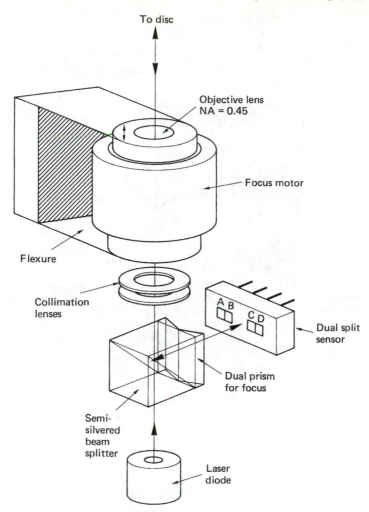

To disc

Objective lens
NA = 0.45

Focus motor

Flexure

Collimation
lenses

A B
C D

Dual split
sensor

Dual prism
for focus

Semi-
silvered
beam
splitter

Laser
diode

Figure 11.16 Philips laser head showing semisilvered prism for beam splitting. Focus error is derived from dual-prism method using split sensors. Focus error $(A + D) - (B + C)$ is used to drive focus motor which moves objective lens on parallel action flexure. Radial differential tracking error is derived from split sensor $(A + B) - (C + D)$. Tracking error drives entire pickup on radial arm driven by moving coil. Signal output is $(A + B + C + D)$. System includes 600 Hz dither for tracking. (Courtesy *Philips Technical Review*)

moving coil. The assembly is statically balanced to give good resistance to lateral shock.

Figure 11.17 shows a Sony laser head used in early consumer players. The cylindrical-lens focus method is used, requiring a four-quadrant sensor. Since this method has a small capture range, a focus-search mechanism is necessary. When a disk is loaded, the objective lens is ramped up and down looking for a zero crossing in the focus error. The three-spot method is used for tracking. The necessary diffraction grating can be seen adjacent to the laser diode. Tracking error is derived from side-spot sensors (E, F). Since the side-spot

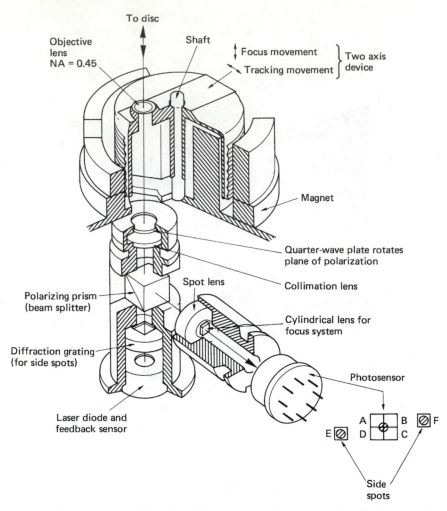

Figure 11.17 Sony laser head showing polarizing prism and quarter-wave plate for beam splitting, and diffraction grating for production of side spots for tracking. The cylindrical lens system is used for focus, with a four-quadrant sensor (A, B, C, D) and two extra sensors E, F for the side spots. Tracking error is E–F; focus error is (A + C) – (B + D). Signal output is (A + B + C + D). The focus and tracking errors drive the two-axis device. (Courtesy *Sony Broadcast*)

system is sensitive to angular error, a parallel-tracking laser head traversing a disk radius is essential. A cost-effective linear motion is obtained by using a rack-and-pinion drive for slow, coarse movements, and a laterally moving lens in the light path for fine rapid movements. The same lens will be moved up and down for focus by the so-called two-axis device, which is a dual-moving coil mechanism. In some players this device is not statically balanced, making the unit sensitive to shock, but this was overcome on later heads designed for portable players. Some designs incorporate a prism to reduce the height of the pickup above the disk.

11.8 CD readout in detail

Many descriptions are simplified to the extent that the light spot is depicted as having a distinct edge of a given diameter. In reality such a neat spot cannot be obtained. It is essential to the commercial success of CD that a useful playing time (75 min max.) should be obtained from a recording of reasonable size (12 cm). The size was determined by the European motor industry as being appropriate for car dashboard-mounted units. It follows that the smaller the spot of light which can be created, the smaller can be the deformities carrying the information, and so more information per unit area. Development of a successful high-density optical recorder requires an intimate knowledge of the behaviour of light focused into small spots. If it is attempted to focus a uniform beam of light to an infinitely small spot on a surface normal to the optical axis, it will be found that it is not possible. This is probably just as well as an infinitely small spot would have infinite intensity and any matter it fell on would not survive. Instead the result of such an attempt is a distribution of light in the area of the focal point having no sharply defined boundary. This is called the Airy distribution[5] (sometimes pattern or disk) after Lord Airy (1835), the then Astronomer Royal.

If a line is considered to pass across the focal plane, through the theoretical focal point, and the intensity of the light is plotted on a graph as a function of the distance along that line, the result is the intensity function shown in Figure 11.18.

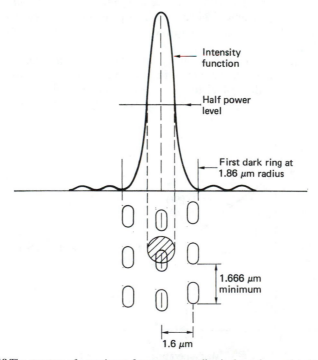

Figure 11.18 The structure of a maximum frequency recording is shown here, related to the intensity function of an objective of 0.45NA with 780 μm light. Note that track spacing puts adjacent tracks in the dark rings, reducing crosstalk. Note also that as the spot has an intensity function it is meaningless to specify the spot diameter without some reference such as an intensity level.

It will be seen that this contains a central sloping peak surrounded by alternating dark rings and light rings of diminishing intensity. These rings will in theory reach to infinity before their intensity becomes zero. The intensity distribution or function described by Airy is due to diffraction effects across the finite aperture of the objective. For a given wavelength, as the aperture of the objective is increased, so the diameter of the features of the Airy pattern reduces. The Airy pattern vanishes to a singularity of infinite intensity with a lens of infinite aperture which of course cannot be made. The approximation of geometric optics is quite unable to predict the Airy pattern. An intensity function does not have a diameter, but for practical purposes an effective diameter typically quoted is that at which the intensity has fallen to some convenient fraction of that at the peak. Thus one could state, for example, the half-power diameter.

With a fixed objective aperture, as the tangential diffraction pattern becomes more oblique, less light passes the aperture and the depth of modulation transmitted by the lens falls. At some spatial frequency, all the diffracted light falls outside the aperture and the modulation depth transmitted by the lens falls to zero. This is known as the spatial cut-off frequency. Thus a graph of depth of modulation versus spatial frequency can be drawn and which is known as the modulation transfer function (MTF). This is a straight line commencing at unity at zero spatial frequency (no detail) and falling to zero at the cut-off spatial frequency (finest detail). Thus one could describe a lens of finite aperture as a form of spatial low-pass filter. The Airy function is no more than the spatial impulse response of the lens, and the concentric rings of the Airy function are the spatial analog of the symmetrical ringing in a phase linear electrical filter. The Airy function and the triangular frequency response form a transform pair[6] as shown in Chapter 3.

When an objective lens is used in a conventional microscope, the MTF will allow the resolution to be predicted in lines per millimetre. However, in a scanning microscope the spatial frequency of the detail in the object multiplied by the scanning velocity gives a temporal frequency measured in Hertz. Thus

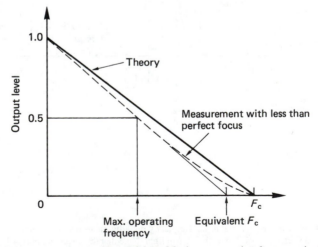

Figure 11.19 Frequency response of laser pickup. Maximum operating frequency is about half of cut-off frequency F_c.

lines per millimetre multiplied by millimetres per second gives lines per second. Instead of a straight line MTF falling to the spatial cut-off frequency, Figure 11.19 shows that a scanning microscope has a temporal frequency response falling to zero at the optical cut-off frequency which is given by:

$$F_c = \frac{2NA}{wavelength} \times velocity$$

The minimum linear velocity of CD is 1.2 m/s, giving a cut-off frequency of

$$F_c = \frac{2 \times 0.45 \times 1.2}{780 \times 10^{-9}} = 1.38\,MHz$$

Actual measurements reveal that the optical response is only a little worse than the theory predicts. This characteristic has a large bearing on the type of modulation schemes that can be successfully employed. Clearly, to obtain any noise immunity, the maximum operating frequency must be rather less than the cut-off frequency. The maximum frequency used in CD is 720 kHz, which represents an absolute minimum wavelength of 1.666 μm, or a bump length of 0.833 μm, for the lowest permissible track speed of 1.2 m/s used on the full-length 75 min-playing disks. One-hour-playing disks have a minimum bump length of 0.972 μm at a track velocity of 1.4 m/s. The maximum frequency is the same in both cases. This maximum frequency should not be confused with the bit rate of CD since this is different owing to the channel code used. Figure 11.18 showed a maximum-frequency recording, and the physical relationship of the intensity function to the track dimensions.

The intensity function can be enlarged if the lens used suffers from optical aberrations. This was studied by Maréchal[7] who established criteria for the accuracy to which the optical surfaces of the lens should be made to allow the ideal Airy distribution to be obtained. CD player lenses must meet the Maréchal criterion. With such a lens, the diameter of the distribution function is determined solely by the combination of Numerical Aperture (NA) and the wavelength. When the size of the spot is as small as the NA and wavelength allow, the optical system is said to be diffraction limited. Figure 11.20 shows how Numerical Aperture is defined, and illustrates that the smaller the spot needed, the larger

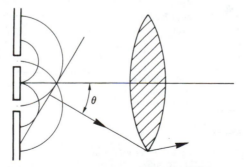

Figure 11.20 Fine detail in an object can only be resolved if the diffracted wavefront due to the highest spatial frequency is collected by the lens. Numerical aperture (NA) = sin θ, and as θ is the diffraction angle it follows that, for a given wavelength, NA determines resolution.

must be the NA. Unfortunately the larger the NA the more obliquely to the normal the light arrives at the focal plane and the smaller the depth of focus will be.

Since the introduction of CD, developments in the technology have continued. The use of a shorter-wavelength laser and a larger numerical aperture allows the spot size to be reduced in DVD. As a result the recording density is increased in comparison with CD.

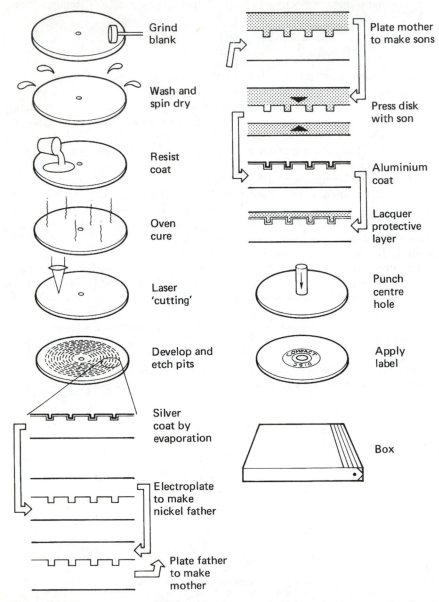

Figure 11.21 The many stages of CD manufacture, most of which require the utmost cleanliness.

11.9 How optical disks are made

The steps used in the production of CDs will next be outlined. Prerecorded MiniDiscs are made in an identical fashion except for detail differences to be noted. MO disks need to be grooved so that the track-following system will work. The grooved substrate is produced in a similar way to a CD master, except that the laser is on continuously instead of being modulated with a signal to be recorded. As stated, CD is replicated by moulding, and the first step is to produce a suitable mould. This mould must carry deformities of the correct depth for the standard wavelength to be used for reading, and as a practical matter these deformities must have slightly sloping sides so that it is possible to release the CD from the mould.

The major steps in CD manufacture are shown in Figure 11.21. The mastering process commences with an optically flat glass disk about 220 mm in diameter and 6 mm thick. The blank is washed first with an alkaline solution, then with a fluorocarbon solvent, and spun dry prior to polishing to optical flatness. A critical cleaning process is then undertaken using a mixture of deionized water and isopropyl alcohol in the presence of ultrasonic vibration, with a final fluorocarbon wash. The blank must now be inspected for any surface irregularities that would cause data errors. This is done by using a laser beam and monitoring the reflection as the blank rotates. Rejected blanks return to the polishing process, those which pass move on, and an adhesive layer is applied followed by a coating of positive photoresist. This is a chemical substance that softens when exposed to an appropriate intensity of light of a certain wavelength, typically ultraviolet. Upon being thus exposed, the softened resist will be washed away by a developing solution down to the glass to form flat-bottomed pits whose depth is equal to the thickness of the undeveloped resist. During development the master is illuminated with laser light of a wavelength to which it is insensitive. The diffraction pattern changes as the pits are formed. Development is arrested when the appropriate diffraction pattern is obtained.[8] The thickness of the resist layer must be accurately controlled, since it affects the height of the bumps on the finished disk, and an optical scanner is used to check that there are no resist defects that would cause data errors or tracking problems in the end product. Blanks passing this test are oven-cured, and are ready for cutting. Failed blanks can be stripped of the resist coating and used again.

The cutting process is shown in simplified form in Figure 11.22. A continuously operating helium cadmium[9] or argon ion[10] laser is focused on the resist coating as the blank revolves. The focus system uses a separate helium neon laser sharing the same optics. The resist is insensitive to the wavelength of the He–Ne laser. The laser intensity is controlled by an acousto-optic modulator driven by the encoder. When the device is in a relaxed state, light can pass through it, but when the surface is excited by high-frequency vibrations, light is scattered. Information is carried in the lengths of time for which the modulator remains on or remains off. The deformities in the resist produced as the disk turns when the modulator allows light to pass are separated by areas unaffected by light when the modulator is shut off. Information is carried solely in the variations of the lengths of these two areas.

The laser makes its way from the inside to the outside as the blank revolves. As the radius of the track increases, the rotational speed is proportionately reduced so that the velocity of the beam over the disk remains constant. This

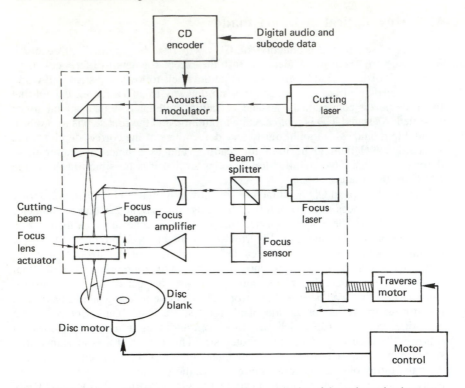

Figure 11.22 CD cuter. The focus subsystem controls the spot size of the main cutting laser on the photosensitive blank. Disc and traverse motors are coordinated to give constant track pitch and velocity. Note that the power of the focus laser is insufficient to expose the photoresist.

constant linear velocity (CLV) results in rather longer playing time than would be obtained with a constant speed of rotation. Owing to the minute dimensions of the track structure, the cutter has to be constructed to extremely high accuracy. Air bearings are used in the spindle and the laser head, and the whole machine is resiliently supported to prevent vibrations from the building from affecting the track pattern.

As the player is a phase contrast microscope, it must produce an intensity function that straddles the deformities. As a consequence the intensity function that produces the deformities in the photoresist must be smaller in diameter than that in the reader. This is conveniently achieved by using a shorter wavelength of 400–500 nm from a helium–cadmium or argon–ion laser combined with a larger lens aperture of 0.9. These are expensive, but only needed for the mastering process.

The master recording process has produced a phase structure in relatively delicate resist, and this cannot be used for moulding directly. Instead a thin metallic silver layer is sprayed onto the resist to render it electrically conductive so that electroplating can be used to make robust copies of the relief structure.

The electrically conductive resist master is then used as the cathode of an electroplating process where a first layer of metal is laid down over the resist, conforming in every detail to the relief structure thereon. This metal layer can then be separated from the glass and the resist is dissolved away and the silver is recovered leaving a laterally inverted phase structure on the surface of the metal, in which the pits in the photoresist have become bumps in the metal. From this point on, the production of CD is virtually identical to the replication process used for vinyl disks, save only that a good deal more precision and cleanliness is needed.

This first metal layer could itself be used to mould disks, or it could be used as a robust submaster from which many stampers could be made by pairs of plating steps. The first metal phase structure can itself be used as a cathode in a further electroplating process in which a second metal layer is formed having a mirror image of the first. A third such plating step results in a stamper. The decision to use the master or substampers will be based on the number of disks and the production rate required.

The master is placed in a moulding machine, opposite a flat plate. A suitable quantity of molten plastic is injected between, and the plate and the master are forced together. The flat plate renders one side of the disk smooth, and the bumps in the metal stamper produce pits in the other surface of the disk. The surface containing the pits is next metallized, with any good electrically conductive material, typically aluminium. This metallization is then covered with a lacquer for protection. In the case of CD, the label is printed on the lacquer. In the case of a prerecorded MiniDisc, the ferrous hub needs to be applied prior to fitting the cartridge around the disk.

As CD and prerecorded MDs are simply data disks, they do not need to be mastered in real time. Raising the speed of the mastering process increases the throughput of the expensive equipment. Pressing plants have been using computer tape streamers or hard disk drives in order to supply the cutter with higher data rates.

11.10 How recordable MiniDiscs are made

Recordable MiniDiscs make the recording as flux patterns in a magnetic layer. However, the disks need to be pregrooved so that the tracking systems can operate. The grooves have the same pitch as CD and the prerecorded MD, but the tracks are the same width as the laser spot: about 1.1 µm. The grooves are not a perfect spiral, but have a sinusoidal waviness at a fixed wavelength. Like CD, MD uses constant track linear velocity, not constant speed of rotation. When recording on a blank disk, the recorder needs to know how fast to turn the spindle to get the track speed correct. The wavy grooves will be followed by the tracking servo and the frequency of the tracking error will be proportional to the disk speed. The recorder simply turns the spindle at whatever speed makes the grooves wave at the correct frequency. The groove frequency is 75 Hz; the same as the data sector rate. Thus a zero crossing in the groove signal can also be used to indicate where to start recording. The grooves are particularly important when a chequer-boarded recording is being replayed. On a CLV disk, each seek to a new track radius results in a different track speed. The wavy grooves allow the track velocity to be corrected as soon as a new track is reached.

The pregrooves are moulded into the plastics body of the disk when it is made. The mould is made in a similar manner to a prerecorded disk master, except that the laser is not modulated and the spot is larger. The track velocity is held constant by slowing down the resist master as the radius increases, and the waviness is created by injecting 75 Hz into the lens radial positioner. The master is developed and electroplated as normal in order to make stampers. The stampers make pregrooved disks that are then coated by vacuum deposition with the MO layer, sandwiched between dielectric layers. The MO layer can be made less susceptible to corrosion if it is smooth and homogeneous. Layers that contain voids, asperities or residual gases from the coating process present a larger surface area for attack. The life of an MO disk is affected more by the manufacturing process than by the precise composition of the alloy.

Above the sandwich an optically reflective layer is applied, followed by a protective lacquer layer. The ferrous clamping plate is applied to the centre of the disk, which is then fitted in the cartridge. The recordable cartridge has a double-sided shutter to allow the magnetic head access to the back of the disk.

11.11 Channel code of CD and MiniDisc

CD and MiniDisc use the same channel code. This was optimized for the optical readout of CD and prerecorded MiniDisc, but is also used for the recordable version of MiniDisc for simplicity.

The frequency response falling to the optical cut-off frequency is only one of the constraints within which the modulation scheme has to work. There are a number of others. In all players the tracking and focus servos operate by analysing the average amount of light returning to the pickup. If the average amount of light returning to the pickup is affected by the content of the recorded data, then the recording will interfere with the operation of the servos. Debris on the disk surface affects the light intensity and means must be found to prevent this reducing the signal quality excessively. Chapter 6 discussed modulation schemes known as DC-free codes. If such a code is used, the average brightness of the track is constant and independent of the data bits. Figure 11.23(a) shows the replay signal from the pickup being compared with a threshold voltage in order to recover a binary waveform from the analog pickup waveform, a process

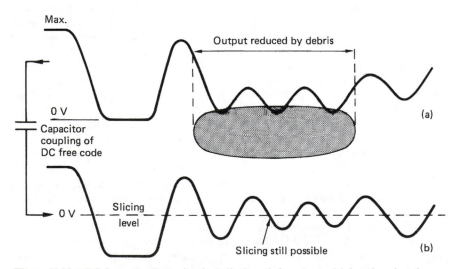

Figure 11.23 A DC-free code allows signal amplitude variations due to debris to be rejected.

known as slicing. If the light beam is partially obstructed by debris, the pickup signal level falls, and the slicing level is no longer correct and errors occur. If, however, the code is DC-free, the waveform from the pickup can be passed through a high pass filter (e.g. a series capacitor) and Figure 11.23(b) shows that this rejects the falling level and converts it to a reduction in amplitude about the slicing level so that the slicer still works properly. This step cannot be performed unless a DC-free code is used. As the frequency response on replay falls linearly to the cut-off frequency determined by the aperture of the lens and the wavelength of light used, the shorter bumps and lands produce less modulation than longer ones. It is a further advantage of a DC-free code that as the length of bumps and lands falls with rising density, the replay waveform simply falls in amplitude but the average voltage remains the same and so the slicer still operates correctly.

CD uses a coding scheme where combinations of the data bits to be recorded are represented by unique waveforms. These waveforms are created by combining various run lengths from $3T$ to $11T$ together to give a channel pattern $14T$ long where T is half a cycle of the master clock.[11] Within the run length limits of $3T$ to $11T$, a waveform $14T$ long can have 267 different patterns. This is slightly more than the 256 combinations of eight data bits and so eight bits are represented by a waveform lasting $14T$. Some of these patterns are shown in Figure 11.24. As stated, these patterns are not polarity conscious and they could be inverted without changing the meaning.

Not all of the $14T$ patterns used are DC-free, some spend more time in one state than the other. The overall DC content of the recorded waveform is rendered DC-free by inserting an extra portion of waveform, known as a packing period, between the $14T$ channel patterns. This packing period is $3T$ long and may or may not contain a transition, which if it is present can be in one of three places. The packing period contains no information, but serves to control the DC content of the overall waveform.[12] The packing waveform is generated in such a way that

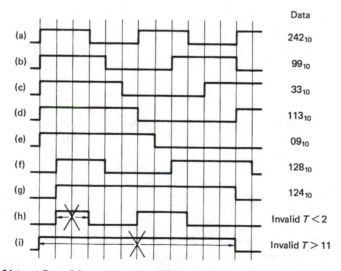

Figure 11.24 (a–g) Part of the codebook for EFM code showing examples of various run lengths from $3T$ to $11T$. (h,i) Invalid patterns which violate the run-length limits.

in the long term the amount of time the channel signal spends in one state is equal to the time it spends in the other state. A packing period is placed between every pair of channel patterns and so the overall length of time needed to record eight bits is 17T.

Thus a group of eight data bits is represented by a code of fourteen channel bits, hence the name of eight to fourteen modulation (EFM). It is a common misconception that the channel bits of a group code are recorded; in fact they are simply a convenient way of synthesizing a coded waveform having uniform time steps. It should be clear that channel bits cannot be recorded as they have a rate of 4.3 Megabits per second whereas the optical cut-off frequency of CD is only 1.4 MHz.

Another common misconception is that channel bits are data. If channel bits were data, all combinations of 14 bits, or 16 384 different values could be used. In fact only 267 combinations produce waveforms that can be recorded. In a practical CD modulator, the eight bit data symbols to be recorded are used as the address of a lookup table which outputs a fourteen-bit channel bit pattern. As the highest usable frequency in CD is 720 kHz, transitions cannot be closer together than 3T and so successive ones in the channel bitstream must have two or more zeros between them. Similarly transitions cannot be further apart than 11T or there will be insufficient clock content. Thus there cannot be more than 10 zeros between channel 1s. Whilst the lookup table can be programmed to prevent code violations within the 14T pattern, they could occur at the junction of two successive patterns. Thus a further function of the packing period is to prevent violation of the run length limits. If the previous pattern ends with a transition and the next begins with one, there will be no packing transition and so the 3T minimum requirement can be met. If the patterns either side have long run lengths, the sum of the two might exceed 11T unless the packing period contained a transition. In fact the minimum run length limit could be met with 2T of packing, but the requirement for DC control dictated 3T of packing.

Decoding the stream of channel bits into data requires that the boundaries between successive 17T periods are identified. This is the process of deserialization or parsing. On the disk one 17T period runs straight into the next; there are no dividing marks. Symbol separation is done by counting channel bit periods and dividing them by 17 starting from a known reference point. The three packing periods are discarded and the remaining 14T symbol is decoded to eight data bits. The reference point is provided by the synchronizing pattern so called because its detection synchronizes the deserialization counter to the replay waveform.

Synchronization has to be as reliable as possible because if it is incorrect all the data will be corrupted up to the next sync pattern. Synchronization is achieved by the detection of an unique waveform periodically recorded on the track at with regular spacing. It must be unique in the strict sense in that nothing else can give rise to it, because the detection of a false sync is just as damaging as failure to detect a correct one. In practice CD synchronizes deserialization with a waveform that is unique in that it is different from any of the 256 waveforms which represent data. For reliability, the sync pattern should have the best signal to noise ratio possible, and this is obtained by making it one complete cycle of the lowest frequency (11T plus 11T) giving it the largest amplitude and also making it DC-free. Upon detection of the $2 \times T_{max}$ waveform, the deserialization counter that divides the channel bit count by 17 is reset. This occurs on the next system clock, which is the reason for the 0 in the sync pattern after the third 1 and before the merging bits.

CD therefore uses forward synchronization and correctly deserialized data are available immediately after the first sync pattern is detected. The sync pattern is longer than the data symbols, and so clearly no data code value can create it, although it would be possible for certain adjacent data symbols to create a false sync pattern by concatenation were it not for the presence of the packing period. It is a further job of the packing period to prevent false sync patterns being generated at the junction of two channel symbols.

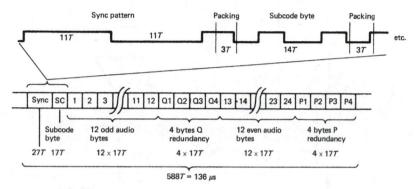

Figure 11.25 One CD data block begins with a unique sync pattern, and one subcode byte, followed by 24 audio bytes and eight redundancy bytes. Note that each byte requires 14T in EFM, with 3T packing between symbols, making 17T.

Each data block or frame in CD and MD, shown in Figure 11.25, consists of 33 symbols 17T each following the preamble, making a total of 588T or 136 μs. Each symbol represents eight data bits. The first symbol in the block is used for subcode, and the remaining 32 bytes represent 24 audio sample bytes and 8 bytes of redundancy for the error-correction system. The subcode byte forms part of a subcode block which is built up over 98 successive data frames.

Figure 11.26 shows an overall block diagram of the record modulation scheme used in CD mastering and the corresponding replay system or data separator. The input to the record channel coder consists of sixteen-bit audio samples that are divided in two to make symbols of eight bits. These symbols are used in the error-correction system that interleaves them and adds redundant symbols. For every twelve audio symbols, there are four symbols of redundancy, but the channel coder is not concerned with the sequence or significance of the symbols and simply records their binary code values.

Symbols are provided to the coder in eight-bit parallel format, with a symbol clock. The symbol clock is obtained by dividing down the 4.3218 MHz T rate clock by a factor of 17. Each symbol is used to address the lookup table that outputs a corresponding fourteen-channel-bit pattern in parallel into a shift register. The T rate clock then shifts the channel bits along the register. The lookup table also outputs data corresponding to the digital sum value (DSV) of the fourteen-bit symbol to the packing generator. The packing generator determines if action is needed between symbols to control DC content. The packing generator checks for run-length violations and potential false sync

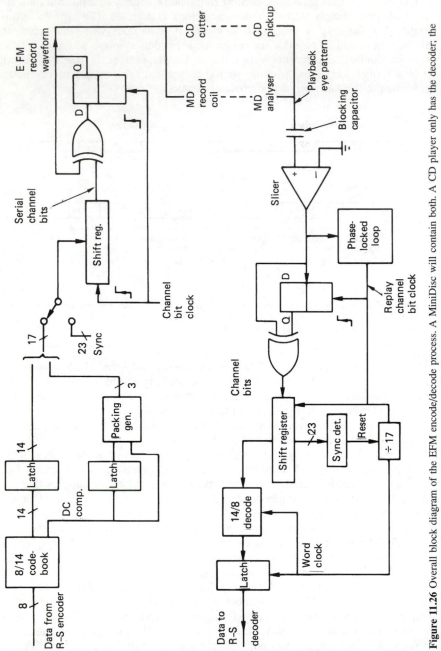

Figure 11.26 Overall block diagram of the EFM encode/decode process. A MiniDisc will contain both. A CD player only has the decoder; the encoding is in the mastering cutter.

patterns. As a result of all the criteria, the packing generator loads three channel bits into the space between the symbols, such that the register then contains fourteen-bit symbols with three bits of packing between them. At the beginning of each frame, the sync pattern is loaded into the register just before the first symbol is looked up in such a way that the packing bits are correctly calculated between the sync pattern and the first symbol.

A channel bit one indicates that a transition should be generated, and so the serial output of the shift register is fed to the JK bistable along with the T rate clock. The output of the JK bistable is the ideal channel coded waveform containing transitions separated by $3T$ to $11T$. It is a self-clocking, run-length-limited waveform. The channel bits and the T rate clock have done their job of changing the state of the JK bistable and do not pass further on. At the output of the JK the sync pattern is simply two $11T$ run lengths in series. At this stage the run-length-limited waveform is used to control the acousto-optic modulator in the cutter.

The resist master is developed and used to create stampers. The resulting disks can then be replayed. The track velocity of a given CD is constant, but the rotational speed depends upon the radius. In order to get into lock, the disk must be spun at roughly the right track speed. This is done using the run-length limits of the recording. The pickup is focused and the tracking is enabled. The replay waveform from the pickup is passed through a high-pass filter to remove level variations due to contamination and sliced to return it to a binary waveform. The slicing level is self-adapting as Figure 11.23 showed so that a 50 per cent duty cycle is obtained. The slicer output is then sampled by the unlocked VCO running at approximately T rate. If the disk is running too slowly, the longest run length on the disk will appear as more than $11T$, whereas if the disk is running too fast, the shortest run length will appear as less than $3T$. As a result the disk speed can be brought to approximately the right speed and the VCO will then be able to lock to the clock content of the EFM waveform from the slicer. Once the VCO is locked, it will be possible to sample the replay waveform at the correct T rate. The output of the sampler is then differentiated and the channel bits reappear and are fed into the shift register. The sync pattern detector will then function to reset the deserialization counter that allows the $14T$ symbols to be identified. The $14T$ symbols are then decoded to eight bits in the reverse coding table.

Figure 11.27 reveals the timing relationships of the CD format. The sampling rate of 44.1 kHz with sixteen-bit words in left and right channels results in an audio data rate of 176.4 kb/s ($k = 1000$ here, not 1024). Since there are 24 audio bytes in a data frame, the frame rate will be:

$$\frac{176.4}{24} \text{ kHz} = 7.35 \text{ kHz}$$

If this frame rate is divided by 98, the number of frames in a subcode block, the subcode block or sector rate of 75 Hz results. This frequency can be divided down to provide a running-time display in the player. Note that this is the frequency of the wavy grooves in recordable MDs.

If the frame rate is multiplied by 588, the number of channel bits in a frame, the master clock-rate of 4.3218 MHz results. From this the maximum and minimum frequencies in the channel, 720 kHz and 196 kHz, can be obtained using the run-length limits of EFM.

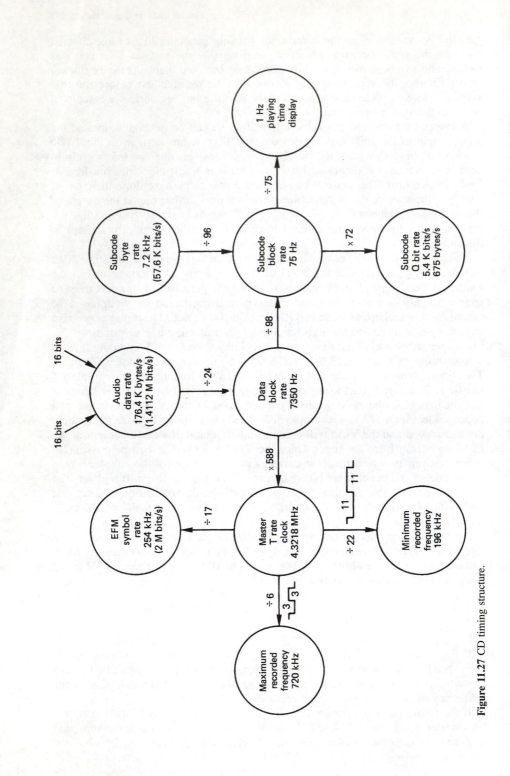

Figure 11.27 CD timing structure.

11.12 Error-correction strategy

This section discusses the track structure of CD in detail. The track structure of MiniDisc is based on that of CD and the differences will be noted in the next section.

Each sync block was seen in Figure 11.25 to contain 24 audio bytes, but these are non-contiguous owing to the extensive interleave.[13-15] There are a number of interleaves used in CD, each of which has a specific purpose. The full interleave structure is shown in Figure 11.28. The first stage of interleave is to introduce a delay between odd and even samples. The effect is that uncorrectable errors cause odd samples and even samples to be destroyed at different times, so that interpolation can be used to conceal the errors, with a reduction in audio bandwidth and a risk of aliasing. The odd/even interleave is performed first in the encoder, since concealment is the last function in the decoder. Figure 11.29 shows that an odd/even delay of two blocks permits interpolation in the case where two uncorrectable blocks leave the error-correction system.

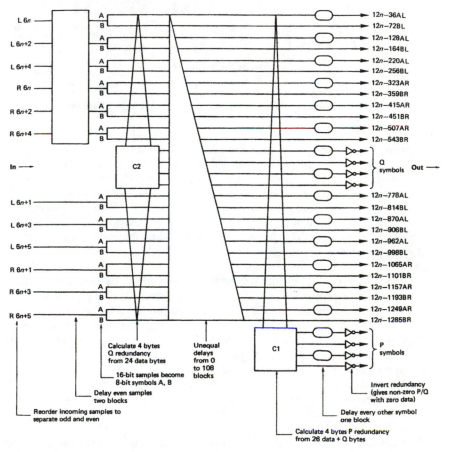

Figure 11.28 CD interleave structure.

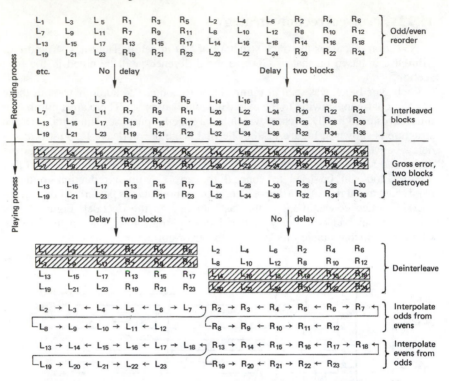

Figure 11.29 Odd/even interleave permits the use of interpolation to conceal uncorrectable errors.

Left and right samples from the same instant form a sample set. As the samples are sixteen bits, each sample set consists of four bytes, AL, BL, AR, BR. Six sample sets form a 24-byte parallel word, and the C2 encoder produces four bytes of redundancy Q. By placing the Q symbols in the centre of the block, the odd/even distance is increased, permitting interpolation over the largest possible error burst. The 28 bytes are now subjected to differing delays, which are integer multiples of four blocks. This produces a convolutional interleave, where one C2 codeword is stored in 28 different blocks, spread over a distance of 109 blocks.

At one instant, the C2 encoder will be presented with 28 bytes that have come from 28 different codewords. The C1 encoder produces a further four bytes of redundancy P. Thus the C1 and C2 codewords are produced by crossing an array in two directions. This is known as crossinterleaving. The final interleave is an odd/even output symbol delay, which causes P codewords to be spread over two blocks on the disk as shown in Figure 11.30. This mechanism prevents small random errors destroying more than one symbol in a P codeword. The choice of eight-bit symbols in EFM assists this strategy. The expressions in Figure 11.28 determine how the interleave is calculated. Figure 11.31 shows an example of the use of these expressions to calculate the contents of a block and to demonstrate the crossinterleave.

The calculation of the P and Q redundancy symbols is made using Reed–Solomon polynomial division. The P redundancy symbols are primarily for

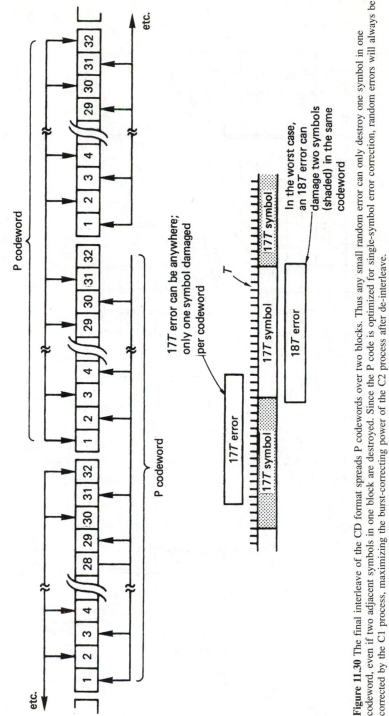

Figure 11.30 The final interleave of the CD format spreads P codewords over two blocks. Thus any small random error can only destroy one symbol in one codeword, even if two adjacent symbols in one block are destroyed. Since the P code is optimized for single-symbol error correction, random errors will always be corrected by the C1 process, maximizing the burst-correcting power of the C2 process after de-interleave.

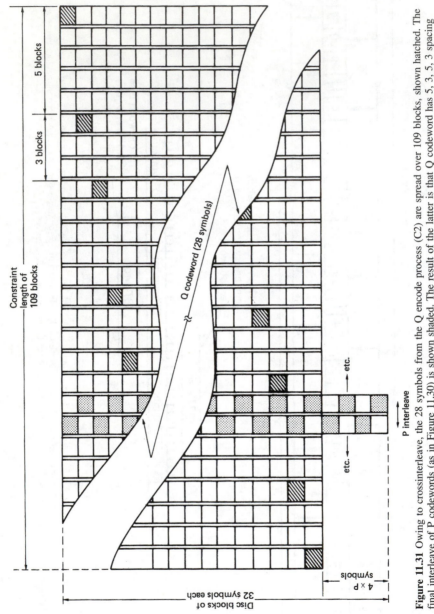

Figure 11.31 Owing to crossinterleave, the 28 symbols from the Q encode process (C2) are spread over 109 blocks, shown hatched. The final interleave of P codewords (as in Figure 11.30) is shown shaded. The result of the latter is that Q codeword has 5, 3, 5, 3 spacing rather than 4, 4.

detecting errors, to act as pointers or error flags for the Q system. The P system can, however, correct single-symbol errors.

11.13 Track layout of MD

MD uses the same channel code and error-correction interleave as CD for simplicity and the sectors are exactly the same size. The interleave of CD is convolutional, which is not a drawback in a continous recording. However, MD uses random access and the recording may be discontinuous. Figure 11.32 shows

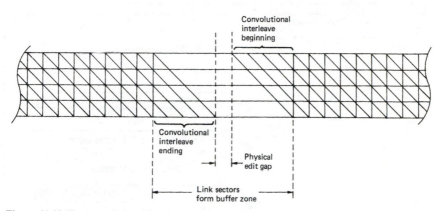

Figure 11.32 The convolutional interleave of CD is retained in MD, but buffer zones are needed to allow the convolution to finish before a new one begins, otherwise editing is impossible.

that the convolutional interleave causes codewords to run between sectors. Re-recording a sector would prevent error correction in the area of the edit. The solution is to use a buffering zone in the area of an edit where the convolution can begin and end. This is the job of the link sectors. Figure 11.33 shows the layout of data on a recordable MD. In each cluster of 36 sectors, 32 are used for encoded audio data. One is used for subcode and the remaining three are link sectors. The cluster is the minimum data quantum which can be recorded and represents just over two seconds of decoded audio. The cluster must be recorded continuously because of the convolutional interleave. Effectively the link sectors form an edit gap which is large enough to absorb both mechanical tolerances and the interleave overrun when a cluster is rewritten. One or more clusters will be assembled in memory before writing to the disk is attempted.

Prerecorded MDs are recorded at one time, and need no link sectors. In order to keep the format consistent between the two types of MiniDisc, three extra subcode sectors are made available. As a result it is not possible to record the entire audio and subcode of a prerecorded MD onto a recordable MD because the link sectors cannot be used to record data.

The ATRAC coder produces what are known as sound groups. Figure 11.33 shows that these contain 212 bytes for each of the two audio channels and are the equivalent of 11.6 milliseconds of real-time audio. Eleven of these sound groups will fit into two standard CD sectors with 20 bytes to spare. The 32 audio data sectors in a cluster thus contain a total of $16 \times 11 = 176$ sound groups.

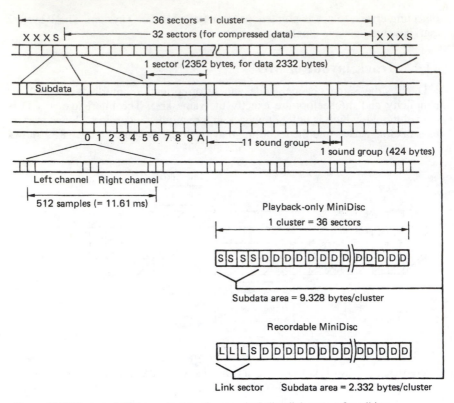

Figure 11.33 Format of MD uses clusters of sectors including link sectors for editing. Prerecorded MDs do not need link sectors, so more subcode capacity is available. The ATRAC coder of MD produces the sound groups shown here.

11.14 Player structure

The physics of the manufacturing process and the readout mechanism have been described, along with the format on the disk. Here, the details of actual CD and MD players will be explained. One of the design constraints of the CD and MD formats was that the construction of players should be straightforward, since they were to be mass-produced. Figure 11.34 shows the block diagram of a typical CD player, and illustrates the essential components. The most natural division within the block diagram is into the control/servo system and the data path. The control system provides the interface between the user and the servo-mechanisms, and performs the logical interlocking required for safety and the correct sequence of operation.

The servo systems include any power-operated loading drawer and chucking mechanism, the spindle-drive servo, and the focus and tracking servos already described. Power loading is usually implemented on players where the disk is placed in a drawer. Once the drawer has been pulled into the machine, the disk is lowered onto the drive spindle, and clamped at the centre, a process known as chucking. In the simpler top-loading machines, the disk is placed on the spindle by hand, and the clamp is attached to the lid so that it operates as the lid is closed.

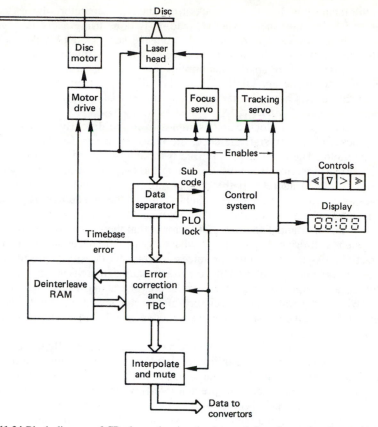

Figure 11.34 Block diagram of CD player showing the data path (broad arrow) and control/servo systems.

The lid or drawer mechanisms have a safety switch which prevents the laser operating if the machine is open. This is to ensure that there can be no conceivable hazard to the user. In actuality there is very little hazard in a CD pickup. This is because the beam is focused a few millimetres away from the objective lens, and beyond the focal point the beam diverges and the intensity falls rapidly. It is almost impossible to position the eye at the focal point when the pickup is mounted in the player, but it would be foolhardy to attempt to disprove this.

The data path consists of the data separator, timebase correction and the de-interleaving and error-correction process followed by the error-concealment mechanism. This results in a sample stream that is fed to the convertors. The data separator that converts the readout waveform into data was detailed in the description of the CD channel code. The separated output from both of these consists of subcode bytes, audio samples, redundancy and a clock. The data stream and the clock will contain speed variations due to disk runout and chucking tolerances, and these have to be removed by a timebase corrector.

The timebase corrector is a memory addressed by counters arranged to overflow, giving the memory a ring structure as described in Chapter 3. Writing into the memory is done using clocks from the data separator whose frequency rises and falls with runout, whereas reading is done using a crystal-controlled clock, which removes speed variations from the samples, and makes wow and flutter unmeasurable. The timebase-corrector will only function properly if the two addresses are kept apart. This implies that the long-term data rate from the disk must equal the crystal-clock rate. The disk speed must be controlled to ensure that this is always true, and there are several ways in which it can be done.

The data-separator clock counts samples from the disk. By phase-comparing this clock with the crystal reference, the phase error can be used to drive the spindle motor. The alternative approach is to analyse the address relationship of the timebase corrector. If the disk is turning too fast, the write address will move towards the read address; if the disk is turning too slowly, the write address moves away from the read address. Subtraction of the two addresses produces an error signal which can be fed to the motor. In these systems, the speed of the motor is unimportant. The important factor is that the sample rate is correct, and the system will drive the spindle at whatever speed is necessary to achieve the correct rate. As the disk cutter produces constant bit density along the track by reducing the rate of rotation as the track radius increases, the player will automatically duplicate that speed reduction. The actual linear velocity of the track will be the same as the velocity of the cutter, and although this will be constant for a given disk, it can vary between 1.2 and 1.4 m/s on different disks.

An alternative method is used in more recent drives, especially those with anti-shock mechanisms or the capability to play DVDs as well. Here the disk speed is high and poorly controlled. A large buffer memory is used and this soon fills up owing to the high speed. To prevent overflow, the player pickup is made to jump back one track per revolution so that the data flow effectively ceases. Once the memory has emptied enough, the skipping will stop and replay of the disk track will continue from exactly the right place. An undetectable edit of the replay data takes place in the memory.

Owing to the use of constant linear velocity, the disk speed will be wrong if the pickup is suddenly made to jump to a different radius using manual search controls. This may force the data separator out of lock, and the player will mute briefly until the correct track speed has been restored, allowing the PLO to lock again. This can be demonstrated with most players, since it follows from the format.

Following data separation and timebase correction, the error-correction and de-interleave processes take place. Because of the crossinterleave system, there are two opportunities for correction, first, using the C1 redundancy prior to deinterleaving, and second, using the C2 redundancy after de-interleaving. In Chapter 6 it was shown that interleaving is designed to spread the effects of burst errors among many different codewords, so that the errors in each are reduced. However, the process can be impaired if a small random error, due perhaps to an imperfection in manufacture, occurs close to a burst error caused by surface contamination. The function of the C1 redundancy is to correct single-symbol errors, so that the power of interleaving to handle bursts is undiminished, and to generate error flags for the C2 system when a gross error is encountered.

The EFM coding is a group code which means that a small defect that changes one channel pattern into another will have corrupted up to eight data bits. In the worst case, if the small defect is on the boundary between two channel patterns, two successive bytes could be corrupted. However, the final odd/even interleave on encoding ensures that the two bytes damaged will be in different C1 codewords; thus a random error can never corrupt two bytes in one C1 codeword, and random errors are therefore always correctable by C1. From this it follows that the maximum size of a defect considered random is $17T$ or $3.9\,\mu m$. This corresponds to about a $5\,\mu s$ length of the track. Errors of greater size are, by definition, burst errors.

The de-interleave process is achieved by writing sequentially into a memory and reading out using a sequencer. The RAM can perform the function of the timebase-corrector as well. The size of memory necessary follows from the format; the amount of interleave used is a compromise between the resistance to burst errors and the cost of the de-interleave memory. The maximum delay is 108 blocks of 28 bytes, and the minimum delay is negligible. It follows that a memory capacity of at least $54 \times 28 = 1512$ bytes is necessary. Allowing a little extra for timebase error, odd/even interleave and error flags transmitted from C1 to C2, the convenient capacity of 2048 bytes is reached. Players with a shock-proof mechanism will naturally require much more memory.

The C2 decoder is designed to locate and correct a single-symbol error, or to correct two symbols whose locations are known. The former case occurs very infrequently, as it implies that the C1 decoder has miscorrected. However, the C1 decoder works before de-interleave, and there is no control over the burst-error size that it sees. There is a small but finite probability that random data in a large burst could produce the same syndrome as a single error in good data. This would cause C1 to miscorrect, and no error flag would accompany the miscorrected symbols. Following de-interleave, the C2 decode could detect and correct the miscorrected symbols as they would now be single-symbol errors in many codewords. The overall miscorrection probability of the system is thus quite minute. Where C1 detects burst errors, error flags will be attached to all symbols in the failing C1 codeword. After de-interleave in the memory, these flags will be used by the C2 decoder to correct up to two corrupt symbols in one C2 codeword. Should more than two flags appear in one C2 codeword, the errors are uncorrectable, and C2 flags the entire codeword bad, and the interpolator will have to be used. The final odd/even sample de-interleave makes interpolation possible because it displaces the odd corrupt samples relative to the even corrupt samples.

If the rate of bad C2 codewords is excessive, the correction system is being overwhelmed, and the output must be muted to prevent unpleasant noise. Unfortunately the audio cannot be muted simply by switching the sample values to zero, as this would produce a click. It is necessary to fade down to the mute condition gradually by multiplying sample values by descending coefficients, usually in the form of a half-cycle of a cosine wave. This gradual fadeout requires some advance warning, in order to be able to fade out before the errors arrive. This is achieved by feeding the fader through a delay. The mute status bypasses the delay, and allows the fadeout to begin sufficiently in advance of the error. The final output samples of this system will be correct, interpolated or muted, and these can then be sent to the convertors in the player.

The power of the CD error correction is such that damage to the disk generally results in mistracking before the correction limit is reached. There is thus no point in making it more powerful. CD players vary tremendously in their ability to track imperfect disks and expensive models are not automatically better. It is generally a good idea when selecting a new player to take along some marginal disks to assess tracking performance.

The control system of a CD player is inevitably microprocessor-based, and as such does not differ greatly in hardware terms from any other microprocessor-controlled device. Operator controls will simply interface to processor input ports and the various servo systems will be enabled or overridden by output ports. Software, or more correctly firmware, connects the two. The necessary controls are Play and Eject, with the addition in most players of at least Pause and some buttons which allow rapid skipping through the program material.

Although machines vary in detail, the flowchart of Figure 11.35 shows the logic flow of a simple player, from Start being pressed to sound emerging. At the beginning, the emphasis is on bringing the various servos into operation. Towards the end, the disk subcode is read in order to locate the beginning of the first section of the program material. When track-following, the tracking-error feedback loop is closed, but for track crossing, in order to locate a piece of music, the loop is opened, and a microprocessor signal forces the laser head to move. The tracking error becomes an approximate sinusoid as tracks are crossed. The cycles of tracking error can be counted as feedback to determine when the correct number of tracks have been crossed. The 'mirror' signal obtained when the readout spot is half a track away from target is used to brake pickup motion and re-enable the track-following feedback.

The control system of a professional player for broadcast use will be more complex because of the requirement for accurate cueing. Professional machines will make extensive use of subcode for rapid access, and in addition are fitted with a hand-operated rotor which simulates turning a vinyl disk by hand. In this mode the disk constantly repeats the same track by performing a single track-jump once every revolution. Turning the rotor moves the jump point to allow a cue point to be located. The machine will commence normal play from the cue point when the start button is depressed or from a switch on the audio fader. An interlock is usually fitted to prevent the rather staccato cueing sound from being broadcast.

Another variation of the CD player is the so-called Karaoke system, which is essentially a CD jukebox. The literal translation of Karaoke is 'empty orchestra'; well-known songs are recorded minus vocals, and one can sing along to the disk oneself. This is a popular pastime in Japan, where Karaoke machines are installed in clubs and bars. Consumer machines are beginning to follow this trend, with machines becoming available which can accept several disks at once and play them all without any action on the part of the user. The sequence of playing can be programmed beforehand.

CD changers running from 12 volts are available for remote installation in cars. These can be fitted out of sight in the luggage trunk and controlled from the dashboard. The RAM buffering principle can be employed to overcome skipping caused by road shocks. Personal portable CD players are available, but these have not displaced the personal analog cassette in the youth market. This may be due to the cost of player and disks relative to Compact Cassette. Personal CD players are more of a niche market, being popular with professionals who are more likely to have a quality audio system and CD collection. The same CDs can then be enjoyed whilst travelling. There has been a significant development of such devices which now incorporate anti-shock memory as well as very low

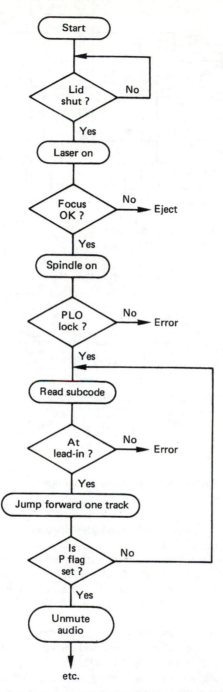

Figure 11.35 Simple flowchart for control system, focuses, starts disk, and reads subcode to locate first item of program material.

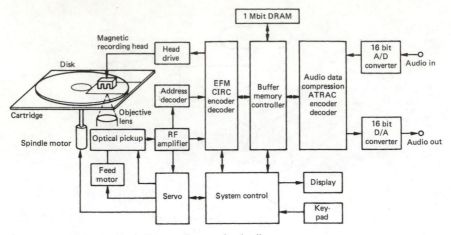

Figure 11.36 MiniDisc block diagram. See text for details.

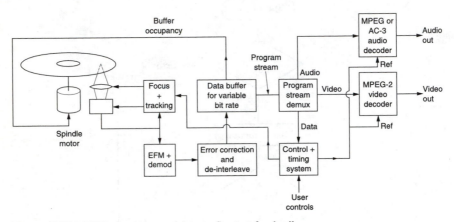

Figure 11.37 A DVD player's essential parts. See text for details.

power logic which combines with recent developments in battery technology to give remarkable running times.

Figure 11.36 shows the block diagram of an MD player. There is a great deal of similarity with a conventional CD player in the general arrangement. Focus, tracking and spindle servos are basically the same, as is the EFM and Reed–Solomon replay circuitry. A combined CD and MD player is easy because of this commonality.

The main difference is the presence of recording circuitry connected to the magnetic head, the large buffer memory and the compression codec. Whilst MD machines are capable of accepting 44.1 kHz PCM or analog audio in real time, there is no reason why a twin-spindle machine should not be made which can dub at four to five times normal speed.

In many respects the audio channel of a DVD player is similar in concept to that of MD. Figure 11.37 shows that the DVD bitstream emerging from the error-correction system is a multiplex of audio and video data. These are routed to appropriate decoders. The audio bitstream on a DVD may be compressed according to AC-3 or MPEG Layer II standards. Audio-only DVDs may also be uncompressed or use lossless compression, either of which will offer an improved sound quality when compared with the use of lossy compression.

References

1. Bouwhuis, G. *et al.*, *Principles of Optical Disk Systems*, Bristol: Adam Hilger (1985)
2. Mee, C.D. and Daniel, E.D. (eds) *Magnetic Recording*, Vol.III, Chapter 6, New York: McGraw-Hill (1987)
3. Goldberg, N., A high density magneto-optic memory. *IEEE Trans. Magn.*, **MAG-3**, 605 (1967)
4. Various authors, *Philips Tech. Rev.*, **40**, 149–180 (1982)
5. Airy, G.B., *Trans. Camb. Phil. Soc.*, **5**, 283 (1835)
6. Ray, S.F., *Applied Photographic Optics*, Chapter 17, Oxford: Focal Press (1988)
7. Maréchal, A., *Rev. d'Optique*, **26**, 257 (1947)
8. Pasman, J.H.T., Optical diffraction methods for analysis and control of pit geometry on optical disks. *J. Audio Eng. Soc.*, **41**, 19–31 (1993)
9. Verkaik, W., Compact Disc (CD) mastering – an industrial process. In *Digital Audio*, edited by B.A. Blesser, B. Locanthi and T.G. Stockham Jr, New York: Audio Engineering Society 189–195 (1983)
10. Miyaoka, S., Manufacturing technology of the Compact Disc. In *Digital Audio, op. cit.*, 196–201
11. Ogawa, H. and Schouhamer Immink, K.A., EFM – the modulation system for the Compact Disc digital audio system. In *Digital Audio, op. cit.*, 117–124
12. Schouhamer Immink, K.A. and Gross, U., Optimization of low-frequency properties of eight-to-fourteen modulation. *Radio Electron. Eng.*, **53**, 63–66 (1983)
13. Peek, J.B.H., Communications aspects of the Compact Disc digital audio system. *IEEE Commun. Mag.*, **23**, 7–15 (1985)
14. Vries, L.B. *et al.*, The digital Compact Disc – modulation and error correction. Presented at the 67th Audio Engineering Society Convention (New York, 1980), Preprint 1674
15. Vries, L.B. and Odaka, K., CIRC – the error correcting code for the Compact Disc digital audio system. In *Digital Audio, op. cit.*, 178–186

Glossary

Accumulator Logic circuit which adds a series of numbers which are fed to it.

AES/EBU Interface Standardized interface for transmitting digital audio down cable between two devices (*see* Channel status).

Aliasing Beat frequencies produced when sampling rate (q.v.) is not high enough.

Auditory masking Reduced ability to hear one sound in the presence of another.

Azimuth recording Magnetic recording technique which reduces crosstalk between adjacent tracks.

Bit Abbreviation for Binary Digit.

Byte Group or word of bits (q.v.), generally eight.

Channel coding Method of expressing data as a waveform which can be recorded or transmitted.

Channel status Additional information sent with AES/EBU (q.v.) audio signal.

CLV Constant linear velocity. In disks, the rotational speed is controlled by the radius to keep the track speed constant.

Codeword Entity used in error correction which has constant testable characteristic.

Coefficient Pretentious word for a binary number used to control a multiplier.

Coercivity Measure of the erasure difficulty, hence replay energy of a magnetic recording.

Companding Abbreviation of compression and expanding; increases dynamic range of a system.

Concealment Means of rendering uncorrectable errors less audible; e.g. interpolation.

Critical band Band of frequencies in which the ear analyses sound.

Crossinterleaving Method of coding data in two dimensions to increase power of error correction.

Crosstalk Unwanted signal breaking through from adjacent wiring or track on recording.

Curie temperature Temperature at which magnetic materials demagnetize.

Cylinder In disks, set of tracks having same radius.

Decimation Reduction of sampling rate by omitting samples.

Dither Noise added to analog signal to linearize quantizer.

DSP Digital signal processor; computer processor optimized for audio use.

EDL Edit decision list; used to control editing process with timecode.

EFM Eight to fourteen modulation; channel code (q.v.) of Compact Disc.

Entropy The useful information in a signal.

Faraday effect Rotation of plane of polarization of light by magnetic field.

Ferrite Hard non-conductive magnetic material used for tape heads and transformers.

Flash convertor High-speed ADC technology used with oversampling.

Fourier transform Frequency domain or spectral representation of a signal.

Galois field Mathematical entity on which Reed–Solomon coding (q.v.) is based.

Gibb's phenomenon Shortcoming of digital filters causing ripple in frequency response.

Hamming distance Number of bits different between two words.

Headroom Area between normal operating level and clipping.

Interleaving Reordering data on recording medium to reduce effect of defects.

Interpolation Replacing missing sample with the average of those either side.

Jitter Time instability, similar to flutter in analog.

Kerr effect *see* Faraday effect.

Limit cycle Unwanted oscillation mode entered by digital filter.

MTF Modulation transfer function; measure of the resolving ability of a lens.

Non-monotonicity Convertor defect which causes distortion.

Oversampling Using a sampling rate which is higher than necessary.

Phase linear Describes circuit which has constant delay at all frequencies.

Product code Combination of two one-dimensional error-correcting codes in an array.

Pseudo-random code Number sequence which is sufficiently random for practical purposes but which is repeatable.

Reconstruction Creating continuous analog signal from samples.

Reed–Solomon code Error-correcting code which is popular because it is as powerful as theory allows.

Requantizing Shortening sample wordlength.

Sampling rate Rate at which samples of audio waveform are taken (*see* Aliasing).

SDIF-2 Digital audio interface for consumer use.

Seek Moving the heads on a disk drive.

Subcode Additional non-audio data stored on recording media.

Wordlength Number of bits describing sample.

Index

www.focalpress.com

Join Focal Press on-line

As a member you will enjoy the following benefits:

- an email bulletin with **information on new books**
- a regular **Focal Press Newsletter**:
    - o featuring a selection of new titles
    - o keeps you informed of **special offers, discounts and freebies**
    - o alerts you to **Focal Press news and events** such as author signings and seminars
- complete access to **free content** and reference material on the focalpress site, such as the focalXtra articles and commentary from our authors
- a **Sneak Preview** of selected titles (sample chapters) *before* they publish
- a chance to have your say on our **discussion boards** and **review books** for other Focal readers

Focal Club Members are invited to give us feedback on our products and services.
Email: worldmarketing@focalpress.com – we want to hear your views!

Membership is **FREE**. To join, visit our website and register. If you require any further information regarding the on-line club please contact:

Lucy Lomas-Walker
Email: lucy.lomas-walker@repp.co.uk
Tel: +44 (0) 1865 314438
Fax: +44 (0)1865 314572
Address: Focal Press, Linacre House,
Jordan Hill, Oxford, UK, OX2 8DP

Catalogue

For information on all Focal Press titles, our full catalogue is available online at www.focalpress.com and all titles can be purchased here via secure online ordering, or contact us for a free printed version:

USA
Email: christine.degon@bhusa.com
Tel: +1 781 904 2607

Europe and rest of world
Email: jo.blackford@repp.co.uk
Tel: +44 (0)1865 314220

Potential authors

If you have an idea for a book, please get in touch:

USA
editors@focalpress.com

Europe and rest of world
focal.press@repp.co.uk